KNOWLEDGE BASED HYBRID SYSTEMS

IFIP Transactions B: Applications in Technology

International Federation for Information Processing

Technical Committee 5:

Computer Applications in Technology

IFIP Transactions Abstracted/Indexed in:
INSPEC Information Services
Index to Scientific & Technical Proceedings®
CompuMath Citation Index®, Research Alert™, and SciSearch®

B-11

KNOWLEDGE BASED HYBRID SYSTEMS

Proceedings of the IFIP TC5/WG5.3/IFAC International Working Conference on Knowledge Based Hybrid Systems in Engineering and Manufacturing, KNOWHSEM '93, Budapest, Hungary, 20-22 April, 1993

Edited by

ISTVÁN MEZGÁR
CIM Research Laboratory
Computer and Automation Research Institute
Hungarian Academy of Sciences
Budapest, Hungary

PETER BERTÓK
Department of Computer Science
Royal Melbourne Institute of Technology
Melbourne, Australia

1993

NORTH-HOLLAND
AMSTERDAM • LONDON • NEW YORK • TOKYO

ELSEVIER SCIENCE PUBLISHERS B.V.
Sara Burgerhartstraat 25
P.O. Box 211, 1000 AE Amsterdam, The Netherlands

Library of Congress Cataloging-in-Publication Data

IFIP TC5/WG5.3/IFAC International Working Conference on Knowledge Based Hybrid Systems in Engineering and Manufacturing (1993 : Budapest, Hungary)
Knowledge based hybrid systems : proceedings of the IFIP TC5/WG5.3/IFAC International Working Conference on Knowledge Based Hybrid Systems in Engineering and Manufacturing, KNOWHSEM '93, Budapest, Hungary, 20-22 April 1993 / edited by István Mezgár, Péter Bertók.
p. cm. -- (IFIP transactions. B, Applications in technology, ISSN 0926-5481 ; B-11)
Includes bibliographical references and index.
ISBN 0-444-81484-1
1. Expert systems (Computer science)--Congresses. 2. Computer-aided engineering--Congresses. 3. Computer integrated manufacturing systems--Congresses. I. Mezgár, István, 1951- . II. Bertók, Péter, 1952- . III. Title. IV. Series.
QA76.76.E95I387 1993
006.3'3--dc20 93-11626
CIP

Keywords are chosen from the ACM Computing Reviews Classification System, ©1991, with permission.
Details of the full classification system are available from
ACM, 11 West 42nd St., New York, NY 10036, USA.

ISBN: 0 444 81484 1
ISSN: 0926-5481

This book is printed on acid-free paper.

Printed in The Netherlands

PREFACE

Knowledge based systems have emerged as efficient solution for complex problems in many areas, including manufacturing and engineering. There are many problems, however, when neither symbolic procedures nor algorithmic routines alone can provide solution. A balanced mixture of causal and heuristic processing modules offers more efficient and comprehensive approach in complex cases. We call the different architecture types using the combination of these techniques "knowledge based hybrid systems" (KBHS).

The International Working Conference on "Knowledge Based Hybrid Systems in Engineering and Manufacturing" wanted to provide a forum for presenting interesting results using these approaches.

Seventeen papers from eleven countries were selected for presentation, and three keynote papers were invited. A fairly comprehensive and in-depth discussion of the main subjects was aimed at by the organizers.

The papers were grouped into seven sessions, each covering a separate topic. The sessions included both theoretical and practical papers. Hybrid Systems for CIM Applications, Knowledge Based Product Design, Coupling Learning with Other Paradigms, Problems of Manufacturing Automation, Hybrid Systems in Scheduling, Knowledge Based Simulation, Hybrid Techniques in Monitoring and Maintenance were selected as being of special interest. Formal and informal sessions were held to exchange ideas.

Some presentations were accompanied or followed by demonstrations. So called mini-symposia were held in conjunction with these demonstrations to provide opportunities for deep and detailed discussions, and to make the conference really "working".

The event was sponsored by the International Federation for Information Processing (IFIP) TC5/WG 5.3 (Computer Aided Manufacturing) and co-sponsored by the International Federation of Automatic Control (IFAC) Manufacturing Technology Committee.

We wish to thank all the Authors, Referees, and Program Committee Members for their efforts, that made this Working Conference possible. Their commitment and enthusiasm were among the main driving forces while organizing this conference.

On behalf of the National Organizing Committee we also thank the Hungarian Academy of Sciences and the OMFB (National Committee for Technological Development) for their support.

Special thanks are due to Digital Equipment Hungary Ltd., IBM Hungary Ltd., ICON Ltd. (Authorized Reseller of SUN Microsystems, Inc.) and SILWARE Ltd. (Authorized Distributor of Silicon Graphics, Inc.) for providing appropriate computing environment for the demonstrations.

May, 1993

The Organizing Committee and
the Editors

INTERNATIONAL PROGRAM COMMITTEE

Bertók, P., Hungary
Browne, J., Ireland
Buzacott, J., Canada
Deng, Z., China
Desrochers, A., USA
Doumeingts, G., France
Horváth, M., Hungary
Jacucci, G., Italy
Kaltwasser, J., FRG
Kimura, F., Japan
Kochan, D., FRG
Kovács, G., Hungary
Kowalik, J. S., USA
Kusiak, A., USA, Chairman
Lu, S. C-Y., USA
Márkus, A., Hungary
Menga, G., Italy
Mezgár, I., Hungary, Vice-Chairman
Michelini, R.C., Italy
Nemes, L., Australia
Ponomaryov, V.M., Russia
Stecke, K.E., USA
Stringa, L., Italy
Talavage, J., USA
Tamura, H., Japan
Tomljanovich, M., Italy

NATIONAL ORGANIZING COMMITTEE

Balatoni, Gy.
Erdélyi, F.
Hencsey, G.
Hermann, Gy.
Horváth, I.
Mezgár, I., Chairman
Lehoczki, G.
Salamon, M.
Somló, J.
Straub, E.
Tóth, T.

International Working Conference on Knowledge Based Hybrid Systems in Engineering and Manufacturing KNOWHSEM '93

ORGANIZED BY

Computer and Automation Research Institute,
Hungarian Academy of Sciences

SPONSORED BY

IFIP/TC5/WG 5.3 (International Federation for Information Processing, Working Group 5.3)

CO-SPONSORED BY

IFAC (International Federation of Automatic Control) Manufacturing Technology Committee

SUPPORTED BY

Digital Equipment Hungary Ltd., Hungarian Academy of Sciences, IBM Hungary Ltd., ICON Ltd. (Authorized Reseller of SUN Microsystems, Inc.), National Committee for Technological Development, SILWARE Ltd. (Authorized Distributor of Silicon Graphics, Inc.)

Contents

COUPLING LEARNING WITH OTHER PARADIGMS

PROBLEMS OF MANUFACTURING AUTOMATION

HYBRID SYSTEMS IN SCHEDULING

KNOWLEDGE BASED SIMULATION

HYBRID TECHNIQUES IN MONITORING AND MAINTENANCE

MINI-SYMPOSIA
(Demonstrations, discussions)

MINI-SYMPOSIUM I.

Hybrid Tools for CIM System Development

MINI-SYMPOSIUM II.

Hybrid Systems for Product Design

MINI-SYMPOSIUM III.

Layout Design

MINI-SYMPOSIUM IV.

Management of Production Systems

KEYNOTE PAPERS

Knowledge Based Hybrid Systems (B-11)
I. Mezgár and P. Bertók (Editors)
Elsevier Science Publishers B.V. (North–Holland)

Hard Problems of Soft Knowledge

T. Vámos

George Mason University
Center for Artificial Intelligence, Fairfax, VA
Computer and Automation Institute
Hungarian Academy of Sciences, Budapest

Abstract

Methods of Artificial Intelligence were mostly concentrated on reasoning in the domain of well structured information. Introduction of different procedures in uncertainty calculations were based on similar schemes. Human knowledge, due to our philogenic and ontogenic development, is more pattern-like. Logic and several other generic reasoning methods are metapatterns, patterns of pattern relations. Patterns, i.e. sets of some kinds of coherent phenomena, are given representations of unstructured or weakly structured knowledge. Methods of visual pattern recognition can not be used in a straightforward way, because the decision spaces of knowledge patterns are usually nonmetric. This fact is due to the gravitational distortion effected by the hidden, tacit knowledge. The paper reviews recent efforts and results in management of these difficulties, gives more details on those man-machine symbiotic systems, which are promising in a better representation and decision support.

Keyword Codes: I.2.4; I.2.8; I.5.0
Keywords: Knowledge Representation, Formalisms and Methods; Problem solving; Pattern Recognition, General

1. Introduction: the soft deadlock

Some people agree with me that Artificial Intelligence is in some respects in a deadlock. I start with this provocative statement not only to turn on you attention, but in order to explain the paths of my thoughts regarding next agenda.

Let us review what could AI achieve and what not. In this quest we can have a bookkeeping based on the original claims. That is important because year by year we meet the renewal of these claims advertising some novel or only renamed ideas.

The most successful systems are those which handle vast amounts of well-structured knowledge, i.e. data, which have certain, well defined definitions and clear conceptual structures, mostly ordered in hierarchies. Intelligent data bases with flexible logical structures do similar tasks, such systems can find (let us not use the more ambitious word: discover) relations, new common features in the database. These features could have been hidden, but in some way explicit by the data definitions.

These systems can include or even learn rules, logical relations among sentence type statements, created by the defined or found dependencies. If the problem concerned contains as much information as it can not be overlooked by the traditional ways, but the combinatorial multitude of possible relations is still computable, we get really useful results. Typical examples: configuration of computer systems; finding appropriate combinations of tools and manufacturing tasks; complicated cases of taxation; search in chemical knowledge.

Relevant results can be reported in related fields, in tasks around the kernel of Artificial Intelligence. Programming languages and advanced tools for knowledge representation; pattern recognition; sophisticated hypertext systems, which help in understanding linguistic texts, getting closer to natural language representation; visual representation of objects, relations are the most important achievements. Several of these are born in the non-academic environments. This small remark indicates the drive of pragmatic trends, somehow deviating from the speculative theoretical developments. Theory was probably the best in fighting complexity of big graph structures, helping heuristic search. One of these results was the idea of genetic algorithms. We come back to problems of heuristics later.

Not too much was achieved of the original claims, imitating high level human creativity, brain activity by computers, i.e. solving the general problems of machine translation, diagnostic and recognition systems which reach or even surpass the capacity of the best experts.

The real problems start with uncertainty and uncertainty related logical complications of not well structured knowledge. This is a long story, much-much longer than of computer science and Artificial Intelligence. The philosophical aspects started with the Greek Classics, but the entire history of probability and related fields, philosophical and mathematical logics reflect the unsolvable problems of uncertainty. I refer here to my earlier publications and the great predecessors quoted in these.

The fundamental result is the final unsolvability of the really uncertain problems, and the dubious approximation of them. This fact is mostly due to the complexity syndrome, complexity of combinations and branchings in all aspects of behavior, i.e. interactions, propagation and combination of effects, time variance, nonlinear responses, etc. Mathematics, cognitive science, epistemology converge in the same conclusion, related to uncertainty and complexity. Latest results in compexity theory assert new evidences of non-computabilities.

These considerations don't speak against the use of all achievements of mathematics, computer science, system theory, all these provided sometimes powerful instruments for model building. The major conclusion is a critical one: we have to analyze the intrinsic relations between the model and method applied, the relations between our model and the real investigated problem, taking into the account the limits of the modelling methods and the limitations of the different types of representations. The loss of an ideal certainty of the naive rationalism stimulates further critical research in problem solving.

2. Patterns: representation of soft knowledge

The idea of using patterns is not new, may be it is older than logic and surely older than any kind of probability-uncertainty consideration. The notion received different names in different contexts, depending on time and the nomenclature of the field. We use the wording pattern, because of several reasons. The first is its visual connotation, which will be discussed a little bit more in details later. The second is the relation to pattern recognition methods of computer science, and the third is a hint to the seminal work of Polya: *Patterns of Plausible Reasoning*.

Pattern is the representation of unstructured or weekly structured knowledge. This representation can include well structured knowledge as well, their typical representation patterns are graphs. All other relations are represented by sets, assemblages of informations which have any kind of coherence and these are just our patterns.

The pattern representation is a result of hundreds of millennia of philogenetic development, starting much earlier than any kind of more or less intelligent beings existed. The survival system of any animal is based on the memorizing of information which are advantageous or detrimental, collecting these vital informations in bunches according their happenings, storing those reactions of their own, which were advantageous or detrimental in connection of these vital information patterns. On a very primitive level the unicellular does the same: it has the ability to respond to combinations of two input informations (e.g. direction of light and flow of water), and on an extremely high level managing fantastic complexity, we do the same.

In our everyday activity we are supported by many millions of situation- and response-patterns. The mere difference in the view of these patterns was the opinion, if these patterns are "only" governing our automatic, lower level activities, or they reflect the general picture of all our performance. Cognitive science proved the rather trivial truth, that the highest levels of creative human practice is similarly pattern driven. Skill, taste, invention, associative ability, phantasy, ingenuity are all based on very large storages of patterns (i.e. cases, concurrences, information bunches) and adroitness to establish similarities and/or contrasting. The quoted book of Polya proves that the creative work of a mathematician or physicist are not too much results of strong deductive reasoning, but just of the described pattern recognition intuition.

I add an other idea: those intellectual activities, which were considered as being of different nature and separating us from the animal, like instinct, logical reasoning, the use of professional deductive and inductive methods, are metapatterns, i.e. patterns of pattern behavior, dynamics, interrelations. If we analyze the origins of logic we find the empirical pattern of some returning consequences (*if* men, *then* die; *if* raining, *then* get wet) and a generalization of this consequence rule pattern, i.e. metapattern. Several other abstractions of pattern behavior help us in the forms of metapatterns. It should be mentioned that the usual way, how this scheme works, is a preserved close relation between abstraction and concrete items. In most cases we have in mind a prototype like concrete pattern behind the abstract notion (a chair or a wardrobe behind the conceptual furniture, a familiar figure and face behind the conceptual other sex, a well remembered story behind an abstract moral, etc).

An other typical pattern of ways of thinking is the manner how we treat conceptual patterns. Numbers are excellent examples. Number mystic and mathematical number theory are equally interested in patterns of numbers, which numbers are themselves conceptual patterns, ciphers of certain groups of objects. The same scheme returns in mathematical thinking about similarities of formulae, chemical conjectures, based on structural-formular representations, i.e. special metapatterns are created for generic pattern processing.

Science uses metaphoric metapatterns, where relations can be interpreted by some logical considerations, but these are sometimes much weaker, than the problems concerned. Entropy is a typical one, used first in thermodynamics, than in information theory, but tried to be applied in other fields too. The Hamiltonian formulation is an other generic one. We should not forget, that these are all representations in the brain, which has certain relations to the reality, but the mode of representation, which is now our concern, is an adaptation to the structures of our brain, which was developed for other purposes. We know the difficulties of expressing something in a different language, which has a different cultural history and therefore different in vocabulary and ways of expression. The difficulty of our brain, in understanding relations, which were not the subjects of our survival-developmental history are somehow similar. The main metaphors of human thinking are related to vision and within the visual perception to geometry. Symmetry, linearity and other "basic" forms play a key role not only in treating visual objects, but also in very far, metaphoric relations.

Analyzing roots of representation has relevant practical consequences. The first is epistemic: having in mind the repre- sentation- transformation problems we get a hint to a critical view of any modelling processes. The second is methodological: accepting the pattern view we get an other way of modell-representation and creating more intelligent man-machine systems.

3. Pattern recognition and metrics

Pattern recognition discipline developed several well applicable methods for evaluation of complex patterns, i.e. providing discrimination-separation techniques among patterns, even in the case of some overlappings. All are based on some metrics, the visual field itself is a metric one, and because of its fundamental role in brain development, it is the essential internal metaphor of any kind of thinking. The trouble in our complex problems starts with this metric.

The decision space of highly complex problems, containing lots of uncertainties, both in estimates of individual data and in their relations, is usually non-metric. The relations are not transitive, not commutative, triangulation can not be applied. This space is more difficult to be managed than the space of irreversible thermodynamic processes, although we can find some metaphoric similarities in having problems where the closure of a trajectory doesn't result in the same value as the start had, and the value of a function integrated on the trajectory is dependent on the direction.

What is the reason of this strange behavior? Logically, in a well defined, closed world this would be not the case, the principle (or dogma) of excluded middle would lead to unambiguous trajectories of reasoning and real consequences, independent of direction. The metric of the decision space of a classic probability distribution would be similar, but more dispersed. Our hypothesis for the reason of the irregularity phenomena is the presence of hidden, tacit knowledge. Using a simple metaphor we coined this phenomenon the gravitational distortion of the decision space, because a hidden big mass of different density material causes distortions on the theoretically even planar fields of gravitation. The effect of the tacit knowledge (experience, skill, instinct) of a human expert is similar, if we could extract all relevant knowledge about the problem concerned, the space of decision would be a metric one and methods of pattern recognition could be transferred for solution.

The pattern metaphor (we are generally thinking in metaphors, this is the due to the view on representations) suggests, that if we could create infinite dimension representations, the metric problem would disappear, we look at the infinite problem space (infinite number of influencing data and relations among them) from a limited dimension point of view. This means, that if we would like to improve the decision techniques, we have to look for the information causing the distortion in metric. Unfortunately, in most cases, this is not a mathematically treatable task, if it would be, the problem of metrication, and by that the evaluation of hidden knowledge would be solved. The criterion of an appropriate metrication is the success condition of pattern recognition: are the patterns (cases, situations) well separable by that pattern recognition way, is this separation process in some extent (for some practical purposes) generic, does the separation lead to a required level of good decision results? As it can be seen from this enumeration of criteria, these are very soft conditions, as soft as the whole problem complex and this is the reason of the hard difficulties.

4. Context dependency of metrics

Let us see two minor examples. The first can be taken from medical diagnosis. Most of the data contain some uncertainty. Body temperature, blood analysis, EKG, EEG data, etc. all are defined in some ranges, these are time by time revised due to new knowledge. This rating is due to individual data. The second step is a well definable environment, e.g. values related to ages, body weights, sexes, physiological cycles. Both can be well interpreted within the conceptual range of structured machine learning. The next step is more pattern oriented with all its difficulties. Some irregular phenomena are treated as physic, i.e. related to a constitutional specificity of the person and being not pathological. Some are irrelevant in the view of one disease but very meaningful in the context of an other. The same data have different meaning in the incubation period of a malady and in the recovery period. The experienced physician does no logical structuring of data, but has a decision looking at the entire pattern in a parallel way, not omitting the information received from the two previous steps of processing structured data.

The same can be followed in economy. Data are uncertain per se, because they are collected in a predefined conceptual structure. An endless dispute between economists and statisticians refer to this contradiction, the statistician would like to have a "reliable", comparable data base, i.e. lasting definitions, the economist is driven by his most challenging problem: the change in structures. We all feel, that data are nonlinear and context dependent. Prices, GNP/capita, growth rate, living standard, having the same numerical value, are not the same, depending on location, age, social status and environment, season, period, etc. Relevant are the differences in short range and long range view. State of infrastructure, future and relevance of savings, personal and community investment in education or health care can dramatically change priorities, i.e. metric. All judgements in micro- or macro-economics should take these complete patterns in consideration.

This picture is trivial and we have to emphasize, that it is similar in any complex relation. The seemingly well defined problems of engineering design presents similar problems. Economy of production and of use, reliability, maintenance, durability, user friendliness, environment protection, esthetics, fashion, market peculiarities, life-style related requirements play the same role in creation uncertain, complex decision patterns like the others in biology or economy. The trivial picture was unattended by many people working in AI, because it is an argument against well developed algorithmic methods. It was considered by some, who claimed to solve these problems as well, by their generic procedures.

Similarity of patterns is viewed in two relations: similarities to certain "ideal" cases , conceptual entities (objects), and similarities among equally treated ones. Examples for the first are patients' medical patterns, related to canonical, textbook descriptions of maladies, or legal cases, related to vices defined by Statute Law. The second type is characterized by the case based Common Law practice or by economical comparison studies.

5. Need for combination of hard and soft methods

Similarly to many other problems -most of them are somehow related to this one- a pragmatic compromise is advisable, using all, what is available for a machine supported decision system, delineate their limitations, and try to combine machine methods with others.

The following step is the application of some appropriate clustering method and evaluation procedure. In most cases the objective is somehow different to the visual pattern recognition: usually the clear clustering among patterns is less important in this case, some points can belong to different patterns, the overall evaluation is primary. Clustering can help in this respect as well, but it can overdo its job, filtering or at least weakening some relevant information. This small experience on clustering indicates, how context sensitive the procedure itself is, how relevant is the experimental issue.

We pursue two objectives: get an overall value, which contrasts the pattern, looked from any other. This is a generalization of the signature concept. The second is a view of the pattern, which helps in finding similarities, analogies, associations.

6. The task of pattern identification

The first task is accomplished by an appropriate combination of weights and data, standardization included. Our experience in medical diagnosis proved the feasibility of linear combinations. The emphasis was on weights, expressing context dependent relevance. This scale of weights is really the most delicate issue. Our first experience, and this was later similar one of the MYCIN group, that affirmation should be taken weaker, than negative informations, which push the evaluation away of from some hypothetical patterns. The scale, according to cognitive experience, is different in refinement, at least 3, at most 10 grades. This is also a decision, due to experimental results.

Human estimate does linearization in a spontaneous way, this linearization is the natural pshysiological-psychological operation of any sensory signal processing (Weber-Fechner, Stevens-Law). The typical sigmoid characteristic of evolutionary processes (incubation, quasilinear or exponential growth, saturation), must be checked for time varying processes. This characteristics is not so much imbedded in the spontaneous human response. Other filtering, like noise, cyclic phenomena, are separate issues, they should be taken care of in the data acquisition phase, not to be let for data evaluation in the pattern identification process. The reason lies in the earlier remark: filtering can hide relevant informations, if it is done in the data acquisition process. The importance of these lost, or corrected values is considered one by one and not in a general picture.

The use of linear combination of data and weights is not exclusive, our system contains a free choice for any other combination algorithms (e.g. using momenta of higher order). According to our experience the fine setting of weights is primer.

The domain expert and the end user should get separate and user friendly means to this fine tuning, adjustment of weights for getting an appropriate separating evaluation.

Theoretically this could be done by an automatic process, using some least squares or maximum likelihood methods on the data bases, but this usually immense computer work would conceal the essential objective of the process: mining (and several times deepening) the hidden knowledge of the domain expert. Knowledge mining and improving is just one part of the issue, which we had pointed out. The metric is expressed by the weights used, all these linearization, separation of affirmative and negative, the possible use of momenta and this fine tuning are measures for getting a more adequate metric. This ideal metric expresses the real meaning behind the insufficient data and the insufficient extent of well defined knowledge. We come back to the relations of the strange and ideal metric once more.

7. The similarity issue

The other step, looking for a general view of patterns in a search for similarities, is even more complex.. This is the point, where every artificial intelligence system halts and not because of the intellectual weakness of the creators. The very hard NP problem is this one. We quoted several times the metaphor of a space of infinite dimensions and a combinatorially more infinite number of interrelations. Points of each pattern are of that kind, having so many possible parameters, including their own ones, and the interrelated dynamics. Most probably, this is one reason of the incredible evolutionary diversity of the brain, its neurons, synaptic and nonsynaptic relations. No other biological organization developed in such a diversity. This unique production evolved just because of the necessity, to cope with that part of the real world's pattern infinity, which was vital for survival. Hundreds million years of evolution created this pattern reception and response abilities, which are patterns themselves (the metapatterns). The human brain was a late and highest success of the process, but we follow many admirable abilities of central neural systems in much less sophisticated creatures. The genetic learning process of many-many billions of individuals, many millions of generations and practically infinite and variational teaching feedback actions of reality is much beyond of any algorithmically computable limit.

We should not drop this result of Nature because of a technological hype. According to our view, one of the main tasks of computer science is the undertaking, to find the optimal ways of combining this psychological-intellectual abilities with existing and future technologies. The technology itself should be developed in this direction. The practical demand feels this, most of the last decade's new products are characterized by this trend, research does it in much less extent. It is therefore easy to find many products on the market, which are ready made helpful supports for our purposes.

The other side of the man-machine relation is not less important. The patterns and metapatterns of behavior, thinking are deposits of earlier experience, received sometimes in completely different environment. They can be as harmful prejudices as they are blessings of talents. The man-machine symbiosis, if it is well designed, can balance good and bad features of both side, provide a steady feedback control for human guesses, hypotheses.

8. Visual representation, a key for man-machine symbiotic systems

Accepting the lessons of the admirable abilities of the brain, its origins and roles, our agenda is much more clear. We have to get closer to representation interface of the mind and create the computer interface in a sense, which should be the most convenient for both partners. Undoubtedly, this is the visual representation, which is the most important and most rich one of human perception. Additionally it permits a parallel representation of information both for the computer and the human. After the popularity success of connectivism, no argument is needed for parallelism.

Regularity and deformation are the key issues of this kind of perception. Not by chance, art of any kind applies these effects of normal, satisfactory status and of abnormality, which should evoke attention. This means, that we have to use the most regular shapes and restrained colors for regular patterns and sharp geometrical irregularities, bright colors for any disturbing information. This is a flat triviality, but can not be omitted.

The representation should include dynamics, the figure should be able to move in time and let pictorial tracks of this movement (e.g. shadows). The need for the last type of representation is a user dependent issue. Some people have excellent memory for patterns of movement and the abundance of instant pictures can be distracting for those. Others would require this quasiparallel view of dynamics, as a support for identification. (For the first: Mozart could hear the whole composition simultaneously)

Symmetry and asymmetry, relations of size, shapes, expressing individual excessive values or some returns in regular or irregular ripples are subjects of design, dependent on the number and nature of the pattern components, attitudes of the users. The mentioned wide variety of graphic products makes possible a user friendly and rather individual design for each purpose. The problem is fortunately not the graphic software, but the psychology of the project concerned.

9. Two versions of visual representation

In this respect we meet two basic alternatives. The first is a representation by as simple, familiar and easily receptive figures as possible, the circle and its deformations are ideal ones. The second requires more phantasy and more psychological experimentation. It is a pictorial representation of the subject, the relational distortions and dynamics included. In our medical project we use sketches of babies representing typical or atypical dysfunctions, but the same idea can be used in a project on world economy, using maps or any other usual or unusual representations of activity characteristics. The latter way is not only a challenging one for design creativity, but also a practical way of representation much more information, than it can be accomplished by simple diagrams. Much work was directed in this line, a reference can be given to the beautiful books of *Tufte*. Statisticians explore the possibilities of advanced computer graphics in representation of multidimensional data.

A practice of these highly meaningful representations can support the easy way of data input. Instead of filling long, boring queries, the picture can be accepted, modified. In our medical project the data input by long menus was one of the major obstacle of usage, the practitioners hated to do this apparently needless work, instead of the usual tacit, hidden communication with the tacit, hidden knowledge of the user and among the practitioners. The regular looking parts of the figure can be a check for all noninteresting data, one or a few related figure parts, as pointed, can evoke all usually related others, call for attention and checking. A primitive version of this kind of representation was used for a long time in car repair documentations.

A witty intermediate representation idea was the use of *Chernoff*-faces, human face characteristics representing deviations.

The graphic tools for this representation are similarly available, but the special design of this input-output representation can be a very captivating challenge of knowledge acquisition cooperation between the domain expert and the knowledge engineer.

It is apparent that these kinds of graphic representations offer an indirect way of exploring the metric of the decision space, just that indirect way, which on the other hand is the most direct one, looking from the human point of view. This latter is most important, because metric itself is a human representation, especially those metrics, which are related to any kind of human decision.

10. Mining for hidden knowledge

We omitted an other possibly important aspect: mining for missing information (dimensions in the decision space). The easier case is the one, if the missing information is present in the data base (in one or an other form), but was neglected or not used in an appropriate way. The outlined experimentation procedure is a helpful way to lift them onto the surface. Missing relevant information, which doesn't exist in the data base is always a research task, but the visualized experimentation on the data base can give useful hints, where and how to look for the hidden information, not discovered other objects. The same is valid for the suggested graphic representation. which is a kind of portrayal of metric, the gravitational distortion metaphor returns here! It suggests an analogous procedure to the discovery of remote celestial objects, based on the trajectory anomalies of the well observed.

11. Conclusion

We tried to outline a renewed trend in computer science: abandoning of the promises of a general problem solver, but continuing efforts towards a pragmatic, quasioptimal man-machine symbiosis.

Knowledge Based Hybrid Systems (B-11)
I. Mezgár and P. Bertók (Editors)
Elsevier Science Publishers B.V. (North-Holland)

METHODOLOGIES AND APPLICATIONS OF KNOWLEDGE-BASED HYBRID SYSTEMS

Axel Lehmann
Institut für Technische Informatik
Universität der Bundeswehr München
Germany

Abstract

This paper presents a summary of methodological approaches and techniques often used for the representation and processing of information in knowledge-based systems (KBS). Starting from major types of problem solving, it compares the essential characteristics of conventional programming techniques with those often used for the realization of knowledge-based systems. Regarding strengths and weaknesses of these techniques, different approaches and tools for the implementation of knowledge-based hybrid systems (KBHS) are discussed. Finally, examples and experiences with the application of hybrid systems in the problem domain of simulation and diagnosis are presented.

Keyword Codes: I.2.1; I.2.4; I.2.8
Keywords: Artificial Intelligence, Applications and Expert Systems; Knowledge Representation Formalisms and Methods; Problem Solving, Control Methods and Search.

1. Introduction

Following the definition of Feigenbaum, Waterman et. al. [HaK87]

> a "knowledge-based system (or expert system) is .... an intelligent computer program that uses knowledge and inference procedures to solve problems that are difficult enough to require significant human expertise for their solution. Knowledge .... plus the inference procedures used, can be thought of as a model of the expertise (of the best practitioners) of the field".

That means that the realization and application of a knowledge-based system (KBS) always concerns the implementation and installation of a domain specific problem solver. Regarding major problem categories, we can distinguish different types of problem solving, e.g.:

- analysis
  (diagnosis, selection, classification, advise)
- synthesis
  (construction, configuration, design, planning)
- simulation
  (prediction, evaluation)

For performing those problem solving capabilities efficiently and effectively, different categories of information, knowledge or data, respectively, have to be distinguished and represented adequately. In that context, three major categories of knowledge are ([SEV87]):

- facts expressing valid propositions,
- beliefs expressing plausible propositions and
- heuristics expressing rules of good judgement in situations where valid algorithms do not exist.

This paper gives a summary of state of the art approaches and techniques for the realization of such problem solvers. In the first section, major methods, techniques and tools for adequate knowledge representation and processing are presented. The presentation focuses on possibilities of combining and integrating different programming paradigms, forms of knowledge representation and inferencing mechanisms for a single application - thus forming knowledge-based hybrid systems. In that context, a survey of state of the art tools and their main characteristics are also analysed. In the second section, some examples and experiences are summarized concerning the realization of simulation and diagnosis applications by the use of hybrid tools. The final conclusions analyse state of the art and future perspectives of knowledge-based hybrid systems concerning research and industrial applications, as well.

2. Methodologies, Techniques and Tools

Over the past decades, most problem solving applications of computers have been realized on data processing systems (DPS). Major differences of those programming systems and of knowledge-based systems (KBS) concern:

- information representation, e.g.:
  - DPS: numerical,
    e.g. by numbers, alphanumerical characters;
  - KBS: symbolical or numerical,
    e.g. by patterns, pictures, character strings, numbers;

- representation of vagueness, e.g.
  - DPS: statistics, probability theory;
  - KBS: certainty factors, fuzzy logic/sets;

- specification of problem solving processes, e.g.:
  - DPS: procedural, by algorithms;
  - KBS: declarative and procedural, by non-deterministic reasoning and by methods;

- program controlling mechanisms, influenced in:
  - DPS: by sequence of instructions in a program, calculations of quantitative values/formulas;
  - KBS: by inferencing strategies;

- program dynamics, e.g.:
  DPS: control flow-oriented;
  KBS: data flow-oriented.

According to the classification of techniques applied for the implementation of knowledge-based systems as given in [SEV87], we have to distinguish between basic techniques (e.g. symbolic programming, proportional logic) and those derived from the basic ones (e.g. constraint satisfaction, production rules) or derived from the first level ones (e.g. optimization, explanation facilities). The interdependencies and derivations of common used blocks of implementation techniques are demonstrated in Figure 1.

In this paper, we will concentrate on the analysis of different formalisms and techniques for knowledge representation. The importance of an efficient realization of the designer and of the user interfaces of a knowledge-based system as well as different formalisms and techniques for their realization are not neglected. But these techniques are not taken into account in the context of this paper when talking about hybrid concepts.

The most widely used formalisms and techniques for knowledge representation include:
- logical calculus (modal, temporal),
- production rules,
- objects, classes etc.,
- semantic nets,
- constraints,
- case based reasoning or pattern matching, respectively,
- fuzzy sets, fuzzy logic,
- certainty factors,
- probability theory.

Detailed descriptions and examples for their application are widely discussed in the literature (see e.g. [HaK87], [SEV87], [HLW83]. The goal of this section is to discuss and analyse those tools offering alternative techniques for the implementation of knowledge bases and inference engines by an homogeneous programming environment or by a single tool. For that reason, we have also to distinguish between three major categories of programming environments for the implementation of KBS.

- extensions of programming languages (LE),
- general development environments (DE) and
- shells (SH).

Table 1 presents a list of those products which will be further analysed in this section. At first, a classification according to these 3 categories is presented. It can be recognized that about half of these products can be classified as hybrid development tools. It can also be seen, that most of these products are not only used by AI-specialists but mostly also by specialiced software development teams and by application domain experts. Most applications concern the problem solving types of diagnosis, classification and advisory systems, as well.

A survey of Tables 2 indicates, that most products offer at least as knowledge representation technique production rules. Table 2a demonstrates, how a single rule and rule sets, as well, can be structured and assembled. Table 2b shows that forward and backward chaining are widely used as inferencing mechanisms; most tools offer in addition the combined application of both mechanisms.

Different formalisms and concepts for an object oriented representation of knowledge is summarized in Table 3. Objects are often represented by frames or scripts (in case of dynamic process representations). In this table, at first, different formalisms for object-oriented structuring of knowledge are presented, such as classes, hierarchies or object-attribut tupels. In addition, in this table, tools are classified according to their possibilities of inheritance and their types of attributes offered for characterizing the properties of objects.

Table 4 contains information about different formalisms and techniques for the representation of vague knowledge. In Table 4a, numerical and logical techniques for the representation of uncomplete, of uncertain and of inconsistent information are presented. Table 4b illustrates, how vagueness can be implemented in the selected products. As a result we can deduct that several products do not offer special techniques. Others offer mostly mechanisms like certainty factors or confidence values.

Finally, Table 5 indicates the knowledge acquisition facilities of the analysed products. More than half of those tools offering acquisition support are offering editors or generators for production rules or frames (object specifications).

3. Applications and Experiences

In this section, the basic concept and the realization of two different types of problem solving systems - a simulation and a diagnosis application - are discussed. Both concepts could be implemented and tested by prototype versions realized by hybrid development systems.

3.1 Discrete Event Simulation

The general objective of this research project was directed towards supporting users and developers of discrete event simulation models in the different steps of model construction and its experimental application. Supporting users and developers means to offer them various kinds of knowledge about modelling theories, techniques, tools, statistical analysis etc. which can be efficiently represented by using different data and knowledge representation techniques.

Regarding their specific task, taxonomy and accessibility by the user as analyst, at least three different categories of combined applications of expert systems and simulation models have to be distinguished [Leh87]. The implementation of those architectures requires a hybrid programming environment:

- Simulation models embedded in expert systems
  Simulation models as integral part of an expert system can be used to represent time-dependent knowledge or time-dependent inferencing techniques (Fig. 3a).

- Expert systems as integral part of a simulation model (Fig. 3b and c)
  The global goal of an expert-system-integrated modelling environment can be seen as an approach to express and to apply fragmentary or uncertain knowledge in a model by means of an expert system, e.g. concerning the non-deterministic job-flow or user behaviour in a system. Besides this form of application in quantitative simulation, new approaches like qualitative simulation or inductive modelling are based on such combinations.
  Regarding the degree of integration of expert system and coded model, a distinction can be made between:
  * an interactive cooperation of a parallel developed expert system and an executable model (Fig. 3b, both could be realized in a different programming environment), and
  * an expert system embedded in the coded model (Fig. 3c) and mostly realized in the same programming environment as the executable model, e.g. in PROLOG/TPROLOG.

- Expert systems as supporting framework for a goal-directed application of modelling tools
  This category of expert systems is developed for application by unexperienced users in different stages of a modelling process, like for the selection of a problem-adapted modelling method or for the construction and experimental application of an executable model with respect to cost-benefit considerations. In contrast to the second category, these expert systems are directly accessible by the analyst. Regarding the function of these expert systems, we have to distinguish between:
  * decision support systems (Fig. 3d) applicable as classification systems (giving the analyst final, weighted recommendations) or as advisory systems (leading the analyst in a step-by-step consultation, indicating alternative solutions weighted e.g. by certainty factors) [Leh87];
  * intelligent front-ends (Fig. 3e), which offer the analyst several application-dependent interfaces to a single modelling tool or to a simulation language; this class of expert systems is used to bridge over the gap between problem domains and a modelling tool by internal mapping of the meaning and description of objects, items

etc. of a specific application domain in terms used by the modelling tool.

The global goal of our research and tutorial project INT^3 concerns the implementation of a highly interactive, user-adaptable and knowledge-based modelling facility on a standard PC and its primary application by graduate students [LeS87]. There we used RESQ and M.1/S.1 as hybrid programming environment and implemented a classification and an advisory system (according Fig. 3d), as well. With MEDUSA we also implemented architectures according Fig. 3c and d on Sun workstations with SIMKIT/KEE as hybrid programming environments. In the later implementation we used frames, production rules, semantic net and probabilistic formalisms for knowledge representation. In the first PC project, we applied production rules, certainty factors and probabilistic functions.

3.2 Fault Diagnosis

For the application domain of fault diagnosis and maintenance support for public switching systems, we implemented a knowledge-based hybrid system under KEE on Sun-workstations [BFK89]. The general types of problem solving and its capabilities were:

- classification (of event messages)
- diagnosis (reasoning about symptoms)
- advisory (given advise how to maintain a switch)

For its realization, techniques like frames, production rules, certainty factors, Mycin-mechanism, forward and backward chaining were applied in a single knowledge-based diagnosis system.

Currently we are working in a research cooperation for adding case-based reasoning formalisms for the correlation of alarm streams in such an environment.

4. Conclusions

As indicated by Figures 4 and 5, knowledge-based development systems are widely used [HaK87]. After the first wave and enthusiasm of small systems, a wider progress and application of hybrid systems running on powerful workstations can be observed. But, nevertheless, the real industrial or practical application suffers from:

- missing support of <u>knowledge maintenance techniques</u> applicable in practice,
- missing support in the <u>adequate selection and efficient application</u> of KBS-techniques in hybrid environments (real specialists are required!),
- missing techniques for wide application of <u>machine learning</u> and
- limited <u>flexibility,</u> efficiency and validation support.

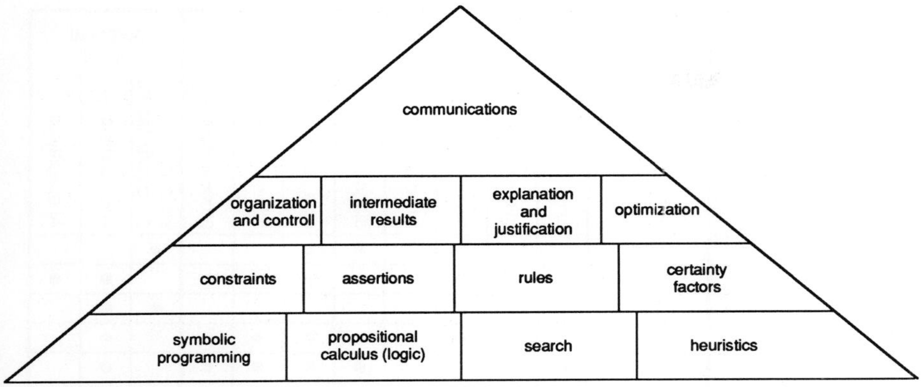

Figure 1: Building blocks for knowledge-based systems [SEV 87]

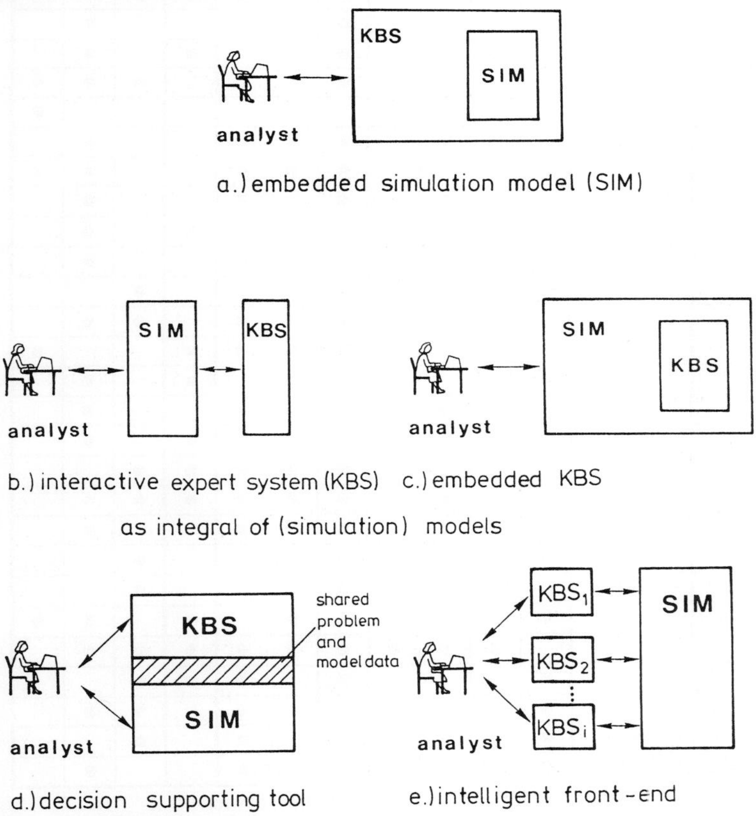

Figure 3: Architectures of knowledge-based modeling systems

Product	programming environment	Paradigms				problem solving types				potential users		
		rules	rule + object.	hybrid systems	inductive	diagn., class.	conf., plan.	advisory	process control	special. developm. teams	appl. domain	AI specialists
ADEPT	SH			●		●	●			●		
ADS	DE			●		●	●	●	●		●	●
Babylon	DE			●						●	●	
COPERNICUS	SH		●	●		●	●	●	●		●	
DB Expert	SH	●				●	●	●	●		●	
E´G´E´R´i´A	DE		●			●	●	●	●		●	●
ESE	SH	●				●		●		●		
Goldworks II	DE		●	●		●	●				●	
GS Real-Time	SH		●			●	●		●		●	
Guru	DE	●				●	●	●		●	●	●
HUMBLE	SH		●									●
IPW	LE Prolog			●		●	●	●			●	
Kappa/PC	SH	●	●			●	●	●			●	●
KBMS	SH			●		●	●	●			●	
KEE	DE			●		●	●	●	●		●	
KES	DE		●			●	●	●			●	
KET	DE	●			●	●		●		●		
Knossos	SH		●			●	●	●		●	●	
Knowledge Craft	DE			●		●	●	●	●			●
Knowledge Tool	LE PL/I	●	●				●		●		●	
Kool	DE		●	●		●	●	●			●	
Leonardo	DE		●	●		●		●	●		●	
Level 5 Object	SH		●	●		●	●	●	●	●		
Nexpert Object	SH	●	●	●		●	●	●	●	●	●	
ProKappa	LE C			●		●	●	●	●		●	
Tablo	DE	●				●	●	●		●		
Testbench	SH		●			●				●		
TIRS	DE	●	●				●		●		●	●
TWAICE	DE		●	●	●	●	●	●	●	●	●	
VAX Dec. Exp.	SH	●				●		●	●	●		
Xi Plus	SH	●				●	●	●		●	●	
Zeno Expert	SH	●			●	●		●		●		
Zeno Professional	LE C	●				●	●	●			●	

SH: shells LE: language extensions of ... DE: development environments

Table 1: Overview of programming environments for knowledge-based systems [BSW 91]

Product	structuring of rules			premises							conclusions		
	rule packages	meta rules	demons	relations	OR	NOT	arithm. oper.	set oper.	pattern matching	variables	i/o-oper.	calculations	subroutine calls
ADEPT	●			●	●	●	●	●		●	●	●	●
ADS	●	●	●	●	●	●	●	●	●	●	●	●	●
Babylon	●		●	●	●	●	●	●	●	●	●	●	●
COPERNICUS	●		●	●	●	●	●	●		●	●	●	●
DB Expert	●			●	●	●	●	●		●	●	●	●
E´G´E´R´i´A	●	●	●	●	●	●	●	●		●			
ESE	●	●	●	●	●	●	●	●	●		●	●	●
Goldworks II	●	●	●	●	●	●	●	●	●	●	●	●	●
GS Real-Time	●	●	●	●		●	●	●	●	●	●	●	●
Guru	●	●	●	●	●	●	●	●	●	●	●	●	●
HUMBLE					●	●				●	●	●	●
IPW	●	●	●	●	●	●	●	●	●	●	●	●	●
Kappa/PC	●	●	●	●	●	●	●	●	●	●	●	●	●
KBMS	●	●	●	●	●	●	●	●	●	●	●	●	●
KEE	●	●	●	●	●	●	●	●	●	●	●	●	●
KES			●	●	●	●	●		●	●	●	●	
KET	●										●	●	●
Knossos	●	●		●	●	●	●	●		●	●	●	●
Knowledge Craft	●	●	●	●	●	●	●	●	●	●	●	●	●
Knowledge Tool	●	●	●	●	●	●	●	●	●	●	●	●	●
Kool	●	●	●	●	●	●	●	●	●	●	●	●	●
Leonardo	●				●	●	●	●			●	●	●
Level 5 Object	●	●	●	●	●	●	●	●	●	●	●	●	●
Nexpert Object	●	●	●			●	●	●	●		●	●	●
ProKappa	●	●	●	●	●	●	●	●	●	●	●	●	●
Tablo	●				●	●	●			●	●	●	●
Testbench					●	●					●		
TIRS			●	●	●	●	●	●	●	●	●	●	●
TWAICE	●	●	●	●	●	●	●	●			●	●	●
VAX Dec. Exp.	●	●		●	●	●	●		●		●	●	●
Xi Plus	●	●	●	●	●	●	●	●	●	●	●	●	●
Zeno Expert	●				●						●		●
Zeno Professional	●	●		●		●	●			●	●	●	●

Table 2a: Representation of production rules [BSW 91]

Product	inference mechanisms						sequence of fixing rules
	forward chaining	backward chaining	combined strategies	backtracking (→ conclus.)	time-dep. execution	programmable conflict solution	default methods
ADEPT	●	●	●	●		●	priority
ADS	●	●	●	●		●	ADS-special algorithm
Babylon	●	●	●	●		●	textual sequence
COPERNICUS	●	●	●	●			sequence of rules
DB Expert	●	●	●				
E´G´E´R´i´A	●	●	●	●		●	
ESE	●	●	●			●	special algorithm; CF desc.
Goldworks II	●	●	●				priority-factors
GS Real-Time	●	●	●	●	●	●	rule priority levels
Guru	●	●	●			●	
HUMBLE	●	●	●				
IPW	●	●	●	●		●	
Kappa/PC	●	●	●			●	
KBMS	●	●	●	●		●	priorities
KEE	●	●	●		●	●	agenda (Fifo)
KES	●	●	●	●			
KET	●			●			
Knossos	●	●	●				
Knowledge Craft	●	●	●	●	●	●	special algorithm
Knowledge Tool	●				●	●	priority; "new before old"
Kool	●	●	●	●	●	●	LIFO-mechanisms
Leonardo	●	●	●				
Level 5 Object	●	●	●	●	●	●	dynamic CF
Nexpert Object	●	●	●	●		●	stat. und dyn. categories
ProKappa	●	●	●			●	priorities
Tablo	●						
Testbench							
TIRS	●	●	●	●			
TWAICE	●	●	●	●		●	evidence
VAX Dec. Exp.	●	●	●	●		●	value analysis
Xi Plus	●	●	●	●	●	●	"first rule"
Zeno Expert	●					●	best solution
Zeno Professional	●			●		●	deterministic

Table 2b: Representation of production rules [BSW 91]

Product	formalisms			types of attributes				inheritance properties						
	O-A-tupels	hierarchies	classes, instances	simple data types	strings	enumerations	object instances	not possible	static	dynamic	multiple	changeable	defin. by methods	message passing
ADEPT	•	•	•	•	•	•	•			•			•	
ADS	•	•	•	•	•	•	•		•			•	•	•
Babylon	•	•	•	•	•	•	•				•	•	•	•
COPERNICUS	•	•	•	•	•	•			•				•	
DB Expert														
E´G´E´R´i´A	•	•	•	•	•	•	•			•	•	•		•
ESE	•	•		•	•	•		•						
Goldworks II	•	•	•	•	•	•	•				•		•	•
GS Real-Time	•	•	•	•	•	•	•		•					
Guru														
HUMBLE	•	•	•	•	•	•	•	•					•	•
IPW	•	•	•	•	•	•			•		•	•	•	•
Kappa/PC	•	•	•	•	•	•	•		•			•	•	•
KBMS	•	•	•	•	•				•		•		•	
KEE	•	•	•	•	•	•	•			•	•	•	•	•
KES	•	•	•	•	•	•			•					
KET														
Knossos	•	•	•	•	•	•	•			•	•			
Knowledge Craft	•	•	•	•	•	•	•			•	•	•	•	•
Knowledge Tool	•	•	•	•	•		•			•	•		•	•
Kool	•	•	•	•	•	•	•			•		•	•	•
Leonardo	•	•	•	•	•		•			•				
Level 5 Object	•	•	•	•	•	•	•		•		•		•	•
Nexpert Object	•	•	•	•	•				•	•	•	•	•	•
ProKappa	•	•	•	•	•	•	•		•	•	•	•	•	•
Tablo														
Testbench		•	•	•	•	•	•			•				
TIRS	•	•	•	•	•		•			•	•		•	
TWAICE	•	•	•	•	•	•			•	•		•	•	
VAX Dec. Exp.														
Xi Plus														
Zeno Expert														
Zeno Professional	•			•	•		•		•					

Table 3: Representation of objects [BSW 91]

Knowledge \ Concept	numerical	logical
uncomplete		- non monotonic logic - modal logic
uncertain	- probability theory - certainty factors CF - fuzzy sets	- multi value logic - fuzzy logic
inconsistent	- individual programming	- multiple worlds - individual programming

Table 4a: Techniques for representation of vague knowledge [BSW 91]

Product	processing of vague knowledge: uncomplete reasoning	processing of vague knowledge: uncertain reasoning	propagation of vagueness	programmable strategies
ADEPT	discrediting	fuzzy-sets		yes
ADS		numerical, CF	Mycin-model, threshold values, priorities	yes
Babylon				
COPERNICUS	default values, "unknown"-values	CF	probab. theory	yes
DB Expert	CF, fuzzy IF	fuzzy IF		
E´G´E´R´i´A	logical, numerical	logical, numerical		
ESE	default-reasoning	evidence,confidence values, fuzzy logic	Emycin model	no
Goldworks II				
GS Real-Time				
Guru	CF,"unknown"values fuzzy variables	CF, fuzzy variables	probab. theory	yes
HUMBLE		CF (mycin model)	Mycin model	yes
IPW				
Kappa/PC				
KBMS				
KEE	priorities			
KES	OF	OF	defaults, CF	no
KET	confidence			yes
Knossos		CF, mod. predicates	Mycin model	no
Knowledge Craft				
Knowledge Tool				
Kool				
Leonardo		CF, bayes theorem		no
Level 5 Object		product-space confidence, numerical	confidence values	yes
Nexpert Object	3 value logic			
ProKappa	priorities			
Tablo				
Testbench				
TIRS				
TWAICE	CF	programmable	programmable	yes
VAX Dec. Exp.	3 value logic	3 value logic	variable logic	no
Xi Plus				
Zeno Expert	first all, first best		best sol. propag.	yes
Zeno Professional	quasihomorphol. Inference	quasihomorphol. Inference	best solution propagation	yes

Table 4b: Representation of vague knowledge [BSW 91]

Product	KA support	tool compatibility	supported methods
ADEPT	yes	yes	rule-, frame- generators
ADS	no		
Babylon	yes	yes	rule-, frame- generators
COPERNICUS	yes	no	GRAPES (Graphical description)
DB Expert	no		
E´G´E´R´i´A	yes	yes	rule generator
ESE	no		
Goldworks II	no		
GS Real-Time	yes	yes	rules, objects, frames, graphics, etc.
Guru	yes	yes	rule editor
HUMBLE	K.A.		
IPW	yes		
Kappa/PC	no		
KBMS	yes	no	inductive rule generator
KEE	no		
KES	no		
KET	yes	yes	
Knossos	yes	yes	rules, frames, objects
Knowledge Craft	yes	yes	rule-, frame generator
Knowledge Tool	no		
Kool	yes	yes	rule-, frame generator
Leonardo	yes	yes	frame generator
Level 5 Object	yes	yes	rule-, object editor; K. tree
Nexpert Object	yes	yes	Nextra
ProKappa	yes	yes	spreadsheet
Tablo	yes		
Testbench	yes	yes	frame generator
TIRS	no		
TWAICE	yes	yes	rule-, formula generator
VAX Dec. Exp.	yes	yes	rule tab., flow diagr., rule gen., graphics
Xi Plus	yes	yes	
Zeno Expert	yes	yes	rule-decision tree generation
Zeno Professional	yes	yes	rule generation

Table 5: Knowledge Acquisition (K.A.) [BSW 91]

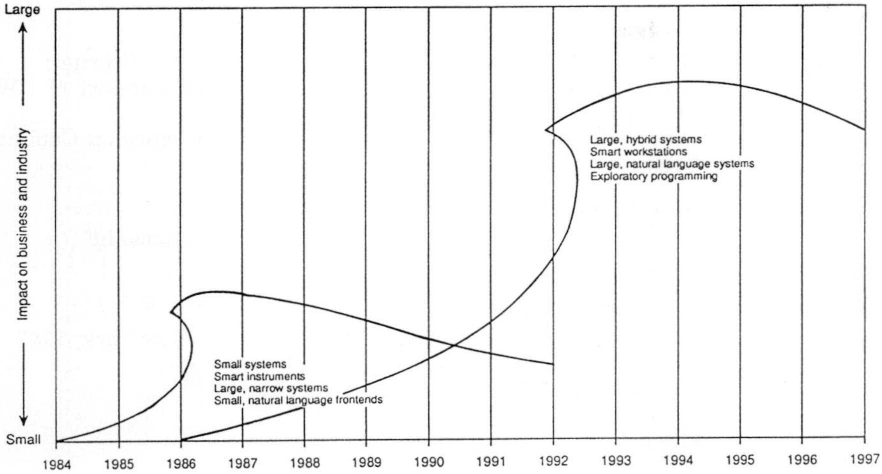

Figure 4: Industrial (partical) application of KBS [Hak 87]

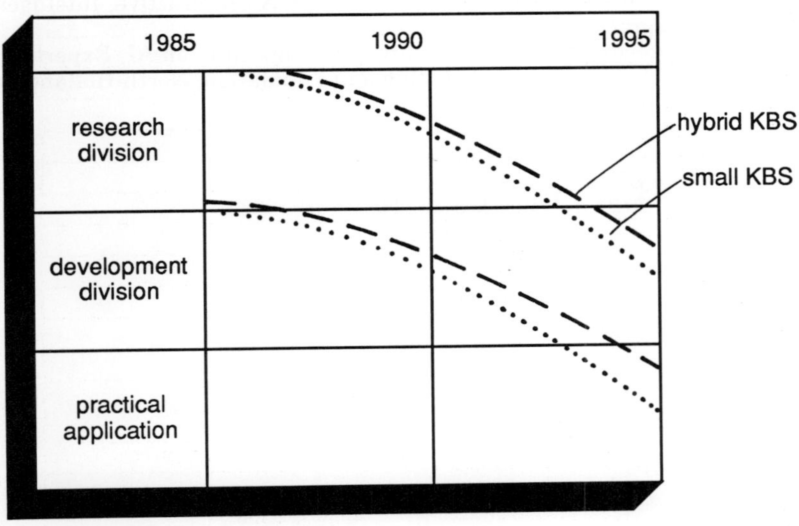

Figure 5: Practical application of hybrid KBS [Hak 87]

References

[BFK89] Berberich; Feicht; Kwee-Christoph; Lauer; Lehmann; Stürner: "An Expert System Approach for Maintenance Support of EWSD Switches" in: Proceedings of ICC '89 - International Conference on Communications; IEEE, Boston, June 1989

[BSW91] v. Bechtolsheim; Schweichhart; Winand: "Expertensystemwerkzeuge - Produkte, Aufbau, Auswahl" Vieweg-Verlag, Braunschweig, 1991

[HaK87] Harmon; King: "Expert Systems"; 2nd Edition, J. Wiley & Sons, New York, 1987

[HLW83] Hayes-Roth; Lenat; Waterman: "Building Expert Systems"; Addison-Wesley Publ. Comp., 1983

[Leh87] Lehmann: "Taxonomy and Application of Expert Systems in Simulation" Proceedings of the International Symposium on AI, Expert Systems and Languages in Modelling and Simulation; North-Holland Publish Comp., 1987

[LeS87] Lehmann; Szczerbicka: "Goal-Directed Modelling with INT^3: An Interactive, Intelligent and Integrated Modelling Environment". Proceedings of the International Symposium on AI, Expert Systems and Languages in Modelling and Simulation; North-Holland Publish Comp., 1987

[SEV87] Shapiro; Eckroth; Vallasi: "Encyclopedia of Artificial Intelligence" (Vol 1+2); J. Wiley & Sons, New York, 1987

Knowledge Based Hybrid Systems (B-11)
I. Mezgár and P. Bertók (Editors)
Elsevier Science Publishers B.V. (North-Holland)

CONCURRENT ENGINEERING: MODELS AND SOLUTION APPROACHES

Andrew Kusiak

Intelligent Systems Laboratory, Department of Industrial Engineering, The University of Iowa, Iowa City, IA 52242 - 1527, USA

Abstract

Products and components are frequently traditionally designed without considering constraints imposed by a manufacturing system. In a concurrent engineering approach, design and manufacturing engineering are viewed as an integrated area. Perhaps the first and the most visible relationship between design and manufacturing has been explored by the design-for-automated assembly methods. Most of the design-for-manufacturing literature does not consider constraints related to manufacturing systems, but rather deal with manufacturing processes. Although the area of design-for-manufacturing processes is still evolving and remains an important effort within concurrent engineering, there is a need to look at other aspects of manufacturing. An aspect of manufacturing that offers a high potential for improvement of productivity is concerned with design-for-manufacturing operations. In fact the cost of operating a manufacturing system can be significantly reduced if enough consideration is given to design of products and components. Improvements in design processes should be accompanied by effective design of manufacturing systems. The design-for-manufacturing approach for each of the first two scenarios implies that a product should be designed for a manufacturing system that does not exist. Based on the principles of concurrent engineering, a product should be designed for a manufacturing system, as much as the manufacturing system should be designed for the product. Compromising the design of a product and design of a manufacturing system should improve the competitiveness of the industry. Manufacturing systems and processes should not just be viewed as passive constraints in design of products and components but remain active.

Key Codes:I.2.1; F.1.1;J.6
Keywords:Applications of Expert Systems, Models of Computations, Computer Aided Engineering, Concurrent Engineering

1. INTRODUCTION

Products and components are frequently designed without considering constraints imposed by a manufacturing system. With the introduction of concurrent engineering, design and manufacturing engineering are viewed as an integrated area. Perhaps the first and the most visible relationship between design and manufacturing has been manifested by the design-for-automated assembly (see for example, Boothroyd et al. 1982). Most of the literature on design-for-manufacturing does not consider constraints related to manufacturing systems, but rather deal with manufacturing processes. Although the area of design-for-manufacturing processes is still evolving and remains an important effort within concurrent engineering, there is a need to look at other aspects of manufacturing.

An aspect of manufacturing that offers a high potential for improvement of productivity is concerned with design-for-manufacturing operations. In fact the cost of operating a manufacturing system can be significantly reduced if enough consideration is given to design of products and components. Improvements in design processes should be accompanied by effective design of manufacturing systems.

2. LITERATURE REVIEW

The literature concerned with concurrent engineering is rather extensive, however, the author is not aware of any references discussing to a meaningful extend the research outlined in this paper. Numerous references either mention the topics of this proposal or are indirectly related to this proposal. Some of the representative papers are reviewed next.

Corbet (1986) summarized the main issues involved in design for economic manufacturing as follows: establish the specification and target cost; identify the likely high cost areas; supply cost target information to design, manufacturing and purchasing; prepare preliminary design, with inputs from manufacturing engineering; supply target cost to appropriate manufacturing and purchasing areas; and continue development, ensuring that costs incurred by modifications do not increase the target costs.

Stoll (1990) identified the following two major goals of concurrent design: parallel development of the product and the manufacturing process, and evaluation of multiple product and process design alternatives. He defined concurrent design as a five-step process: step l: definition of functional requirements and constraints of the design problem and development of manufacturability design goals: step 2: optimization of current product design methods using the principles of design-for-manufacturing and methodologies such as parts reduction assessment and robustness assessment; step 3: identification of innovative new product and process concepts that are easy to manufacture; step 4: evaluation, combination, and refinement of the results of step 3 into one or more 'best' concepts: and step 5: selection of one or more 'best' designs from the list preferred concepts developed in step 4. A concurrent design process is supported by a number of different methodologies and tools that include design axioms, design-for-manufacturing guidelines, manufacturing process design rules, group technology, failure mode and effects analysis, and value analysis.

Foreman (1990) identified four major categories of tools that make a concurrent design process works: total quality control (Taguchi method, quality function deployment, process control planning), computer integrated manufacturing (simulation and analysis, electronic data interchange, networks and data communications, group technology, value engineering, solid modeling), just-in-time productivity improvement (design-for-manufacturing / assembly, synchronous manufacturing, continuous improvement programs) and human systems (project management, communications, problem solving).

El-Gizaway and Hwang (1989) developed a strategy for concurrent product and process design for aerospace components. This strategy involves the following activities: development of an initial design concept, definition of product constraints, knowledge-based manufacturability analysis, tentative process selection, product redesign for manufacture and modular modeling of products and processes. The product design is evaluated with respect to productivity and cost objectives.

Cutkosky and Tenenbaum (1990) described a methodology for concurrent product and process design and its implementation in a computational framework. The methodology assumes simultaneous design of a part or assembly and the process for producing it. The computational framework actually resembles a concurrent engineering team. It consists of a number of knowledge sources (human experts and expert systems) for simultaneous execution of various life-cycle activities of the product.

Assemblability is one of the most important life-cycle values of a product. The most popular approaches for design-for-assembly are the Hitachi assemblability evaluation method (Miyakawa et al. 1986) and the design-for-assembly (Boothroyd et al. 1982).

The assemblability evaluation method (AEM) was developed to improve the design quality for better assembly. The AEM identifies weak points in the design by assessing the difficulty of assembly operations through an assemblability evaluation score, and calculating the estimated assembly cost. The complete assembly operation of a design is a combination of elementary operations that are represented by symbols. There are about 20 elementary assembly operations, each assigned a penalty score is a function of penalty scores of all elementary assembly operations involved. The estimated assembly cost is evaluated to verify the improvement of the design quality.

Design-for-assembly (DFA) is a method for quantitative assembly evaluation. The method involves two major steps: determine whether a part can be eliminated from the design in order to minimize the number of parts, and estimate the handling and assembly costs for each part to determine whether proposed design changes save in assembly cost. The handling and assembly cost is based on the type of the assembly process used-manual, robotic or high-speed automatic.

Brown et al. (1991) proposed an integrated approach to quality engineering in support of design-for-manufacturing. The approach is based on the use of QFD for specification of customer attributes, product features and functions, component characteristics, process characteristics and production operations. The QFD cascade allows for carrying new, difficult to meet and important requirements from one stage to the next with more detailed investigation performed at each subsequent stage.

Rehg et al. (1988) described a concurrent design system that integrates various life-cycle values of a product into the design process. The system framework consists of three types of design knowledge sources: design agents, program modules for performing design activities: design critics that are used for analysis of an involving design from different aspects: and design translators, responsible for mapping a design representation into another.

Ishii et al. (1993) proposed a general framework for the development of knowledge-based systems that allow for applying concurrent design approach in the area of mechanical design. The underlying concept of this framework is called design compatibility analysis. The concept allows for analysis of the compatibility of a proposed design with design requirements and constraints and analysis of the proposed design elements compatibility within the system. It also provides a justification for results of compatibility analysis and suggest design improvements.

Development of a prototype system for concurrent design based on the principle of cooperative work of specialists was described in Altenhof et al. (1989). The system, which allows for incorporation of reliability, maintainability and supportability requirements in a concurrent design process, consists of five elements: a human-element (an individual or a team' of experts that interacts with the system through a user interface), a CAD/CAE system for design of products, an estimator/predictor elements which receives the design feature and determines values of design dependent parameters, an evaluator element for evaluation of the current state of the product design and a database of economic and physical factors.

A computer-aided environment for concurrent design was presented by Subramanyan and Lu (1989). The environment, based on the model-based reasoning concept, provides capabilities for feature-based product modeling, manufacturability facility modeling, process design and manufacturability analysis.

Thompson and Lu (1989) developed a methodology for representing and implementing the design rationale as a concept that enables concurrent execution of product and process design activities. The concept is illustrated by designing a prototype computer-based environment for concurrent design. The design rationale is represented in terms of design plans for future product and process activities and design constraints detected during the design process. The design rationale supports the four most important concepts in the system framework: product specifications, product designs, process designs, and manufacturing resources.

Molloy et al. (1991) presented a rule-based approach for representation of design-for-assembly knowledge. The feature-based part representation is used for DFA analysis at the part level, ensuring that the part is easily handled and oriented, and at the product level for generation of the assembly sequence. The assembly sequence is determined by generating its disassembly sequence. The disassembly sequence is generated by identification of movable directions of each part.

Knight and Kim (1991) described a system for assisting human experts in specific domains in performing concurrent design. The system applies the QFD method for product and process design. Four basic modules of the system perform the concept selection, component design, manufacturing process selection and process parameter selection activities. Each of these activities is divided into three phases: specification of the problem, generation of possible solutions and selections of the best alterative.

The most recent coverage of many topics in concurrent engineering is presented in Kusiak (1992, 1993).

3. HYBRID-BASED SYSTEMS

Optimization approaches have been used for solving manufacturing problems, however, they are not always easy to apply because:

- the data required by the optimization algorithms may not be easily available
- their scope of applicability may be limited
- human expertise might be required

- the algorithms are often not able to provide optimal solutions for industrial problems due to their complexity.

On the other hand, knowledge-based systems have typically been used to solve problems that are either too complex for mathematical formulation or too difficult to solve using optimization approaches. Some of the difficulties encountered when using optimization techniques may be offset by combining them with knowledge-based systems. Traditionally, human experts have solved problems using the optimization approach. If knowledge-based systems are to be used successfully, they should be used in such a way that they replace human experts rather then optimization approach.

Based on their operational mode, two classes of knowledge-based systems are identified (Kusiak 1990):

- stand-alone knowledge-based systems, and
- tandem knowledge-based systems.

Stand-Alone Knowledge-Based System

A knowledge-based system in the stand-alone mode uses data and constraints pertinent to the problem and solves it using procedures similar to those used by human experts. It does not use the optimization approach which involves modeling the given problem and solving the model using an optimal or a heuristic algorithm. Many existing knowledge-based systems fall into the stand-alone class.

Tandem-Based System

The tandem knowledge-based system, on the other hand, combines the optimization approach with the knowledge-based system approach for problem solving. It can be thought of as a knowledge-based subsystem linked to a model and algorithm base. The basic approach utilized in a tandem knowledge-based system is as follows: A suitable model is either selected or built for the given problem. To solve the model, an optimal or heuristic algorithm (available in the algorithm base) is selected or built. There are four classes of the tandem knowledge-based systems, namely:

- data modifying system
- model based system, and
- model building system
- algorithm building system.

The basic approach in all the four classes of the tandem knowledge-based system is the same, i.e. they use optimization approach to solve the problem considered. However, the actual problem solving approach in each class is different. The main function in a data modifying system, is to modify, i.e. generate or reduce data (from among available data) as required by the model which is selected by the system. In a model based system, a suitable model and algorithm is selected for the given problem. The model building system attempts to construct an appropriate model or to modify the selected model, for example, add or delete constraints. In the algorithm building system, the system constructs an algorithm for solving the problem.

In each of the above mentioned classes of the tandem knowledge-based system, a suitable algorithm for solving the model is selected or built (in the case of algorithm building system).

Data Modifying System

In a data modifying system, first a model is selected for the problem considered. If the model selected requires:

- additional data to be generated, or
- a subset of data to be extracted from the available set of data

then the knowledge-based system performs the function required, i.e. data generation or data reduction. The resulting data is then used in the model. Next, an appropriate algorithm is used for solving the model.

Model-Based System

The model based system is suitable for problems which can be formulated using a number of models, each model being suitable for a particular situation. For a given problem, the knowledge-based system first selects an appropriate model and then an algorithm. The problem is solved by the algorithm and the solution produced is evaluated. If the solution is implementable, the knowledge-based system accepts it. For example, in the case of the machine layout problem, the solution (layout) is implementable if space constraints are satisfied and adjacency requirements are met in the layout produced by the knowledge-based system (Kusiak 1990). If the solution is not implementable, then the system may take one of the following actions:

- modify certain parameters in the algorithm (if possible) and apply the algorithm again to the problem in order to generate new solution, check whether it is implementable and repeat the above procedure until an implementable solution is obtained,
- modify the solution in order to make it implementable.

Of course, alternative (i) may not be applicable to all algorithms. Even if it is applicable to a particular algorithm, the corresponding parameter can be modified only to a certain extent, beyond which any modification fails to produce acceptable solutions. In such a case, i.e. when the parameter(s) in the algorithm cannot be modified any further, and if the solutions produced thus far are not implementable, the knowledge-based system adopts alternative (ii) mentioned above. Note that the system may also use alternative (i) to improve the current solution. The knowledge-based system for machine layout KBML discussed in Kusiak (1990) uses alternative (i) to improve the current solution and alternative (ii) to make a solution implementable (if necessary).

Model Building System

Knowledge-based systems belonging to this class are, in general, more difficult to develop because they require a careful consideration of the knowledge base. The task of building or at least modifying a model to suit the problem is a difficult one even for human experts, since a number of factors such as data availability, ease of applying algorithms, constraints, etc. have to be examined. One should realize that it is difficult to develop a knowledge-based system, which like a human expert is capable of building or modifying a model to suit the problem considered.

The model building approach involves two steps. In the first step, the knowledge-based system either constructs an appropriate model for the given problem or selects one from the available set of models. In the second step, it determines whether there are any algorithms available to solve the model. If no appropriate algorithms are found, then the system either:

- modifies the current model such that a suitable algorithm for solving it is available;
- constructs another model for the problem.

If the knowledge-based system is unable to do either of the two above, then it considers modifying the algorithm or parameters within the algorithm, so that the problem can be solved. Of course, modification of the solution generated by the algorithm, in order to make it implementable or to incorporate qualitative factors, is done, if necessary.
In the data modifying and model based systems, the models are simply selected from the model base and are not manipulated. Hence their representation is not an important issue. But, in a model modifying system, since models are constructed or modified, their representation should be given careful consideration.

Algorithm Building System

The algorithm building knowledge-based system should have the ability to build an algorithm for solving the problem. During the algorithm execution process, the algorithm and the knowledge-based subsystem closely interact. The knowledge-based subsystem may, for example generate search directions for the algorithm.

Most of the knowledge-based systems presented later in this book fall into more than one of the four classes of tandem systems. For example the knowledge-based system for group technology (KBGT) presented in Kusiak (1990) is using the model based as well as algorithm building approach. Although, at the current stage of development KBGT none of the algorithms is built by the system, however one of them closely interact with the knowledge-based subsystem.

4. CONCURRENCY ANALYSIS

Consider the design process of an electronic product that involves thousands of activities (the example is taken from an industrial company). The entire design process is divided into seven phases. Phase 3, product definition/requirements allocation, is used to illustrate the concepts presented in the paper.

The design activities involved in Phase 3 are represented as the process graph shown in Figure 1. The shaded nodes in Figure 1 are the review activities. The matrix representation of the design process is shown in Figure 2. Applying the concurrency algorithm to the matrix in Figure 2, the matrix in Figure 3 is obtained.

Five groups of activities are identified: G-1 = {1, 2, 3, 4, C1, 14, 20, 25}, G-2 = {6, C2, 15, 28, 31}, G-3 = {7, 13, 22, 30}, G-4 = {16, 17, 29}, G-5 = {C3, 21, 32}, and G-6 = {23}, where C1 = {12, 24}, C2 = {8, 5, 9, 10, 11, 26}, and C3 = {18, 19, 27} are coupled activities. It is observed that the dependencies among groups of activities are indicated by elements outside the block diagonal matrices. For example, activities 2 and 4 in group G-1 should precede activities 6 and 9 in group G-2 respectively. Similarly, activities 12 and 24 in group G-1 have to precede activities 15, 28, and 31 in group G-2. One has to complete the activities 2, 4, 12, and 24 as soon as possible in order to begin the downstream activities in G-2. The groups of activities identified can be performed in parallel, if their predecessors have been finished.

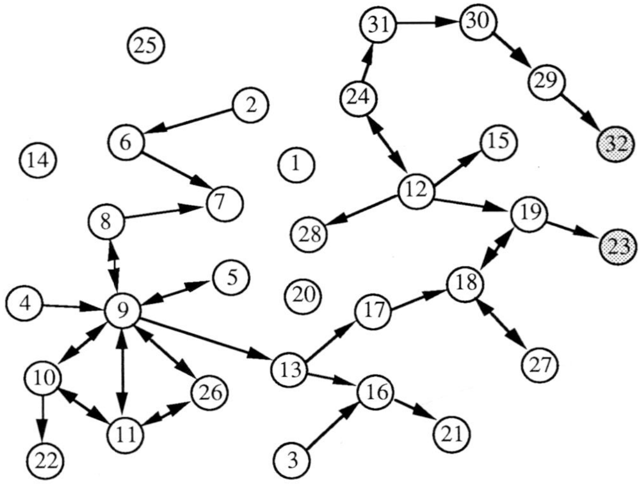

Figure 1. The process graph of Phase 3

The six groups of activities are represented as a process tree (Figure 4). A link between any two groups (levels) represents dependency between the corresponding groups. It is important to show activities in the upper level that impact the lower-level activities. For example, there are some dependencies between level 1 (G-1) and level 2 (G-2) activities. In order to begin the activities in level 2, activities 2, 4, 3, 13, 12 and 24 in level 1 have to be completed as they impact the level 2 activities. A manager should monitor these activities in order to avoid delay in information transfer to the lower-level activities.

Note that the temporal relationship among groups of activities leads to the Gantt chart in Figure 5. Some activities begin when their predecessors at the upper-level have been completed.

A dependency path is defined as the maximum length chain of activities in the process tree. According to this description, it is observed that the dependency path is "4 - C2 - 13 - 17 - C3 -23". A project manager should focus on monitoring activities on the dependency path.

Concurrency analysis detects potential activities that can be performed in parallel. One may attempt to remove or redefine the dependency among groups of activities in order to increase the degree of concurrency. For example, if the dependency between activities 2 and 6 in Figure 3 is removed, activity 6 can be performed in the first level. The concurrency of the design process increases. Note that the temporal relationship between groups of activities indicate that the groups overlap.

	1	2	3	4	5	6	7	8	9	10	11	12	13	14	15	16	17	18	19	20	21	22	23	24	25	26	27	28	29	30	31	32
1	+																															
2		+																														
3			+																													
4				+																												
5					+				*																							
6		*				+																										
7						*	+	*																								
8								+	*																							
9				*	*			*	+	*	*															*						
10									*	+	*																					
11									*	*	+															*						
12												+												*								
13									*				+																			
14														+																		
15												*			+																	
16			*										*			+																
17													*				+															
18																	*	+	*								*					
19												*						*	+													
20																				+												
21																*					+											
22										*												+										
23																			*				+									
24												*												+								
25																									+							
26									*		*															+						
27																		*									+					
28												*																+				
29																													+	*		
30																														+	*	
31																								*							+	
32																													*			+

Figure 2. The activity-activity incidence matrix corresponding to the process graph in Figure 4

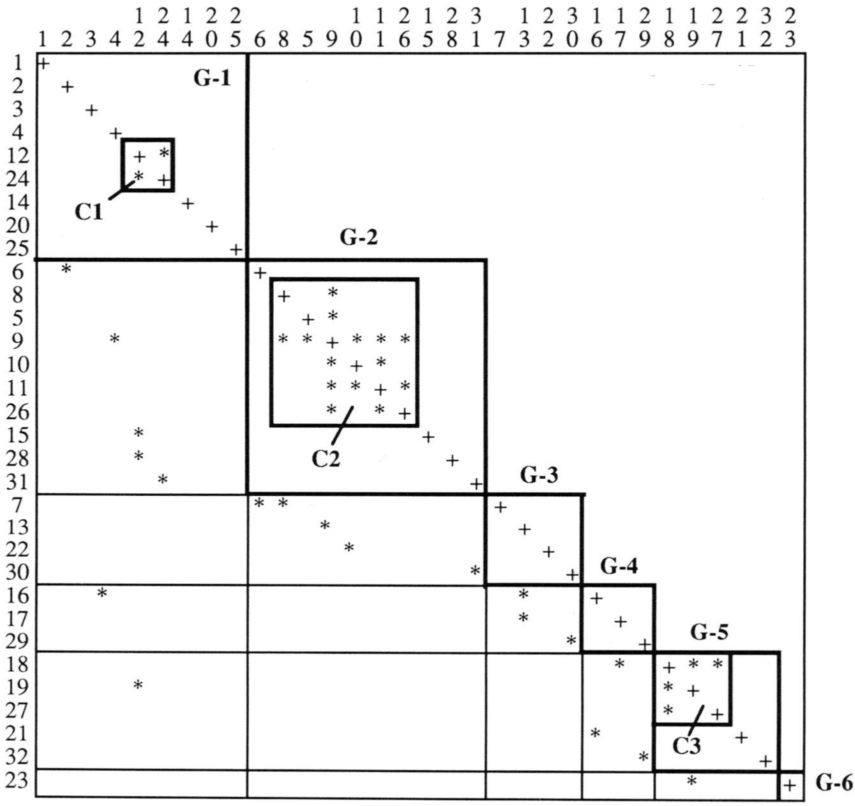

Figure 3. The triangularized activity-activity incidence matrix corresponding to the matrix in Figure 2

The benefits of the concurrency analysis are as follows (Kusiak and Wang 1993):

- Potential groups of activities that can be performed in parallel are determined.
- A critical dependency path is determined without considering the time aspect.
- The duration of product development time may be reduced, due to the increased degree of concurrency of the design process.

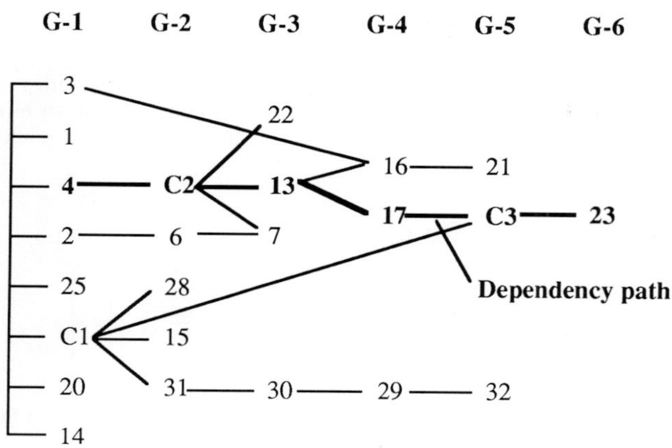

Figure 4. A process tree of Phase 3

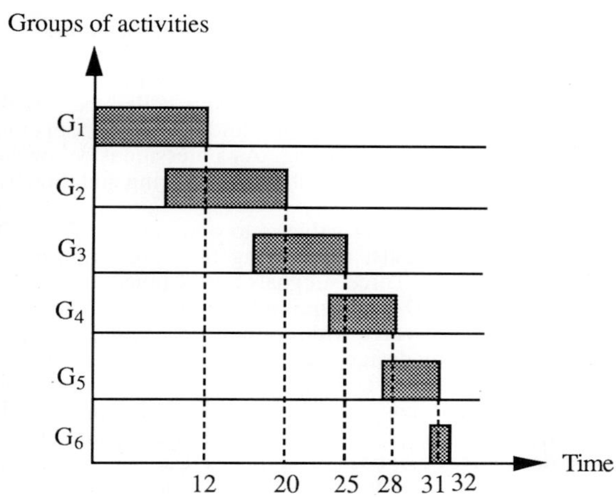

Figure 5. Gantt chart of the design process

5. THE VALVE DESIGN PROBLEM

In this section, the concept of design negotiation is illustrated with the example of a poppet relief valve (Lyons 1982). Figure 6 shows the schematic of a poppet relief valve which includes a poppet valve, poppet valve stem, and helical compression spring enclosed in a pipe (Kannapan and Marshek 1993).

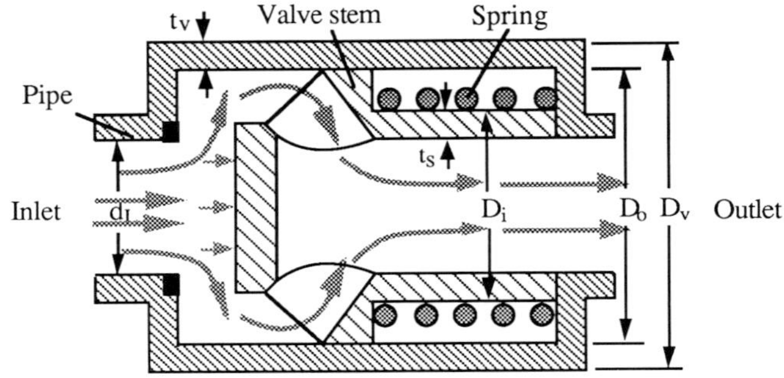

Figure 6. Schematic of a poppet relief valve

The poppet relief valve allows flow of the fluid from the inlet to the outlet when the pressure of the fluid exceeds a certain threshold pressure called the "cracking" pressure. If a pressure is greater than the cracking pressure the fluid opens the poppet valve and holds it in equilibrium against a helical compression spring. As a pressure is below the cracking pressure, the poppet valve is held against a seal due to the helical spring and thereby cuts off fluid flow from the inlet to outlet.

The goal of the design of a relief valve is to optimize the flow. This overall goal is expressed in terms of three subgoals: minimizing flow area, minimizing valve size, and minimizing the volume of spring. Three subgoals are assigned to three design agents (DAs) respectively: valve DA, enclosure DA, and spring DA (Figure 7). The valve DA is responsible to determine the configuration of valve with optimal flow based on the valve requirements. After completing the configuration design, valve DA passes some parameters to the spring DA and the enclosure DA. The spring DA is to design a spring to determine the cracking pressure, the distance the poppet will move, and the stability of the seal in the closed position. The size and the thickness of stem and enclosure are determined by enclosure DA. Note that valve DA dominates the entire design process, because it makes design decisions first and transfer the required information to the agents in downstream design process. In collaborative design, all agents are assumed to be benevolent, i.e., they assist each other to reach a compromise solution.

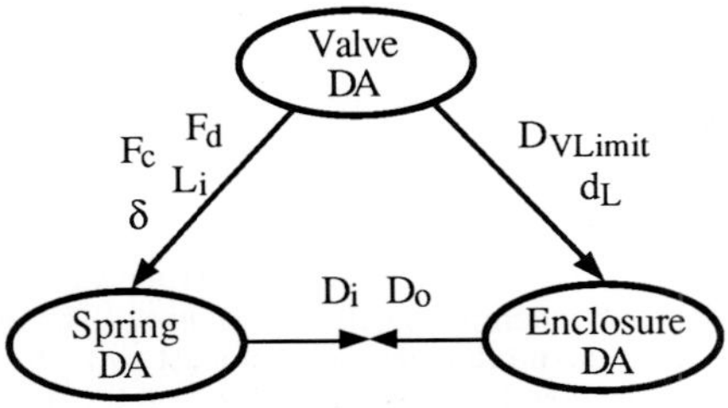

Figure 7. Collaboration among valve DA, spring DA, and enclosure DA

The variables and goals for each design agent are defined as follows:

Valve DA:

- Decision variables: d_L, C_v, C_f, K, L_i
- Aspiration levels: D_{vLimit}, L_{iLimit}
- Intermediate variables: d_o, F_c, d_{eo}, d, F_d
- Goal: $L_i \leq L_{iLimit}$, $D_v \leq D_{vLimit}$

Enclosure DA:

- Decision variables: D_V, S_p, A_1, A_2, t_s
- Aspiration level: A_{1Limit}, A_{2Limit},
- Intermediate variables: t_v, D_i, D_o
- Performance variable: D_V
- Goal: $D_V \leq D_{VLimit}$

Spring DA:

- Decision variables: D, d, G, r_{s1}, r_{s2}
- Aspiration levels: $r_{s1Limit}$, $r_{s2Limit}$, V_{HLimit}, $B_{SLimit1}$, $B_{SLimit2}$, V_{SLimit}
- Intermediate variables: K, C_s, K_s, S_s, N, L_s, L_f, D_o, D_i, C_{l1}, C_{l2}
- Performance variables: V_S, B_s
- Goal: $V_S \leq V_{SLimit}$, $B_S \leq B_{SLimit}$

Note that D_v is determined by the enclosure DA and it is passed back to the valve agent who determines only an upper-bound of D_v. The aspiration level is defined as an acceptable range of a variable. It is assumed that each agent has its aspiration level that secures its interest.

The constraints and design variables are listed in Kannapan and Marshek (1993). The goal of each agent provides only a guidance for design. Decisions made in the upstream design process have impact on the decisions in the downstream design process. For example, the decisions made by valve DA impact the decisions made by the spring DA and enclosure DA.

6. DESIGN-FOR-RELIABILITY

Manufacturing engineers should design manufacturing systems that are highly reliable. The performance of a manufacturing system should not deviate from its nominal value. One way to increase the reliability of a manufacturing system is by providing duplicate manufacturing capabilities, not necessarily the duplicate machine tools. In order to design a reliable system, various design options need to be carefully evaluated before the final design decision is reached.

To illustrate the reliability issue in design of manufacturing systems assume that the following processes have to be performed in a manufacturing cell (MC).

1. Milling (M)
2. Drilling (D)
3. Turning (T)

Due to the nature of the process plans for the parts designed, the machines have to be arranged in a series: {Milling machine, Drilling machine, Turning machine}. It has been estimated that in order to meet the product demand, at least two copies of each machine have to be purchased. A question arises how to arrange machines on the manufacturing floor.

The following designs are considered:

1. Place one of each milling machine, drilling machine, and turning machine in series in one manufacturing cell and provide two of such cells (see Figure 8).
2. Place two machines of each kind in parallel and form a series combination of these parallel units (see Figure 9).

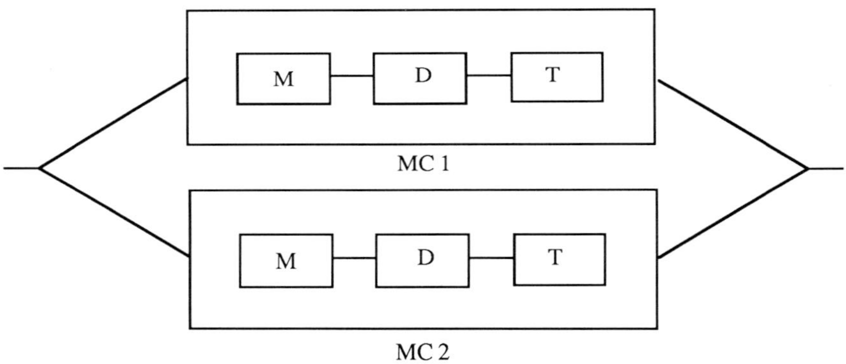

Figure 8. Design 1; each of the two manufacturing cells contains a series of three machines

Assume that the probability that the machines {M, D, T} will function without failure is {0.9, 0.95, 0.8}, respectively.
The two design options can be evaluated on the basis of the reliability theory.

Design 1:
It is required that at least one of the two manufacturing cells functions.
P{at least one of the two cells functions}
= 1 - P{neither of the two cells functions}
$= 1 - [(1 - r(\mathbf{p}))\ (1 - r(\mathbf{p}'))]$
where: p_i (p'_i) is the probability that the machine number i that belongs to machine cell MC1 (MC2) functions

P{at least one of the two cells functions}
= 1 - [(1 - (0.9· 0.95 ·0.8)) (1 - (0.9·0.95 ·0.8))]
= 0.90

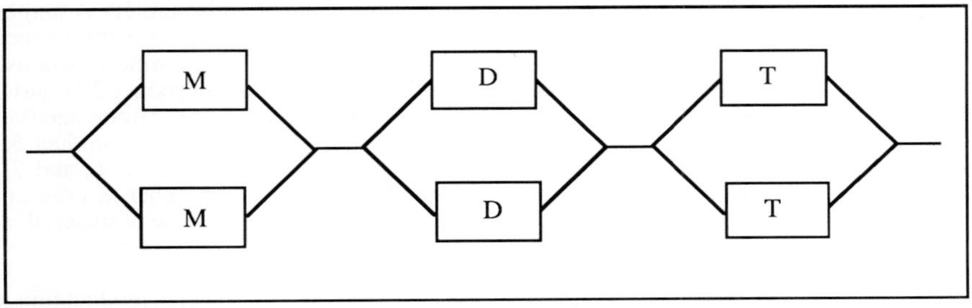

Figure 9. Design 2; arrange two machines of each kind in parallel and form a series of these parallel machines

Design 2:
It is required that at least one of the two machines of each type functions.
P{the machine cell functions}
= r[**1** - (**1** - **p**)(**1** - **p**')]
= (1 - (0.1)(0.1)) (1 - (0.05)(0.05)) (1 - (0.2)(0.2))
= (0.99) (0.9975) (0.96)
= 0.948

From the calculations above, it is clear that replicating machines leads to a system with higher reliability than replicating cells. The design of a manufacturing system presented in Figure 9 is more reliable than the design in Figure 8.

7. DESIGN-FOR-RECONFIGURABILITY, REPEATIBILITY, AND FLEXIBILITY

The ability of a manufacturing system to be reconfigured as the production mix changes is closely related to the type of manufacturing resources (hardware and software) selected and the arrangement of hardware on a manufacturing floor. In order to design a manufacturing system that can be reconfigured to meet the changing production demand, one has to understand the relationship between the structure of products and the structure of a manufacturing system.

The reconfiguration of a manufacturing system may involve relatively minor changes to the system such as, for example, reassigning of operators, retooling of machines, or rearrangement of machine tools. To avoid frequent rearrangement of machine tools (which are expensive, especially, in metal cutting and mechanical assembly processes), manufacturing systems have to be designed to allow for reconfiguration at minimum cost and physical movement of equipment. Also, the reconfigured system when brought back to its original configuration should assure repeatibility of production, i.e., manufacture products or components that meet all specifications regarding quality, standards, and so on.

One way to come up with a reconfigurable system is to design a core manufacturing facility that would be stable in time. Such a facility should be supported by machine tools, stations, and other resources that could be easily rearranged.

As an example, consider the manufacturing system shown in Figure 10. A number of products are manufactured in the core manufacturing facility jointly with the supporting machine tools. The supporting machine tools can be arranged in 8 possible positions (see Figure 10). The layout of the manufacturing floor for product 1 and product 2 is shown in Figure 11 (a). Machines 1, 2, 3, 4, and 5 are used. The manufacturing floor needs to be reconfigured when the product mix is changed to product 3, product 4 and product 5. The machines required for manufacturing the new product mix are machines 1, 6, and 7. The suitable positions for machines 1, 6, and 7 are found according to a number of criteria. The layout after reconfiguration is shown in Figure 11 (b). The direction of flow of material is also changed in the new layout.

The reconfiguration involve assigning machines required for the new product mix to the 8 available locations with certain constraints. The machines and locations can be represented as vertices of a bipartite graph and the problem of assigning machines to these locations can be represented as a bipartite matching problem. Edges are drawn from a machine vertex incident on to the feasible location vertices (see Figure 12). The feasible location for each machine can be determined based on the material handling system used. The weight assigned to each of the edges drawn between machine vertex and the location vertex is determined based on the cost of material movement. The solution of the matching problem provides the minimum cost assignment of machines to the locations.

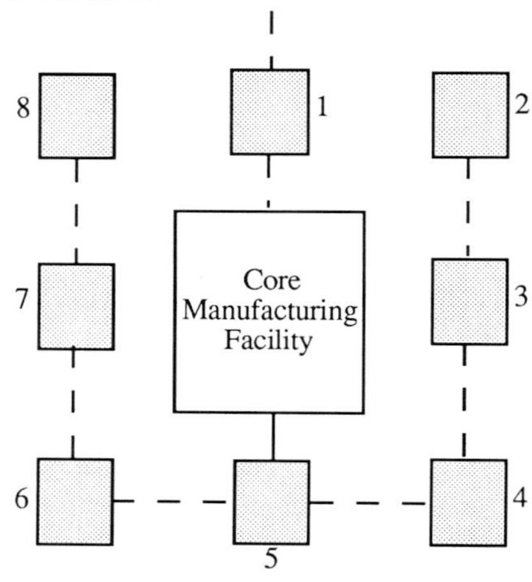

Figure 10. The core manufacturing facility with 8 possible positions for supporting machines

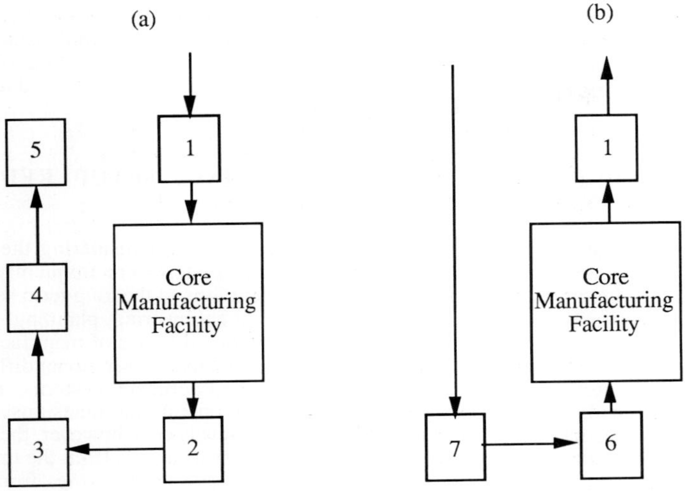

Figure 11: a) Machine layout for product 1 and product 2
b) Machine layout after reconfiguration

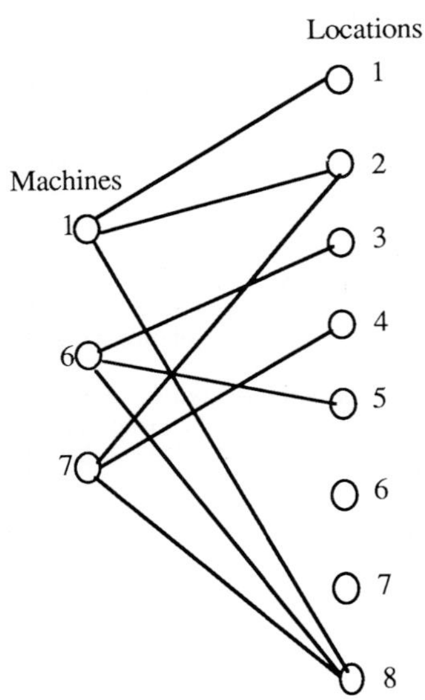

Figure 12. The reconfiguration problem represented as a bipartite matching problem

The approach described above provides only a conceptual framework for dealing with design of products and components for reconfigurability and repeatibility. Industrial scenarios of this problem involve a number of qualitative constraints that have to be considered jointly with a number of quantitative constraints. An artificial intelligence approach can be used to handle the qualitative constraints (see, for example, Kameyama 1993).

8. DESIGN-FOR-PRODUCTIVITY, SCHEDULING, SETUP REDUCTION, AND INVENTORY MANAGEMENT

Products and parts have traditionally been designed without considering the operations aspect of their manufacture. Designs of manufacturing systems are frequently driven by savings of the capital investment without giving considerations to the long term savings from operating those systems. With the advances in concurrent engineering, planning, scheduling, inventory management, setup reduction, and other operational issues of manufacturing have become an important design consideration. It appears that many of operational difficulties can be eliminated by proper design of products and manufacturing systems, rather than development of costly and complex software for making operational decisions. One has to remember that design has a great impact on the product life-cycle cost, however, the design cost is only a fraction of the product life-cycle cost. The short term savings from the investment in the manufacturing resources should not overshadow the long term savings from operating manufacturing systems. The impact of design decisions on the operating cost of manufacturing systems might be difficult or even impossible to reverse without costly redesigns.

9. DESIGN-FOR-PRODUCTIVITY

In order to discuss the impact of design decisions on operational issues, such as the setup cost, material handling cost, and throughput rate, one has to understand the relationship between the structure of products and the structure of a manufacturing system. This relationship is discussed next.

Consider design of a product (assembly) with a number of assembly operations. For each product, a production volume has been determined. Based on the type of operations and the relationship among them, an assembly system is designed. It is important to estimate early in the design process the impact of the structure of products on the structure of an assembly system.
Modifying the structure of a product being designed, one can generate a desired design of an assembly system. In turn, the extent to which the structure of products should be modified is determined by the analysis of the structure of an assembly system. Consider the following example.

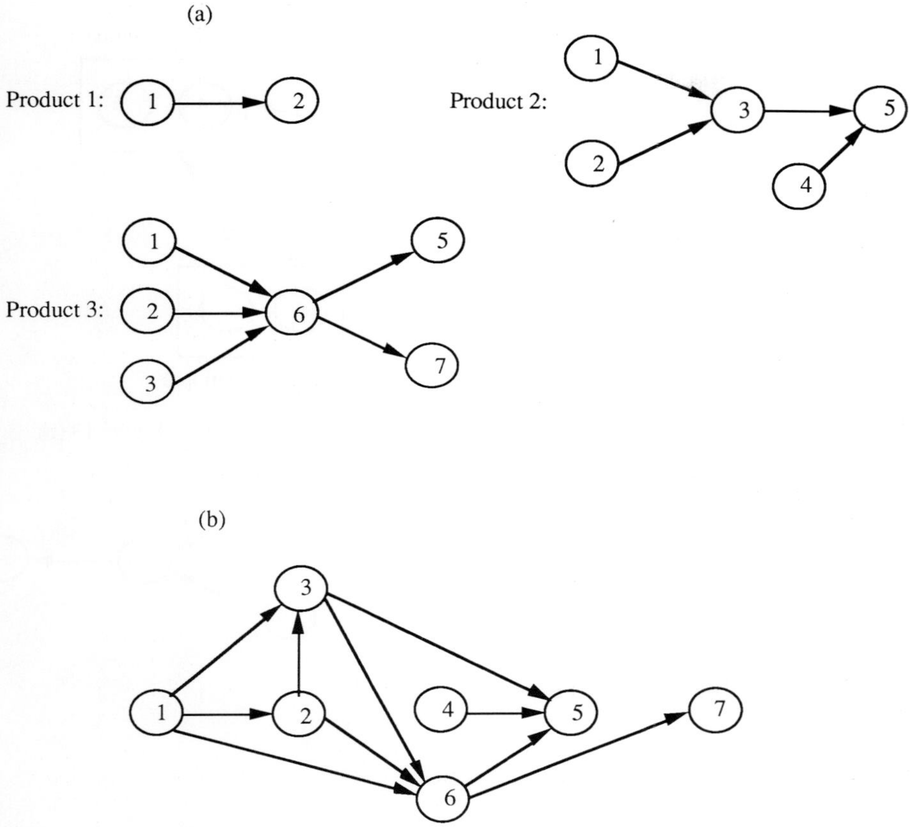

Figure 13: a) Directed graph G_p for each product p = 1, 2, 3, 4
b) The superimposed graph G

Four products with the assembly structure shown in Figure 13 (a) have been designed. The structure of a product p = 1, 2, 3, 4 is represented by a directed graph G_p. Each node in the directed graph G_p represents an assembly operation required by product p and an edge represents the precedence relationship between two assembly operations.

Since the four products are to be assembled in the same assembly system, the graphs in Figure 13 (a) can combined into a superimposed graph G (Figure 13 (b)). Based the superimposed graph G in Figure 13 (b), a design of the assembly system in Figure 14 is obtained. Assume that the assembly system in Figure 14 is balanced, i.e., the capacity constraint at each assembly station is not exceeded.

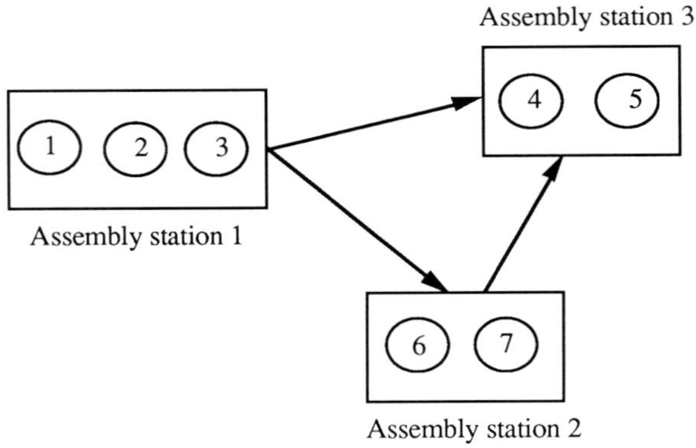

Figure 14. A design of assembly system generated from graph G in Figure 13 (b)

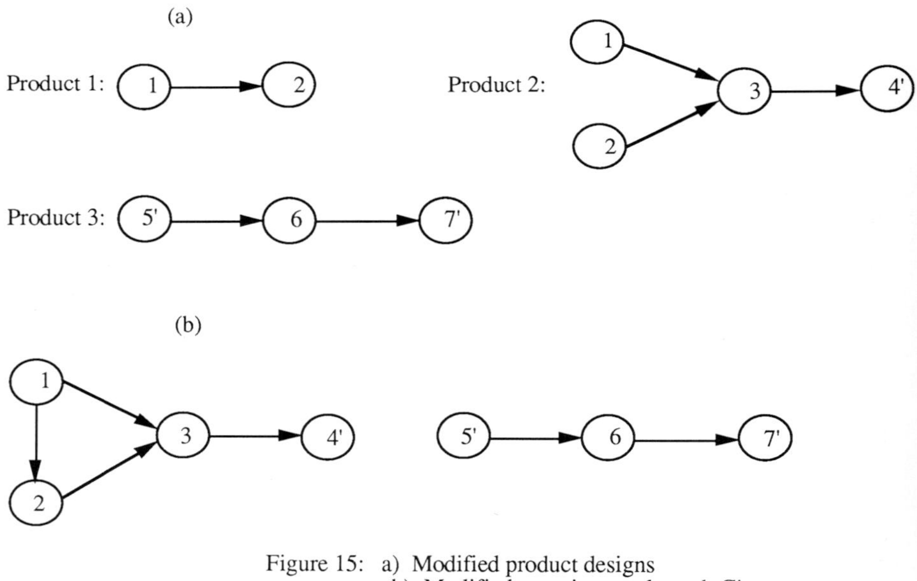

Figure 15: a) Modified product designs
b) Modified superimposed graph G'

Analyzing the design in Figure 14 and the product structures in Figure 13 (a), one can see that the design of assembly system in Figure 14 can be improved by modifying the structure of some products. Assume that the structure of product 2 can be modified by eliminating operations 4 and 5 and incorporating an operation 4', and the structure of product 3 can be modified by eliminating operations 1, 2 and 3, and adding operation 5' and 7'. The modified product designs and the corresponding superimposed graph G' are shown in Figure 15.

The modified design of the assembly system is shown in Figure 16. Since the flow between assembly stations 1, 2, and 3 has been eliminated, the new assembly system includes two stations within some duplicate operations. The operational characteristics of the new system (Figure 16) such as the setup cost, material handling cost, throughput rate, reliability, and so on are far superior than the design in Figure 14.

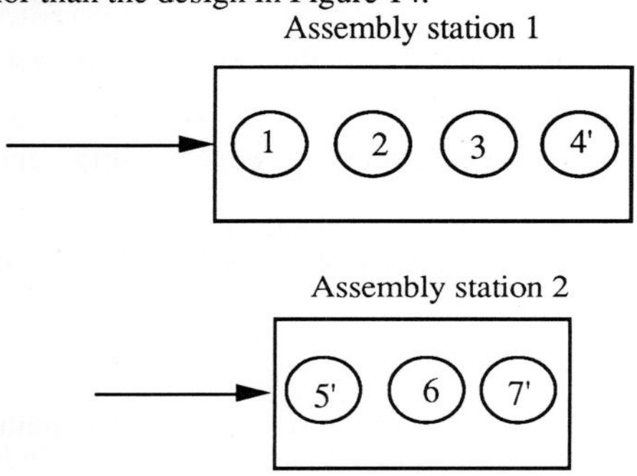

Figure 16. The modified design of the assembly system

Of course, the above discussed considerations apply to processes other than assembly, such as machining, fabrication, and so on.

10. DESIGN-FOR-SCHEDULING

In order to illustrate the impact of design decisions on one of many operational aspects, an example of a design rule aimed at improving schedulability of components is formulated.

Design Rule 1. Assign parts (products) to the existing machining (assembly) cells.

A manufacturing system may consist of a number of group technology cells. In the case where enough production capacity is available, parts should be designed to fit the machine cells (see Figure 17).

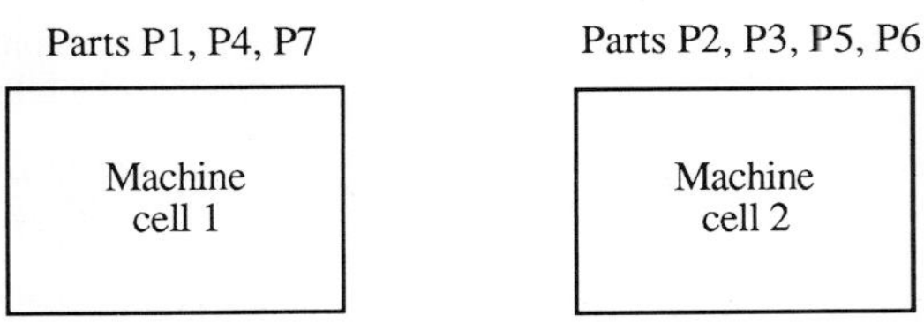

Figure 17. A cellular manufacturing system

Cellular scheduling applies to physical as well as logical (virtual) machine cells (for discussion of cellular manufacturing see Kusiak 1990) The physical cells are obtained by locating machines in the adjacent locations, while the logical cells are visible to the information system only and they are created without physical rearrangement of machines (stations). Details of design for scheduling are discussed in Kusiak and He (1993).

11. CONCLUSION

In this paper, concurrent engineering was viewed from the manufacturing systems perspective. The following issues were discussed in the paper:
1. Modeling the design process.
2. Rules for design-for-reliability, reconfigurability, repeatability, and flexibility of manufacturing systems.
3. Rules for design-for-productivity, design-for-scheduling, design-for-setup reduction, design-for-inventory reduction, and rules considering other operational issues of manufacturing systems.

Acknowledgement

The research presented in this paper has been partially supported by the grant DDM-9215259 from the National Science Foundation and research contracts from three industrial sponsors. Three doctoral students U. Belhe, W. He, and J. Wang have made contributions to some of the ideas presented in the paper.

REFERENCES

Altenhof, J.L., W.J. Fabrycky, M.-F. Feng, and R.G. Mitchiner (1989), Concurrent Mechanical System Design: A Computer-Aided Design Demonstration, in E.J. Haug (Ed), Proceedings of the First Annual Symposium on Mechanical System Design in a Concurrent Engineering, The University of Iowa, Iowa City, pp. 497-512.

Boothroyd, G., C. Poli, and L.E. Murch, Automatic Assembly, Marcell Dekker, New York, 1982.

Brown, A.D., P.R. Hale, and J. Parnaby (1991), An Integrated Approach to Quality Engineering in Support of Design for Manufacture, in J. Corbet, M. Dooner, J. Malleka, and C. Pym (eds), Design for Manufacture: Strategies, Principles and Techniques, Addison-Wesley, Reading, Mass., pp. 146-162.

Corbet, J. (1986), Design for Economic Manufacture", Annals of the CIRP, Vol. 35, No. 1, pp. 93-97.

Cutkosky, M.R. and J.M. Tenenbaum (1990), A Methodology and Computational Framework for Concurrent Product and Process Design, Mechanisms and Machine Theory, Vol. 251, No. 3., pp. 365-381.

El-Gizaway, A.S. and J.-Y. Hwang (1989), A Strategy for Concurrent Design of Aerospace Components, Proceedings of the Winter Annual Meeting of the ASME, San Francisco, CA, December, pp. 143-151.

Foreman, J.W. (1990), Gaining Competitive Advantage by Using Simultaneous Engineering to Integrate Your Engineering, Design, and Manufacturing Resources, in Simultaneous Engineering, Society of manufacturing Engineers, Dearborn, MI, pp. 92-105.

Ishi, K, (1993), Modeling of Concurrent Engineering, in A. Kusiak (Ed.), Concurrent Engineering: Automation, Tools, and Techniques, John Wiley, New York, pp. 19-39.
Kameyama, K.I. (1993), Real-Time Constraint Checking in the Design Process, in A. Kusiak (Ed.), Concurrent Engineering: Automation, Tools, and Techniques", John Wiley, pp. 111-130.
Kannapan, S. M. and Marshek, K. M. (1993). "An Approach to Parametric Machine Design and Negotiation in Concurrent Engineering", A. Kusiak (Ed.), Concurrent Engineering: Automation, Tools, and Techniques, John Wiley, New York, pp..
Knight, T.P. and S.T. Kim (1991), A Knowledge System for Integrated Product Design, Journal of Intelligent Manufacturing, Vol. 2., No. 1, pp. 17-25.
Kusiak, A. (1990), Intelligent Manufacturing Systems, Prentice Hall, Englewood Cliffs, N.J.
Kusiak, A., Ed. (1992), Intelligent Design and Manufacturing, John Wiley, New York, 1992.
Kusiak, A., Ed. (1993), Concurrent Engineering: Automation, Tools, and Techniques, John Wiley, New York.
Kusiak, A. and W. He, Design of Components for Schedulability, European Journal of Operational Research (forthcoming).
Kusiak, A. and J. Wang, Efficient Organizing of Design Activities, International Journal of Production Research, Vol. 31, No. 4, 1993, pp. 753 - 769.
Lyons, J. L. (1982). Lyons' Valve Designer's Handbook, Van Nostrand Reinhold Publishing Co., New York.
Miyakawa, S., T. Ohashi, S. Inoshita, and T. Shigemura (1986), The Hitachi Assemblability Method and Its Applications, Proceedings of the Second International Conference on Product Design for Assembly, pp. 1-6.
Molloy, E., H. Tang, and J. Browne (1991), Design for Assembly within Concurrent Engineering, Annals of the CIRP, Vol. 40, No. 1, pp. 107-110.
Stoll, H.W. (1990), Simultaneous Engineering in the Conceptual Design Phase, Society of Manufacturing Engineering, Dearborn, MI, pp. 165-171.
Subramanyan, S. and S.C.-Y Lu (1989), Computer-Aided Simultaneous Engineering for Components Manufactured in Small and Medium Lot Sizes, in the Proceedings of the Winter Annual Meeting of the ASME, San Francisco, CA, December, pp. 175-184.
Thompson, J.B. and S.C.-Y. Lu (1989), Representing and Using Design Rationale in Concurrent Product and Process Design, in the Proceedings of the Winter Annual Meeting of the SME, San Francisco, CA, December, pp. 109-115.

HYBRID SYSTEMS FOR CIM APPLICATION

Knowledge Based Hybrid Systems (B-11)
I. Mezgár and P. Bertók (Editors)
Elsevier Science Publishers B.V. (North–Holland)

Hybrid Programming Paradigms in CIM-CASE

L.M. Camarinha-Matos; H.J. Pinheiro-Pita and A.L. Osório

Faculdade de Ciências e Tecnologia, Universidade Nova de Lisboa, Quinta da Torre, 2825
Monte Caparica - PORTUGAL

Abstract
The programming paradigms used in CIM-CASE are presented. The concept of CIM-CASE is briefly summarized as a system that will support a systems' designer in deriving particular CIM architectures out of a reference model and resorting to an Electronic Catalog of CIM components. Particular emphasis is put on the illustration and justification of the different paradigms used as well as in a brief presentation of the implementation details.
This work has been developed in the context of CIM-PLATO, an Esprit project of the European Community, involving 14 partners (universities and industry) from 7 countries.

Keyword Codes: D.2.6; H.2.5; I.2.4; J.6
Keywords: Programming Environments; Heterogeneous Databases; Knowledge Representation; Computer-Aided Engineering

1. INTRODUCTION TO CIM-CASE

A software architecture for a Computer Integrated Manufacturing (CIM) system is an extreme case of hybrid programming. A CIM software system can be regarded as a large federation of heterogeneous modules coming from different vendors and running on a distributed and heterogeneous computing infrastructure. These modules have been developed with basis on a variety of programming paradigms/cultures, ranging from a large multiplicity of procedural programming approaches to object oriented or rule based methods.

In order to reduce the integration problems, tools' development is increasingly being driven by requirements of openness and compliance with emerging standards. Nevertheless, topics, such as Information Integration, Integrating Infrastructure and development of Reference Models for Open CIM architectures constitute still important research directions. Integrating issues in advanced Manufacturing Systems are, in a first instance, attempts to build integrated configurations of mostly existing heterogeneous modules or tools. This situation is particularly evident in the scope of international projects, namely in the European Community with ESPRIT and BRITE programmes.

The concept of toolbox as an open library or **Electronic Catalog** of software tools from which subsets, dedicated to particular application domains and specific functional requirements, can be selected and integrated has been emerging in some projects, namely in Esprit CIM-PLATO[13]. In some cases the concept arose as a solution to organize tools developed inside project consortia; in others the motivation was to gather and organize information about the multiplicity of tools available on the market, as the case of the CIMDATA project. The concept of **CIM-CASE** was proposed by us [1] [3] [8] as a successor of the toolbox system, providing functionalities to manage the catalog of tools and support the interactive generation [hierarchical interactive planning] of particular configurations of tools dedicated to specific cases, i.e., particular architectures in the CIM-OSA sense [7].

A **Particular Architecture** is a software configuration for a particular CIM system, comprising a set of tools (selected from the catalog) integrated via an Information System (IS). The assumed integrating infrastructure includes, besides the IS and distributed information access functionalities, the **Control or Administrative Architecture**, i.e., an abstract model of the functionalities of a particular CIM system and desired manufacturing processes, used to control the overall behavior of the integrated system.

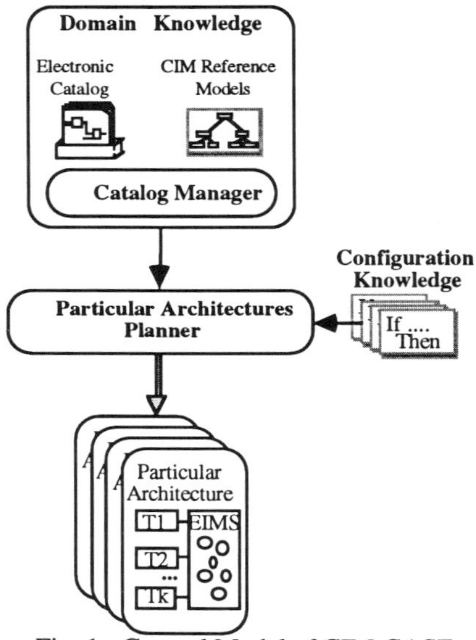

Fig. 1 - General Model of CIM-CASE

In close connection with the catalog of tools, another central component of this system is a metaknowledge base containing knowledge about CIM activities, application domains, reference information concepts and configuration knowledge - **CIM reference models**.

A **Catalog Manager** module is designed to support maintenance, graphical browsing and explanation facilities associated to the catalog and reference models. These functionalities are specially useful due to the large interdisciplinarity involved in CIM models - no one is an expert in all areas of activity needed in CIM. CIM-CASE Catalog Manager has functionalities to consult, insert and modify the Domain Knowledge's structures according to the current user level. Current implementation considers three user levels: normal user, configuration expert and CIM-CASE manager. Figure 2 shows a general aspect of the Catalog Manager interface, browsing over a taxonomy of activities.

A second functional component of the CIM-CASE concept is the **Particular Architectures Planner**, an interactive support system for specification and derivation of particular architectures, including tools' selection and derivation of shared IS concepts.

The CIM-CASE Particular Architecture's Planner has four main phases:

i) Specification and planning of logical structure. The system's designer, using an intelligent SADT/IDEF0 [9] based editor, makes the hierarchical decomposition of the functional architecture according to the requirements. The requirements' specification and the hierarchical planning/decomposition are two parallel processes. As a result, a hierarchy of abstract functionalities, according to the IDEF0 methodology, is obtained.

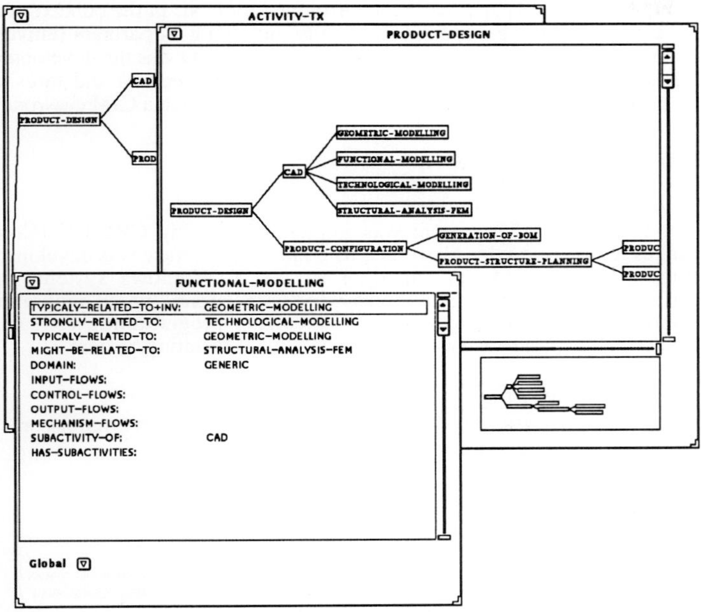

Fig. 2 - General Aspect of Catalog Manager Interface

ii) Building up of realization structure. It might happen that the logical hierarchical structure devised in previous phase is over-detailed regarding the characteristics of tools available in the catalog. Some software packages - the building blocks for our system - are able to perform rather complex tasks. Therefore, from the realization point of view, the hierarchical decomposition could have stop at a level for which tools can be found in the catalog. A kind of "pruning" might therefore be applied at this stage. On the other hand, it might happen that some models in the logical structure are too abstract regarding the level of existing tools. For this case, a recommendation for further decomposition shall be issued.

iii) Evaluation and selection of tools. This is again an interactive process in which a multicriteria-based what-if analysis can be performed. By modifying a set of parameters and filters application order, different solutions can be found and evaluated.

iv) Generation of a particular solution. In this phase a final solution is produced containing the chosen tools as well as a detailed description of the input/output concepts that will fulfill the EIMS - Engineering Information Management System - of the configuration and the requirements in terms of equipment.

The whole process is made in an interactive and flexible way. In all these phases the system uses its Configuration Knowledge to help the user in his decisions or even to propose its own solutions. For a more detailed analysis of CIM-CASE system see [1] [14].

The main objective of this work is to investigate an approach to achieve a kind of specialized **CASE** system to help [interactive planning] in the development of CIM systems. Being a specialized system, the expected degree of automation and help can be higher than in conventional CASE systems. On the other side, by addressing a heterogeneous and distributed context, it has to deal with rather complex problems. Therefore, this work doesn't present a finished system but a first implementation of an emerging concept.

The prototype described in this paper has been developed by UNL in the context of CIM-PLATO, an ESPRIT project of the European Community, involving 14 partners (universities and companies) from 7 countries. The main objective of CIM-PLATO was the development of an industrial toolbox prototype consisting of computer-based procedures and tools which support the design, planning and installation of FMS and FAS systems in a CIM environment.

2. PROGRAMMING PARADIGMS

2.1 General Programming Environment

A prototype of the CIM-CASE system was implemented using CRL/KC (Common Representation Language of Knowledge Craft) and its Graphical Interface was developed in C language resorting to the Xlib library functions of the X Windows system. In this implementation different programming paradigms were used - frame-based programming, reactive programming, rule-based programming, concurrent programming etc.. Current prototype version runs on a SUN Sparc Station using SUN-OS Operating System.

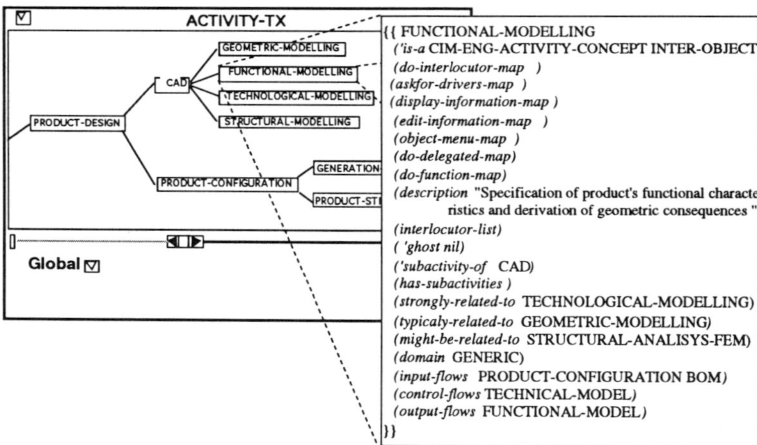

Fig. 3 - Partial view of taxonomy of CIM activities

2.2 Frame Based Programming

Frames are the main knowledge representation mechanism used in CIM-CASE.

The Domain Knowledge part is composed by two information classes: CIM Reference Models and Electronic Catalog. The CIM-CASE Reference Models are organized in hierarchies of objects, namely taxonomies of concepts (taxonomies of CIM activities; CIM Domains; CIM equipments; and a glossary of data terms), being each object represented by a frame. Figure 3 shows an example. The Electronic Catalog consists of a library of models of software tools/ applications available for various CIM activities, and their inter-relationships to the various taxonomies in the Reference Models, being the model of each tool also represented by a frame.

A set of relations were defined trying to model, as accurately as possible, the different interactions among these objects. In figure 3, for instance, slots *Subactivity-of* and *Has-subactivities* are taxonomic relations; *Strongly-related-to*, *Typically-related-to* and *Might-be-related-to* are "horizontal" relations used to define embedded knowledge; *Application-Domain* is a relation that defines the connection between an activity and its application domain (GENERIC by default); and *Input-flows*, *Control-flows* and *Output-flows* are relations to the glossary of data terms. For each relation different inheritance mechanisms are defined.

Associated to each frame there can be methods, and demons, allowing for normal object oriented and reactive programming approaches. Some of these objects are interactive, i.e., an authorized user can dialogue with them. Therefore, a concept of Interactive Object - INTER-OBJECT - was developed. The INTER-OBJECT concept exports all methods needed for this interaction. In point 2.5 a brief explanation of these methods will be done.

2.3 Rule Based Programming

Configuration Knowledge is composed by four rule bases:

i)*Embedded Criticism* and *On-demand Criticism* that, using the CIM reference Models as input, validate and make recommendations about the human expert options;

ii)*Reduction of Focus of Atention* that, based on the successive decisions made by the human expert, reduces the characterization space for the next options;

iii)*Building up the Realization Structure* that, based on the available tools and on their relations with CIM activities, defines the "pruning" of the logical structure or recommends its expansion;

iv) *Selection/ Evaluation* that implements the criteria for driving the derivation of a particular configuration.

Figure 4 shows an example of two rules. Internally a rule is represented by a frame with 3 slots: Test; Then ; and Else. Figure 5 shows the internal structure for the rule *Reducing the space of possible inputs* .The general concept of *rule* contains a method that is the responsible for execution of each rule. Figure 6 shows this method.

Reducing the space of possible activities
IF *(A is an activity attached to box B) &*
(Bk is the set of offspring boxes of B) &
(TA is the tree whose root is A)
THEN *Show nothing but TA when the option attach- activity for each element of Bk is activated.*

Reducing the space of possible inputs
IF *(A is an activity attached to box B) &*
(I is the typical list of inputs for A)
THEN *Show nothing but I when the option connected-input of B is activated.*

Fig. 4 Example of Rules

```
{{Reduc-inputs
  (is-a rule)
  (test (and (attached-activity A B) ))
  (then (display (get-inputs A)))
  (else)
}}
```
where activity A and the box B are instantiated when the rule is fired

Fig. 5 - Internal Representation of a rule

2.4 Reactive Programming

During the specification of the logical structure user decisions should be assessed, in a transparent way (embedded criticism), in order to check if they are in accordance with the Domain Knowledge. Reactive programming is an effective way to implement such critics. Associating a demon to each slot that shall be checked will provoke an automatic firing of checking functions each time the slot is modified. Figure 7 illustrates a demon associated with the slot Activities of the Idef0-box concept. When a configuration already has an attached domain and the user specifies the activities to a particular Idef0-box, the rule C-Check-for-Activities-on-Domain is fired with the intention of checking if all activities belong to the configuration's domain.

2.5 "Advanced" Objects & Behavioral Programming

The process of generating a new CIM Configuration system is a kind of "anarchic" activity that requires a great flexibility in terms of interaction mechanisms. Specification and planning a

Logical Structure is a process for which there is an impossibility of getting a complete requirement's specification at the beginning of the configuration process and, even, at the final stage of the specification phase. Quite often the user starts with a rough idea, and, only when presented with a first sketch of a possible solution, he is able to refine his requirements. From the other side, this process includes a hierarchical decomposition that evolves in parallel with the requirement's specification, i.e., the user shall not be forced to follow a pre-specified path but he shall be able to travel, up and down, in the hierarchical representation. The Interaction Mechanisms described here were developed with this goal in mind.

```
(defun executor-act-mt (Self Slt ctx client &rest
                                             param)
 (let (
     (test (get-value Self 'test))
     (action-T (get-value Self 'then))
     (action-F (get-value Self 'else))
     )
    (if (exec-driver-fn   test client param)
        (exec-driver-fn action-T client param)
        (exec-driver-fn action-F client param)
     )
))
```

Fig. 6 - Method for Rules' Execution

```
{{Activities
   (is-a slot)
   (domain (type instance idef0-box))
   (demon activities-demon)
}}
{{activities-demon
   (instance demon)
   (access new-value)
   (when after)
   (effect side-effect)
   (action activities-demon-fn)
}}
(defun activities-demon-fn (Self slt demon acessor
                                     access-value ctx)
   (call-method 'C-Check-for-Activities-on-Domain
                                     'executor Self))
```

Fig. 7 - Example of a demon

Figure 8 shows the general philosophy behind the Interaction Mechanisms. As a platform to support the interaction a set of parametrizable **interaction forms** like tree-browser, List-Browser, Idef0-browser, filler, etc., was developed. An interaction form is a general script characterizing a graphical presentation and an interaction behavior. The graphical interface is therefore developed using a kind of script-based programming paradigm.

Each application object has a different behavior and appearance depending on the user that interacts with it and based on the kind of interaction form. Therefore, the concepts of Object View (OV) and Object Role (OR) were implemented. An OV is a virtual instantiation of an object where only a particular subset of attributes (structure) and a particular behavior (methods and reactions), available to the current user level, is inherited. An OR is the result of displaying an OV in some interaction form.

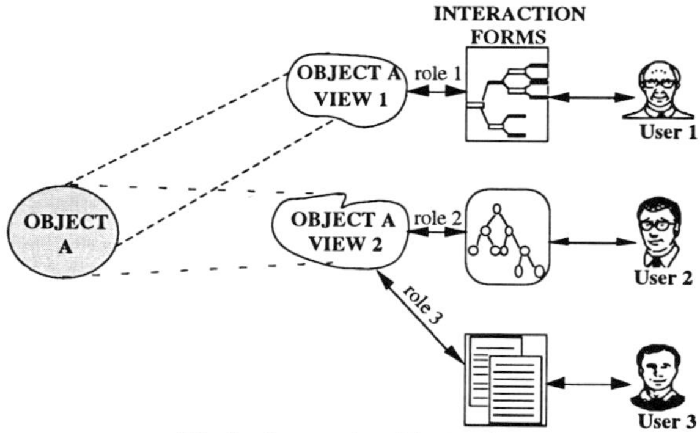

Fig.8 - Interaction Mechanism

Lets consider in more detail the structure of slots and methods of the concept INTER-OBJECT (fig. 9) as well as the general concept of interaction form - INTERLOCUTOR (fig. 10) that are important in this context. When an object receives a message for a self display in a given interaction form, it consults its slot *Do-interlocutor-map* to get the "address" of the selected interaction form. After that it fires its method *AskFor* that creates an OV, available for this interaction. This method uses the information existing in the slot *askfor-drives-map* as well as the value of *current-user-level.* It should also be noted that the attributes inherited by the OV are indicated in the slots *display-information-method* and *edit-information-map* and its behavior, that corresponds to the available menu options, is indicated in the slot *object-menu-map.* Finally it starts the interaction form. The link between the OV created and the instance of the interaction form constitutes the OR for the required interaction.

```
[{ INTER-OBJECT
  (do-interlocutor-map
    ((desk          Desktop)
     (Edit           Filler)
     (Show          Displayer)
     (View          Viewer)
     (Browse       Tree-Browser)
     (List            List-Browser)
     (Idef          IdefBrowser)
     (help            Helper)  ))
   (askfor-drivers-map
    ((Object-Menu          Get-Object-Menu)
     (Window-title          Get-Window-title)
     (Object-Description   Get-Object-Description)
     (Display-Information  Get-Display-
                                         Information)
     (View-Information      Get-View-Information)
     (Edit-Information      Get-Edit-Information)
     (Tree-Information      Get-Tree-Information)
     (List-Information      Get-List-Information)
     (Idef-Information      Get-Idef-Information)
     (Help-Information      Get-help-Information)
    (Desktop-information  Get-Desktop-Inform.) ))
   (Disp.Info.Map ((0 (...))(1 (...))(2 (...))))
   (Edit.Inform.Map ((0 (...))(1 (...))(2 (...))))
   (Obj.Menu.Map ((0 (...))(1 (...))(2 (...))))
   (do-delegated-map)
   (do-function-map)
   (Interlocutor-List)
   (ghost  nil)
   (description)
;Methods
   (DoOption          Object-DoOption-mt)
   (AskFor            Object-AskFor-mt)
   (Event             Object-Event-mt)
   (Close-Inter        Close-Inter-mt)
   (BroadCast         BroadCast-mt)
}}
```

Fig. 9 - Concept of INTER-OBJECT

```
{{INTERLOCUTOR
   (Client-Object)
   (GraphicsObject)
   (Type-of-Graphic-Object)
   (Current-User-level)
   (Current-A-Option-Map)
   (Current-B-option-Map)
   (Current-C-Option-Map)
   (number-of-instances)
   (A-option-Map
        (((0((Close Close-Action)
            (Refresh Refresh-Action)))
          (1((Close Close-Action)
             (Refresh Refresh-Action)))
          (2((Close Close-Action)
             (Refresh Refresh-Action)))
   )   )

   (Inter-new Inter-new-mt)
   (Inter-start Inter-Start-mt)
   (Inter-Xevent Inter-Xevent-mt)
   (Inter-Restart Inter-Restart-mt)
   (Inter-End Inter-End-mt)
   (Send-Specific-Information)
   (Answer-Selected-item)
)
```

Fig. 10 - General Concept of Interlocutor

Sometimes, an object can receive a message to perform an option that does not belong to its internal behavior. In this case the slot *do-delegated-map* is consulted and if it has a delegation of some other object to perform the required method a ghost-object is created. Figure 11 shows an example of this mechanism. It doesn't make sense that an user dialogues with the activity of some other object to perform the required method a ghost-object is created. Figure 11 shows an example of this mechanism. It doesn't make sense that an user dialogues with the activity CAD using an interaction form of type Tree-browser.However this option exists in its Menu. When CAD receives this event it means that the user wants to dialogue with an object taxonomy whose root is the CAD activity itself. In this case, since CAD has delegation of CIM-activity-tx to perform this option (see slot do-delegated-map), a ghost-object is created - *Ghost-CAD-1* - containing the object-menu inherited from CIM-activity-tx but related with CAD trough the relation *Ghost-of* . The final steps, of launching the interaction form (tree-browser), are the same. This ghost-object exists only while the interaction form, through which the dialogue is made, is active.

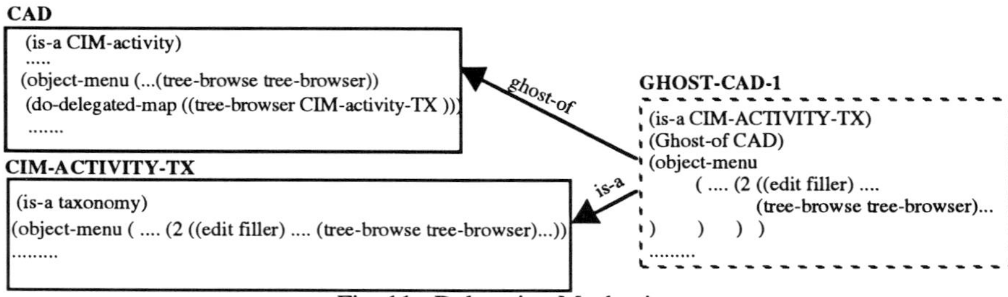

Fig. 11 - Delegation Mechanism

2.6 External Interfaces & Distributed Programming

The graphical interface, as mentioned before, is a very important component to support the interactive planning philosophy of CIM-CASE. The X Windows system was chosen as development platform mainly due to its status as "de facto" standard for UNIX world. The event driven characteristic is another important feature in order to support the required flexibility in terms of user interface. Finally, the "appeal" of its object oriented architecture shall also be mentioned. As the application kernel is developed in LISP/CRL, while the Xlib and specially the high level widgets library (Xview) are available in C, a hybrid architecture arises naturally. Even though, some LISP bindings for Xlib exists, high level widget/toolkits were only available in C (at least when the project started). Therefore, a hybrid structure was necessary for the user interface.

Another important aspect was the decision about a monolithic or distributed architecture.Problems with dynamic linking of Xview library in release 3.2 of KC, due to symbol incompatibilities between SunView (default library that came with this release of KC) and Xview widget sets, forced the separation of LISP/CRL component of CIM-CASE from

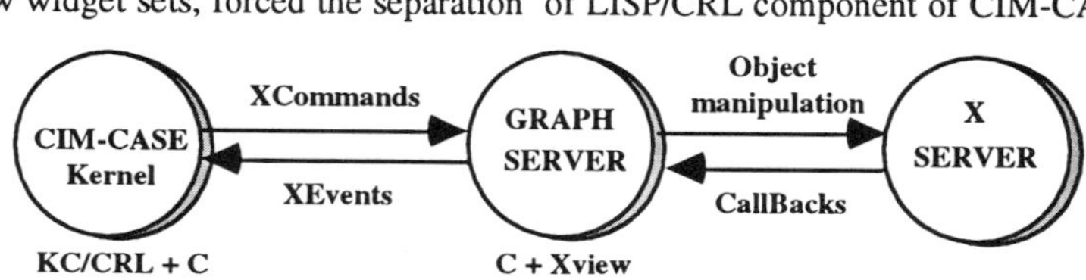

Fig. 12 - CIM-CASE processes distribution

Xview implementation of graphical objects . The big size of KC process (executable binary file about 20 MByte) and difficulties in development phase due to time spent in garbage collection activity, contributed also to this option. The distributed processing philosophy contributed to a faster execution, made possible the use of different machines to run CIM-CASE and in general enforced the modularity of the structure. According to the above constraints, the general implementation structure of CIM-CASE is presented in Fig. 12. The communication mechanisms between CIM-CASE kernel and GraphServer - exchange of XCommands and XEvents messages - were implemented using the Remote Procedure Call (RPC) mechanisms.

•*Reception of X events.* From the CIM-CASE kernel side, the structure illustrated in figure 13 is used to receive an X event coming from the GraphServer.

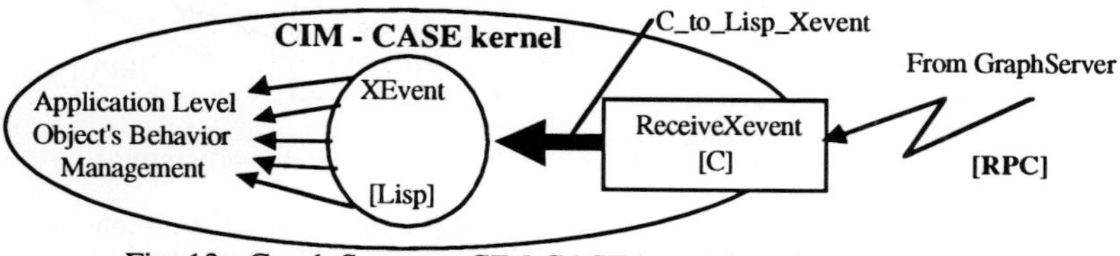

Fig. 13 - Graph Server to CIM-CASE kernel interfacing structure

At lowest level, function Receive_XEvent() supports the connection between Graph Server and CIM-CASE kernel. When Graph Server wants to send an event to CIM-CASE kernel it calls this function through the RPC mechanisms. From the C part, in CIM-CASE kernel, the function LISP_Xevent (in Lisp) is called which does the necessary type translations between the two languages.

```
xevent_status *Receive_XEvent(args)
struct ReceiveXEventArgs *args;
{ rpc_status *aux;
  xevent_status stat;
  . . . /* arguments unpacking */
C_to_lisp_XEvent
      (args->rpc_Interlocutor,
          (Xv_opaque)args->rpc_GO,
          args->rpc_ItemType,args-
>rpc_ItemInfo);
• • •
 return(&stat); }
```

Fig. 14 - C function to receive remote events from Graph Server

```
(def-foreign-callable
    (C_to_LISP_Xevent
        (:name "_C_to_lisp_XEvent")
        (:language :c))
        ( (interlocutor :simple-string)
        (object-id :pointer)
        (item-type :signed-32bit)
        (item-information :simple-string) )
    . . . ; type conversions
     (Xevent interloc object-id
             item-type item-Information)))
```

Fig. 15 - Example of interface functions between C and SunLisp

```
(defun Xevent
    (interlocutor object-id item-type item-
information)
  (case item-type
   (GLOBALMENU
    (call-method interlocutor 'inter-Xevent
                 'Menu item-information))
   (BUTTONOK
    (call-method interlocutor 'inter-Xevent
                 'Button "OK" item-information))
   (BUTTONCANCEL
    (call-method interlocutor 'inter-Xevent
                 'Button item-information))
   (IDEF0DRAG
      (call-method interlocutor 'inter-Xevent
                   'Drag item-information))
. . .
 (return-from Xevent nil)
)
```

Fig. 16 - Event handler in CRL/Lisp called by C interface

The last step is performed by function Xevent which includes the identification of received event type and the dispatch to the corresponding handling method.

•*Sending XCommands.* To send an X Command from CIM-CASE kernel to Graph Server a similar procedure is followed. Figure 17 shows the client side of function XCommand which support the sending of all commands defined to be performed by the Graph Server. This message envelope's includes commands like: object creation, object deletion, sending information to objects, etc..

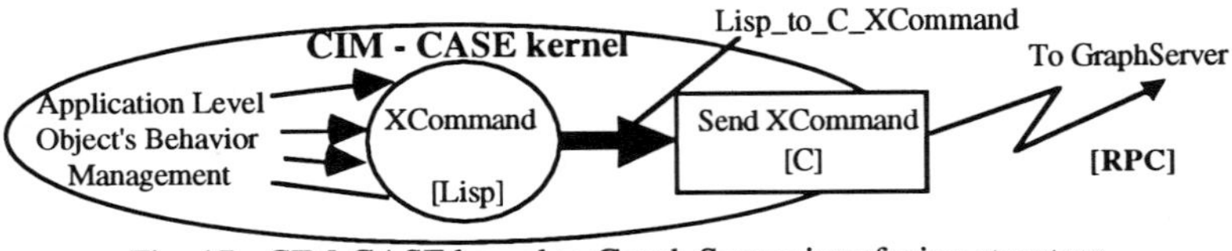

Fig. 17 - CIM-CASE kernel to Graph Server interfacing structure

The function XCommand, according to the requested service, calls the remote function in the Graph Server process. This call is forwarded through the SunLisp external interface to C language, i.e., the function *Lisp_to_C_XCommand* shown in figure 19. The C part has only to pack the arguments using RPC rules and call the remote image of XCommand function.

```
(defun XCommand
 (&key command interlocutor object-id object-type
            title item-type item-information with-
wait)
 • • •
 (Lisp_to_C_XCommand
    command interlocutor object-id object-type
     title item-type item-information with-wait ))
```

Fig. 18 - Skeleton of the function XCommand implemented in SunLisp

```
(def-foreign-function
   (Lisp_to_C_XCommand
      (:return-type :pointer)
      (:name "_Send_XCommand")
      (:language :c))
   (Command :pointer :character)
   (Interlocutor :pointer :character)
   (ObjectId :pointer :character)
   (Title :pointer :character)
   (ItemInformation :pointer :character)
   (ObjectType :signed-32bit)
   (WithWait :signed-32bit))
```

Fig. 19 - Example of interface function between SunLisp and C

```
Send_XCommand( ...)
 . . . /* arguments declaration */
{  IF (Command with immediate answer )
    call the remote Graph Server XCommand
    Wait for the Answer  AND accept RPC services
    ELSE
     call the remote GraphServer XCommand
      . . . /* return status to Lisp */;}
```

Fig. 20 - C implementation of XCommand out port in CIM-CASE kernel side

```
int DispatchNextRPC()
{
   /* declarations  */
 for (;;) {
  readfds = svc_fdset;
  switch (nfd = select(dtblsize, &readfds, NULL,
                                       NULL, NULL))
   {
    case -1: /* instructions */ return(FALSE);
    case 0: continue;
    default:{svc_getreqset(&readfds);
return(TRUE)}
   }
  }
}
```

Fig. 21 - Some aspects of RPC dispatcher

```
Send_XCommand( ...)
{
 IF (Command with immediate answer )
  call the remote Graph Server function XCommand
  Wait for the Answer  // accept RPC services
 ELSE
  call the remote Graph Server function
   XCommand
}

Receive_XEvent(EventType)
{
   IF (previous XCommand is waiting answer )
    signal synchronization mechanism
    Pass information to previous XCommand
   ELSE
    call event handler in Lisp }
```

Fig. 22 - Synchronization between Send_XCommand() and Receive_XEvent()

• *Event handler and synchronization.* Due to the event driven characteristic of CIM-CASE and to the existence of two processes, it was necessary to define synchronization mechanisms in order to schedule events in correct order. After initializations, the CIM-CASE kernel waits for events coming from graph server, in function DispatchNextRPC(). If a remote call is received without any error, the function svc_getreqset() fires the remotely called function. There are two types of XCommands: with and without waiting for answers from the Graph Server. When waiting for an answer from the Graph Server, the CIM-CASE kernel should both wait for an RPC call in DispatchNextRPC() and at the same time it should stay in waiting status for the answer to the previous XCommand. In other words, CIM-CASE kernel can play both the role of client and server regarding the GraphServer. If in waiting status, next XEvent should signal the previous XCommand and pass to it the event received from Graph Server.

•*Brief presentation of RPC's architecture.* Remote Procedure Calls (RPC) mechanism, is available in almost all Unix based machines and its virtualization of low level network communication protocols makes it a good connection support to distributed processing.

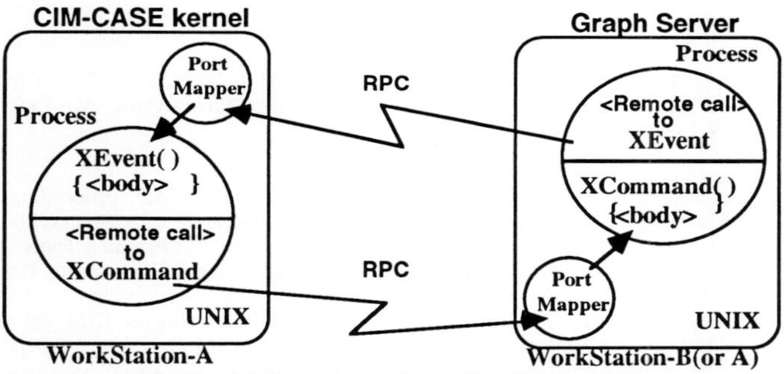

Fig. 23 - RPC services with callback capabilities

Fig. 23 shows the two services XCommand and XEvent implemented in different processes. XCommand, implemented in the Graph Server and called by CIM-CASE kernel and XEvent implemented in CIM-CASE kernel and called by Graph Server.

2.7 Particular Architectures - IMS & Concurrent Access

As described in chapter 1, a particular architecture is a configuration of a set of tools integrated via a common Information System. The generation of such Particular Architectures is the goal of CIM-CASE. The output of the CIM-CASE interactive planning phase includes: List of selected/proposed tools; List of shared concepts to feed the particular IS; and list of required equipment. An Information Management System (IMS) is the integrating kernel for a particular architecture. Another prototype system, developed in conjunction with CIM-CASE is EIMS - Engineering IMS. This system includes both an IMS and an integrating infrastructure for tools running in a distributed environment and accessing the EIMS services via Remote Procedure Calls. The same approach and basic mechanisms developed for the CIM-CASE graphical interface are used in EIMS as well:

i) EIMS graphical interface, that follows the same philosophy described for CIM-CASE interface and provides browsing functionalities over the IS structures;

ii) Tools access. Concurrent access from tools to the EIMS services is allowed through a library of Information Access Methods and resorting to the RPC mechanism. These access methods are "inspired" by STEP and the Express language [12].

A synchronization layer developed on top of LISP/KC to support the RPC services controls the concurrent access of multiple tools running in different nodes. This is an important characteristic not typically found in KBS and that was achieved by a hybrid approach: interfacing the Knowledge Management System (KC) with the Operating System facilities.

The development of a CIM system is a very complex and demanding task. From the software point of view, the emergence of advanced CASE environments, supported by a philosophy of open systems, is a fundamental requirement. In this framework, the prototype described in this paper illustrated the main concepts and pointed directions for future CIM-CASE systems. In spite of several implementation difficulties, the hybrid programming approach, combining several paradigms, proved adequate to fulfill the required levels of programming functionalities and knowledge representation.
Main difficulties are due to the lack of an integrated programming environment combining knowledge representation, object oriented and rule based programming, procedural programming, graphical interfaces development and concurrent/distributed programming. As a consequence, an extra effort was needed to combine different environments, including the access to the operating system/network management services. This implies a group with diversified programming skills and greatly increases the debugging difficulties. Nevertheless, it is our belief that a combination of the above paradigms shall be present in future CIM development platforms.

Acknowledgments

This work was supported in part by EEC [project ESPRIT II 2202, CIM-PLATO], JNICT [project SARPIC and CIM-CASE] and INIC. The authors would like to thank Mr. João Moura-Pires, Mr. Luís Lopes and Dr. Hannu Ahonen for their contributions.

References

1 L.M. Camarinha-Matos; H. Pinheiro-Pita, "Intelligent C.A.S.E. for CIM", Proceedings of IEEE Int. Workshop on Emerging Technologies and Factory Automation, Melbourne, Australia, 11-14 Aug. 92.

2 L.M. Camarinha-Matos and F.Sastron, "Information Integration for CIM Planning Tools", Proceedings of CAPE'91 IFIP Conference on Computer Applications in Production and Engineering, Bordeaux, France, 10-12 Sept. 1991.

3 L.M. Camarinha-Matos and H.J. Pinheiro-Pita "Interactive planning of CIM software systems", Proceedings of 2nd UNIDO Workshop on Industrial Robotics and CIM, Belgrade, 4-6 Sept. 91.

4 A. Steiger-Garção; L.M. Camarinha-Matos - "Design of a knowledge-based information system", in *Integration of Robots into CIM* [Ed. Bernhardt, Dillman, Hörmann, Tierney], Chapman & Hall, 1992 - Cap. 21 & 22.

5 L.M. Camarinha-Matos; A. Steiger-Garção "Knowledge architecture for flexible programming of robotic cells", Proceedings of the 20th International Symposium on Industrial Robotics, Tokyo, 4-6 Out. 1989.

6 L.M. Camarinha-Matos and H.J. Pinheiro-Pita, "CIM-CASE: relatório de uma implementação", Relatório Interno, Março 1992

7 AMICE , Esprit Consortium, [Eds.], Open System Architecture for CIM, Springer-Verlag, 1989.

8 L.M. Camarinha-Matos "Toolbox Manager: Towards a concept of CIM-CASE", Proceedings of the CIM-PLATO Workshop on CIM Planning Tools, University of Karlsruhe, 25-26 Fev 92,

9 Ross, D.T., Application and extensions of SADT, IEEE Computer, Apr 1985.

10 L.M. Camarinha-Matos; H. Pinheiro-Pita; J. Barata; R. Rabelo, "Taxonomy of CIM Activities - 2nd revision", *Esprit CIM-PLATO Document, IS-UNL-04.91/1*, Apr 91.

11 L.M. Camarinha-Matos; J. Moura-Pires; H. Pinheiro-Pita, "CIM-PLATO Common Glossary - 4th revision", *Esprit CIM-PLATO Document IS-UNL-04.91/2*, Apr 91.

12 ISO CD 10303-11, Exchange of Product Model Data - Part II: The Express Language, Jul 1990.

13 Rolf Bernhard "CIM Systems Planning Toolbox - Project Survey and Demonstration", Proceedings of the CIM-PLATO Workshop on CIM Planning Tools, University of Karlsruhe, 25-26 Fev 92,

14 L.M. Camarinha-Matos; H. Pinheiro- Pita; "Interactive Planning in CIM-CASE", to be presented at 1993 IEEE Internationla Conference on Robotics and Automation, Atlanta, Georgia, May 93.

15 L.Camarinha-Matos; A. Luis Osório; "*CIM Information Management System: An Express-Based Integration Platform* ", IFAC Workshop on CIM in Process and Manufacturing Industries, Espoo, Finland, November 23-25, 1992.

Knowledge Based Hybrid Systems (B-11)
I. Mezgár and P. Bertók (Editors)
Elsevier Science Publishers B.V. (North–Holland)

Model Based User Interface Constructor for CIM

Zdeněk Kouba, Jiří Lažanský, Vladimír Mařík, Olga Štěpánková, Tomáš Vlček

Czech Technical University, Faculty of Electrical Engineering,
Technická 2, 166 27 Praha 6, Czech Republic

Abstract

The aim of this paper[1] is to gather the introductory considerations and opinions on the problems connected with the user interface programming for CIM systems. This part of the software developer's activities represents very complicated and tedious work [2]. The interface behaviour and functionality should be specified using some declarative (rather than procedural) means. Such a tool should support significantly the interactive development of the program parts exhibiting the required behaviour of the designed user interface [6,7].

Keyword Codes: D.1.5; H.1.2; H.5.2
Keywords: Object-oriented Programming; User/Machine Systems; User Interfaces

1 INTRODUCTION

The research team has developed several database applications mainly in CIM or related fields during the last ten years. The most important are:

- **TEPRO** - a system for computer aided preparation of **tec**hnological **pro**cedures in an industrial enterprise.
- **PACEB** - a system for full computer support of business and production management in a factory. The system has been implemented relational DBMS FoxPro 2.0 (Fox Software).

All the applications mentioned above can fit into the branch of CIM agendas. All of them are database applications with a closed set of queries. The queries are available to the user through a user-friendly interface.

At the very beginning we have started to work in the environment of the pure operating system and all the job has been done using the Pascal compiler and the generic file management system, only. There has been no data model available and the design of the database was done completely on the physical level. The ratio of the effort devoted to data processing and user interface programming was about 3 : 2. Nowadays we can

1 This research has been carried out by the joint Austrian-Czechoslovak research grant and was prepared during research visits at FAW Linz, Austria.

estimate the ratio about 1 : 2. It means that most of the programming effort is devoted to the user interface design, now. The reason is that the presence of a conceptual data model allows programming on a higher (conceptual) level and makes the data storage and retrieval easier.

Even though the available sophisticated software packages offer many interactive tools for user interface design today, the user interface programming is still very tedious work. The above mentioned tools (e.g. FoxPro user interface development facility called Screen Builder) simplify the design of the geometrical layout of the dialogue screen. However, the procedural background of the dialogue is still to be done by the programmer without any significant support of the development package.

Our experience proves that the quality of the user interface is one of the most important conditions for success of the particular software application. Though many researchers are working in the field of data modelling (relational algebra, object-oriented data models, ...), no significant attention is given to the development of a formal user interface model.

2 SUGGESTED SUPPORTING TOOLS

The final long-term goal of our research can be split into three levels:

1) Methodology of the user interface design. The design of the user interface should have a unified methodology through all the application development. This cannot be achieved in a team composed of several programmers unless they share a common and comprehensive design methodology. It makes sense to develop such a methodology by modification of an existing one (such as the IDEF0 functional modelling tool [5] briefly discussed later).

2) The formal model of the user interface skeleton. We believe that there is a possibility to describe the skeleton of the user interface independently of the particular host programming language.

3) The formal model processor should be able to translate the formal user interface model into the particular target programming language. The skeleton of the user interface dialogue will be generated in this way. The generation process can be accompanied by an interaction with the designer. In such a case the complete user interface screen code would be interactively generated ensuring the uniformity of the user interface behaviour.

The main goal of this study is to summarize critically our experience in user interface programming. We try to specify the basic features requested from an interactive tool supporting development of the user interfaces for various applications. We denote such a program a *model based user interface constructor* - shortly MBUIC.

Many programming environments equipped with an interactive user interface design facility already exist. These facilities usually make it possible to design the geometric layout of the data entries on the screen and to specify the validity conditions for particular data entries. However, they do not usually take into account the mutual dependencies of several entries. In more complicated user interfaces, the mutual dependencies among entries represent the most difficult part of the programmer's work. Our aim is to describe these mutual dependencies by a model or, more precisely, to design a formalism able to express mutual dependencies among entries of a general screen.

Having such a model the MBUIC should be able to generate the skeleton of the user interface program. The principal idea is that the user will interactively hang chunks of code onto the skeleton to achieve the detailed behaviour of the user interface as required.

As mentioned earlier, the user interface programming is a very tedious work. Usually the researchers concentrate on the development of some basic algorithms and the user interface itself is implemented by professional programmers, who transfer the developed algorithms into commercial products.

Those programmers do a very hard work the result of which can very significantly influence the commercial success of the entire project.

Successful development of a MBUIC tool could dramatically reduce the programmers labour costs when designing the final application equipped with a user iterface.

3 THE USER INTERFACE

3.1 Specification

Let us define now what is understood under the term **user interface**. Let us have a software system, the structure of which is introduced by the Fig. 1.

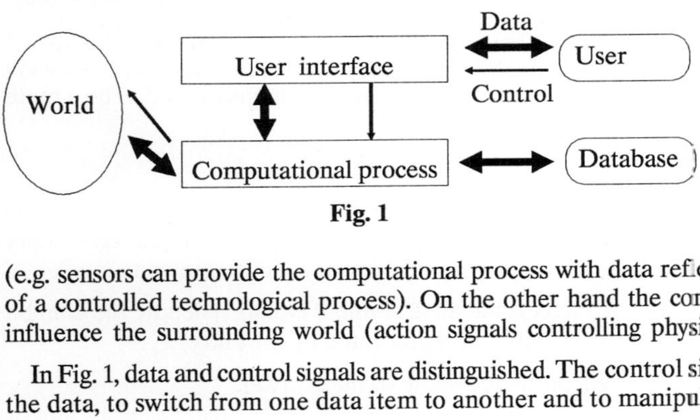

Fig. 1

Fig. 1 represents a computational process including its environment. The computational process may communicate with a database, it receives some data from the surrounding world (e.g. sensors can provide the computational process with data reflecting the current status of a controlled technological process). On the other hand the computational process can influence the surrounding world (action signals controlling physical devices), too.

In Fig. 1, data and control signals are distinguished. The control signals are used to select the data, to switch from one data item to another and to manipulate the data.

Usually, the computational process is not autonomous and it requires control by a human operator. We define the user interface as the software layer responsible for communication with the operator.

The main purpose of the user interface is to visualize the important data being provided by the computational process. Moreover, it enables the operator to enter data to achieve the required behaviour of the computational process.

We can imagine very complicated user interfaces with various graphical facilities. The particular ways how to visualize the data are relevant, but they depend on the particular implementation. Therefore, we restrict our attention to general and substantial features of the user interfaces that are common to every application.

3.2 The Screen

The screen of a computer attracts people usually by graphical features. We will consider the screen on an abstract level not taking into account the graphical features.

Under the term **screen** we will understand a collection of data items. Each item represents a variable (of simple or structured type) of our computational process. We will call the item the (screen) **entry**.

Of course, the entries can be input, output or both. Input entry provides the computational process with a value. Output entry reports on the screen the value of a variable. The input-output entry represents the current value of the variable but the operator is allowed to change this current value, too.

Some entries have special meaning, they are intended to invoke an event which is to be interpreted by the computational process to perform predefined action upon the data. Such entries are sometimes called 'controls'. However, there is no sharp border between data entries and controls and we will not distinguish between them from the modelling point of view.

We will take into account only entries allowing both input and output of data. The editing facility can be simply disabled if we wish to use the entry in the output mode only.

The variable corresponding to an entry can be of various types, it can be an integer or real number, a string or a structured type. Nevertheless, the domain of its values is always defined. In the case of a structured type one value of an entry is considered to be a tuple of simple type values.

3.3 Dependent entries

The formalism used in this section is very similar to that of relational database theory [1, 3]. The main difference is the terminology (entry corresponds to attribute, etc.).

Let us denote e_1 some entry of a screen and let the symbol Dom(e_1) denote the domain of all possible values of the entry (or variable represented by the entry) e_1 .

Let the symbol Val(e_1) denote a particular value of the entry e_1.

It means $\mathrm{Val}(e_1) \in \mathrm{Dom}(e_1)$.

Let $e_1, e_2, \dots, e_n$ be some entries of a considered application. The symbol Dom($e_1, e_2, \dots, e_n$) denotes a relation specifying all allowed tuples

$$<\mathrm{Val}(e_1), \mathrm{Val}(e_2), \dots, \mathrm{Val}(e_n)> \in \mathrm{Dom}(e_1) \times \mathrm{Dom}(e_2) \times \dots \times \mathrm{Dom}(e_n).$$

There is an interesting question how to specify Dom($e_1, e_2, \dots, e_n$). We suppose here that the domain Dom($e_1, e_2, \dots, e_n$) is given by the interface designer and we do not care at the moment by which means he did this task.

The entries $e_1, \ldots, e_n$ are **mutually independent** iff

$$\mathrm{Dom}(e_1, e_2, \ldots, e_n) = \mathrm{Dom}(e_1) \times \mathrm{Dom}(e_2) \times \ldots \times \mathrm{Dom}(e_n).$$

An important property of this notion is that it is downward closed w.r. to subsets. In other words if all the entries from a set $E = \{e_1, \ldots, e_n\}$ are mutually independent, then all the entries of any subset $E_1 \subset E$ are mutually independent, too.

The reverse statement does not hold. Consider the following negative example: Let $<x, y, z>$ be triplet of real number entries, let us define

$$\mathrm{Dom}(x, y, z) = \{<x, y, z>; (x = 0) \vee (y = 0) \vee (z = 0)\}.$$

Obviously[2],

$$\mathrm{Dom}(x) \times \mathrm{Dom}(y) \times \mathrm{Dom}(z) = R \times R \times R.$$
$$\mathrm{Dom}(x, y, z) \neq \mathrm{Dom}(x) \times \mathrm{Dom}(y) \times \mathrm{Dom}(z).$$

However, there holds under the specified conditions

$$\mathrm{Dom}(x,y) = \mathrm{Dom}(y,z) = \mathrm{Dom}(x,z) = R \times R.$$

More generally, suppose the entries from a set $E = \{e_1, \ldots, e_n\}$ are not mutually independent. Nevertheless, there can be a subset $E_1 \subset E$ such that all entries from the set E_1 are mutually independent.

There can exist many other types of independence or dependency. One of them is for example the pairwise independence.

All the entries of a set $E = \{e_1, \ldots, e_n\}$ are pairwise independent iff any pair

$$<e_{i_1}, e_{i_2}>, \quad i_1, i_2 = 1 \ldots n, \; i_1 \neq i_2$$

is independent. For sets of cardinality 2 both notions, pairwise and mutual independence are equivalent; the pairwise and mutual independence means the same.

Notice that pairwise independence of entries in a set E does not guarantee mutual independence of those entries.

3.4 Data dependencies from the user's point of view

We have mentioned the notion of entry dependencies. It seems to be useful to make a remark on dealing with dependencies from a user's point of view.

Let us take into account that the user can fill in the entries sequentially. He can not change more than one entry at once. The problem of data consistency (given by the dependencies) arises here.

The dependency definition through relation (i.e. subset of a Cartesian product of sets) specifies constraints posted on the final values of entries, which appear in the dependency. This definition is static in some sense. It does not describe how to transit from one

---

2 R denotes set of all reals.

point to another, i.e. the feasible sequence of entering the data. The following analysis illustrates the problem in more detail.

Let us have two dependent entries e_1 and e_2.

There are several ways how to solve the dependency of both entries, some of them being introduced here:

1) Postponed consistency check

The user can fill in both entries freely but the change of their values will be accepted after the global check has been done.

The global check tries to verify that $\mathrm{Val}(e_1, e_2) \in \mathrm{Dom}(e_1, e_2)$. If the check is not successful, the user is informed about inconsistency and the data change will be refused[3].

The postponed consistency check is simple to implement but its disadvantage is that until the consistency check (invoked e.g. by pressing a "push button") the inconsistent values of variables may exist.

2) Immediate consistency check

The other possibility of implementation of an entry dependency is to ensure consistent contents of both entries at every moment.

Let the user change the value of the entry e_1 and let the entries e_1, e_2 be dependent. After filling in (or changing) the entry e_1, the conditional domain of the entry e_2 denoted as $\mathrm{Dom}(e_2 \mid e_1 = \mathrm{Val}(e_1))$ is evaluated.

If the current value $\mathrm{Val}(e_2)$ belongs to the conditional domain

$$\mathrm{Dom}(e_2 \mid e_1 = \mathrm{Val}(e_1)),$$

both the entries are consistent. In the opposite case the value of e_2 will be automatically changed to a default value $\mathrm{Val}(e_2) = V_{def,2}(\mathrm{Val}(e_1))$ ensuring

$$\mathrm{Val}(e_2) \in \mathrm{Dom}(e_2 \mid e_1 = \mathrm{Val}(e_1)).$$

The mechanism of immediate consistency check should work in both directions. If we change the value of e_2 so that $\mathrm{Val}(e_2)$ does not belong to $\mathrm{Dom}(e_2 \mid e_1 = \mathrm{Val}(e_1))$, the mechanism evaluates $\mathrm{Dom}(e_1 \mid e_2 = \mathrm{Val}(e_2))$ and it automatically changes the value of e_1 to satisfy the requirement $\mathrm{Val}(e_1) \in \mathrm{Dom}(e_1 \mid e_2 = \mathrm{Val}(e_2))$[4].

3) Unidirectional immediate consistency check

Sometimes the semantics of entries e_1, e_2 allows us to choose e_1 to be fully truly independent while e_2 depends on e_1. Let us denote such a situation by the symbol

3 We do not consider details of refusing the change here.

4 This approach can be generalized to multidimensional dependency, too.

$e_1 \Rightarrow e_2$. A natural way of entering data here is to change e_1 first. In such a case we can implement the dependency of e_1 and e_2 by a mechanism called the unidirectional immediate consistency check.

After the value e_1 has been chosen and the Dom(e_2 | e_1 = Val(e_1)) has been evaluated, the user can choose the value from the set Dom(e_2 | e_1 = Val(e_1)) only. To illustrate this idea, assume that our pair of entries takes the value Val(e_1, e_2) = $<a, b>$. The user's aim is to change the value to $<c, d>$ which is supposed to be consistent. If $d \notin$ Dom($e_2 | e_1 = a$), the user is forced to change the entry e_1 to Val(e_1) = c, first.

Very often the entries on the screen have the "tandem dependency structure". It means that for the entries $e_1, e_2, \ldots, e_n$ holds

$$e_{i_1} \Rightarrow e_{i_2} \Rightarrow \ldots \Rightarrow e_{i_n}\text{, where } i_j = 1, \ldots, n,\ i_j \neq i_k\,.$$

3.5 The implementation of dependencies

Let us analyze how to implement the dependencies, especially the mechanism of immediate consistency check.

The consistency checks are repeated throughout the entire system implementation. To avoid the redundant repetition of the work it would be nice to have a tool enabling the programmer to specify the dependencies in a declarative manner, i.e. to specify the dependency structure, rather than to implement consistency checks themselves.

Our idea is to create a highly modular library of various objects representing entries. Another step is to design a tool making it possible to link those objects together according to a user interface model.

The aim of the research is not to design a new programming language, rather to implement a source code processor able to link different chunks of source code interactively.

The programmer (i.e. the user of our system) will design the screen interactively making use of that tool. At first, he has to make his decision of which components the screen will consists. It means which types of entries there will appear on the screen (push buttons, lists, radio buttons, menu, check boxes, ...). If the type of the entry is a standard one, the appropriate object is present in the library. If the entry were of special type (not yet present in the library), the programmer can also declare its behaviour and store this new entry type to the library.

The new entry type can be a collection of simpler ones being already available in the library. In such a case the user defines the behaviour of the new entry type by redefining how the event signals and the data are transferred among components of the collection. The programmer will define the entry's behaviour making use of some functional modelling formalism (we consider a formalism based on the IDEF0 standard). The new object, which cannot be assembled from the library pre-defined components, is programmed from scratch (e.g. in C+ +).

Let us stress the possible hierarchy of the user interface components. It means that a screen consists of entries but some of those entries can be again screens (i.e. collections of other entries). This hierarchy allows us to work with nested screens making it possible to implement very complex entry structures. The end user can consider the screen on various levels of abstraction. If he is interested in more detailed data, he can zoom an entry into a new screen (collection of entries) at a lower level of abstraction.

4 THE FUNCTIONAL MODELLING

A very important part of our concept is the functional modelling [4]. What are the principal features of the behaviour of entries in a screen? The user selects one of the entries and edits it. When the data editing is finished, different events can be generated. As we have discussed in the description of the immediate consistency check mechanism above, the conditional domain of the dependent entries is evaluated as the response to those events. On success of all the consistency checks the event "data valid" is generated. This fires the next event causing the data to be stored in a variable.

Let us mention now the possible formalism for functional modelling IDEF0.

The IDEF0 methodology has been developed in U.S.A. [5] for the following purposes:

- to enable structured approach to system analysis,
- to describe graphically the functional behaviour of general systems.

The resulting model has the form of a graphical scheme. The interpretation of the scheme elements and its graphical representation follows relatively strict syntax. Therefore the functional scheme is readable for everyone familiar with the IDEF0 methodology.

The basic concept of the IDEF0 is the black box. The box describes a (sub)system having some inputs and outputs (data, material,...) and some input and output control signals (events, messages,...).

The top-down design of the model is the characteristic feature of this methodology. On the highest level the system is represented by a single box with its inputs and outputs. On the lower level the box from the higher level is represented in more detail by a collection of boxes being connected by oriented arcs which represent the information/material flow and/or the control signal transfer among the subsystems on the given level of abstraction. On the lowest level the boxes represent the elementary units of the system with the lowest level of abstraction.

Our idea is to follow this top-down design approach. The behaviour of the object represented by the lowest level boxes can be defined either by assigning the standard object from some library or it can be defined by means of some conventional programming language (C+ +, SQL, ...). Some other formalism (Petri nets, signal flow charts, first order predicate logic, contactor schemes, ...) is possible, too.

5 Object oriented and event driven programming

5.1 The significance of objects

The functional analysis approach introduced above is obviously an object-oriented approach. The correspondence between boxes on the lowest level and the library elements can be easily achieved when following strictly the object-oriented paradigm. It calls for use of some object-oriented programming environment.

The behaviour of any object on any level of the functional scheme can be expressed by the definition of responses to various events. The events can be generated by the user from the keyboard, mouse etc.

The above mentioned library may contain prototypes of objects. The user may use these prototypes and determine their detailed behaviour by adding new features or by overriding the default ones. The concept of inheritance enables to do that in very uniform, modular and simple way.

5.2 The event driven programming paradigm

What do we understand under the term *event*? Let the computational process be able to recognize a set of changes in the surrounding environment (a mouse click, hitting a key of the keyboard, etc.). These significant changes we call (external) events. They are transformed to software signals. When we speak about events we mean usually just these signals. The events represented by software signals may be generated also internally by the computational process itself, usually as a consequence of a user invoked external event[5].

The concept of events has its significance in highly modular systems. A system composed of relatively autonomous components (objects in an OOP sense) uses events as the means for communication among those components. That component which recognizes the external event generates the software signal (an event in its metaphorical meaning) thus informing other components about the new situation. It means the system has to be equipped with a signal distribution mechanism.

To explain the signal flow we have to define the system components hierarchy.

For clear explanation we can consider the application user interface composed of two screens each of them consisting of three entries. This situation may be represented by the hierarchy outlined in Fig. 2.

In Fig. 2 the box **A** represents the entire application, $\mathbf{S}_1$ and $\mathbf{S}_2$ are the screens and $\mathbf{E}_{ij}$ (j = 1, 2, 3) are entries of the screen $\mathbf{S}_i$ (i = 1, 2).

Each box in Fig. 2 will be implemented as an (OOP) object. Notice that some objects are collections of other objects (e.g. $\mathbf{S}_1$ is the collection of $\mathbf{E}_{11}$, $\mathbf{E}_{12}$, $\mathbf{E}_{13}$). We can imagine that there is an event processing routine for each object (object's event handler). The root of the hierarchy tree (object **A**) is the only one which is capable to recognize an external event.

Now we are prepared to discuss the event distribution mechanism in an object hierarchy. At least three event types can be distinguished according to the way of their

---

5 Here, we do not consider events caused by hardware interrupts, time slicing, etc.

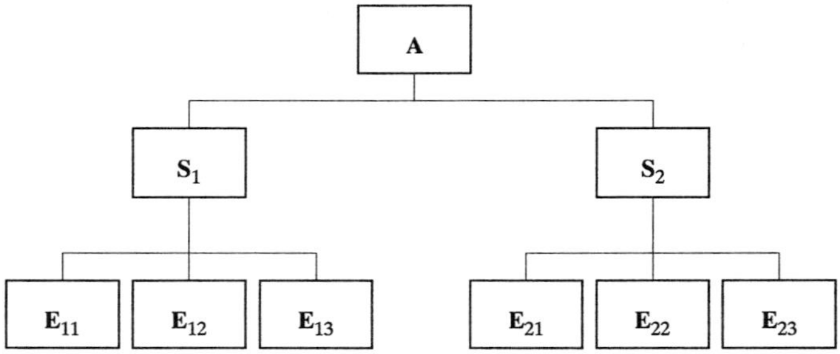

Fig. 2

distribution:

- **broadcasts** are events sent by objects which are collections of other objects to all members of that collection in parallel. Each object can either take an action or ignore the event on its arrival.
- **circulars** are events sent by objects which are again collections of other objects. Collection is organized as a circular list of collection members ordered according to actually valid internal linkage. There is always one *current* member within the collection (e.g. selected by the user). In contrast to the broadcast, the circular is sent to a collection of members sequentially starting with the current member. Each member of the collection may react to the event and one of them may even erase the circular event. If an object erases the event, its successors in the collection will not receive that event at all[6]. Notice, that here the ordering of the members of the collection starts to play the role of a priority chain.
- a **message** is an event sent by one object to another explicitly specified one regardless to their positions in the object hierarchy.

Internally the event is a packet consisting of several sections, namely:

- event distribution type (broadcast, circular, or message)
- sender's identification
- receiver's identification (if applicable)
- data section (carrying the detailed description of the event)

At any moment, the user can fill in at most one entry which is represented by the leaf of the hierarchy tree. This entry will be called the **selected entry**[7]. Consider now only those external events connected with the data entering actions. Thus it is possible to think that the external event (e.g. keyboard hit) is detected by the selected entry. Very often

---

6 The parent object can recognize the erasure.

7 Selected entry is the leaf in the hierarchy tree. There exists a unique path from the root to this leaf. All objects on this path are **selected objects**.

the selected entry itself can execute the external event in which case there is no reason to distribute the event anywhere.

Occasionally, the user presses a special key which should be interpreted as some more global action in the system. The simplest interpretation means that editing of the selected entry has been finished. Usually, this event cannot be processed by the selected entry itself because for instance some consistency checks are to be done. In this case, the selected entry generates an internal event and sends it as a message to the predecessor object in the hierarchy tree.

The addressed object processes the event in an analoguous way. It tries to process the event itself. If the trial is not successful the event is sent higher in the hierarchy, again. If the event arrives at the hierarchy root and the root is not able to process it, the event will be lost (e.g. hitting a special key without any meaning for this application).

When the object recognizes the event, it can be processed in several ways according to the meaning of the event's data section (see above):

- The event can be processed fully within the object, in which case no more events are generated (e.g. screen window resizing).
- If the above statement does not hold, the event is sent as a circular to the object's childern. The selected object (i.e. any object on the path from the root to that one which has started the entire event transaction) always ignores the circular without erasing it.
- If the circular comes back (not being erased by any child) to its sender, the event will be passed upwards in the hierarchy as mentioned earlier. This case may be considered as a part of the test whether the object can process the event or not.

Sometimes it may be advantageous to constrain the upward propagation of certain events. It is possible to declare some object in the hierarchy to be *modal* for a set of events. The modality means that the object is in a special mode which ensures that events from the specified set are never passed higher in the tree. This mechanism can be used to ensure the completeness of user's actions. The alert dialogue window offering several choices may serve as an example. The window cannot be left without selecting one of several admissible answers. Thus the alert dialogue window will be modal for all events excluding the event "correct answer selected".

The above described event handling can be considered as a default one. Since each object has its own event handler, it is possible to change the default behaviour and thus achieve the desired event processing. For instance if some object - a collection member - wants to inform all its peers in the collection, it sends a message to its parent. The parent object converts the message into a broadcast. This is the method how objects on the same hierarchy level can get information about the events within the entire screen.

6 CONCLUSIONS

Event driven programming simplifies the possibility of assembling complex user interfaces from predefined pieces allowing their mutual communication.

An object oriented programming approach coincides with our ideas in the following points:

- high modularity. A library of objects can be prepared in advance. Each predefined object can contain both data and code.
- New objects can be derived from the existing library objects exploating the inheritance property used in OOP.
- Objects usually contain procedural methods. In each of the objects, the event handler can be represented by a method. The inheritance makes it possible to predefine the default event handling of each object. The method overriding facility allowes to redefine the behaviour to the desired one.

The concepts of event driven programming and object oriented programming can form a unified and elegant programming style.

The functional modelling based on IDEF0 methodology is also object oriented in its nature. The arcs in IDEF0 schemes representing the material and control signal flow can be used to model the data and events distribution in a software application. Events correspond to control signals, data to material flow.

The concept of data consistency checks is in agreement with the event driven programming approach. The immediate consistency check may be implemented as a method of the object representing the entry. After entering new data, this method informs other entries about the change by sending a specific event. The dependent entries react to this event by changing their conditional domain and by testing their actual values. Any discovered inconsistecy may result in a new event.

7 REFERENCES

1 Alagič S.: Object-Oriented Database Programming, Springer Verlag, New York, 1988

2 Edmonds E.A., Murray B.S.: Intelligent Presentation and Dialogue Management, in: Advanced Topics in AI (Mařík, Štěpánková, Trappl Eds.), LNAI 617, Springer Verlag, Heidelberg, 1992

3 Gardarin G., Valduriez P.: Relational Databases and Knowledge Bases, Addison-Wesley, 1989

4 IMPPACT - Integrated Modelling of Products and Processes using Advanced Computer Technologies (Bjørke Ø., Myklebust O. Eds.), Tapir Publishers, Trondheim, Norway, 1992

5 Integrated Computer Aided Manufacturing (ICAM) Architecture, Part II, Vol. IV - Functional Modelling Manual (IDEF0), SofTech Inc., U.S.A., 1981

6 Ruland D., Gotthardt H.: Entwicklung von CIM Systemen mit Datenbank Einsatz, Carl Hanser Verlag, München, 1991

7 Pree W.: Object-Oriented Versus Conventional Construction of User Interface Prototyping Tools, VWGÖ, Wien, 1992

Knowledge Based Hybrid Systems (B-11)
I. Mezgár and P. Bertók (Editors)
Elsevier Science Publishers B.V. (North–Holland)

Hybrid Systems for CIM Components' Configuration/ Reconfiguration

István Mezgár and George Kovács

CIM Research Laboratory, Computer and Automation Institute,
Hungarian Academy of Sciences, 1111 Budapest, Kende u. 13-17,
Hungary, e-mail:h724mez@ella.hu

Abstract

The implementation of a Computer Integrated Manufacturing (CIM) system (and even its components/subsystems) is a complex, expensive and long-term investment so it is necessary to design such systems with the most up to date methodologies. The only way to process the design/redesign (or in case of already existing systems configuration/ reconfiguration) task with the proper accuracy is using modeling frameworks that integrate all types of models needed for describing the CIM system in each phase of its life-cycle. An other approach that can assist in CIM design and implementation is the concurrent engineering (CE). The CE technology uses a parallel, interactive, and cooperative team approach to the product and production system design.For the representation of models different programming techniques can be applied. When conventional and artificial intelligence techniques are used in a parallel/integrated way for solving a problem it can be considered as a hybrid way of programming, or application of hybrid systems.After giving a short description of some hybrid experimental systems developed at the CIMRL the paper introduces a knowledge based hybrid prototype system that was developed for configuration/reconfiguration of some CIM components.

Key Codes:I.2.1;J.6
Keywords:Artificial Intelligence, Expert Systems; Simulation,
Concurrent Engineering; Computer Aided Engineering

1. WHAT IS CIM ?

The competition is very strong on the world market, so the high quality and the individualization of the products, the fast reaction of the manufacturer to market demands, coupled with low prices are essential. The market demands can be fulfilled only with the application of different CA (Computer Aided) technologies. In order to raise the efficiency of these systems they have to be integrated into a general system called Computer Integrated Manufacturing (CIM) system.

Most of the different CIM approaches can be classified into two main groups. The first one separates the production related physical activities (manufacturing, assembly, material handling) and the information processing functions (business activities, product design, manufacturing planning, manufacturing control) in the enterprise and declares that automation deals with the physical activities, while CIM deals with automating the four types of information processing functions by using computer systems (E.g.[1]). According to the other approach CIM is an integrated material- and information-processing system, that covers the whole life-cycle of the product, so the inputs of the system are the raw material, energy, information (product-, market- , etc.) and the output is the product(E.g.[2]). According to our opinion the second approach is the one that really covers the essence of CIM, so we use the CIM in this context. The information-, the different design- and planning-, the control-, the manufacturing- and assembly systems are the components of CIM.

There are some important questions in which all approaches agree, namely that implementing CIM is a strategic decision and not only an investment question; in case of CIM systems the integration is the keyword; the role of the human factor is very important.

Usually the most important long-term, strategic reasons for adopting CIM are to maintain/reduce manufacturing costs, maintain/extend market share, integrate information flow, and improve product quality and reliability.

The integration methods are applied in all types of production systems independently of the size of the factory (in SMEs as well as in big companies) and the volume of production (mass vs. one-of-a-kind production, etc.). These fully different production systems need different types of CIM systems.

2. CIM DESCRIPTION WITH MODELS

The only way to create an acceptable, handable process for the stepwise design, planning, etc. and implementation of a CIM system and its subsystems is to describe the system on different levels, from different views in the different stages of its life-cycle by models. In order to solve the task of integration within CIM in a proper and quick way all the systems, subsystems have to be modeled correctly.

A model is the simplified description (from a specific point of view) of the essential properties of an existing or a planned system/object, representing the knowledge about the system/object in a useful form (can be processed, stored, etc.).

During the last few years a lot of efforts were invested in modeling CIM from different aspects. Usually the aims of these efforts were to give a more accurate, general or specific CIM description in order to reduce the time - and the cost - of the design period and to create a "communication and implementation framework" at the same time for the experts who are dealing with these systems in different periods of their life-cycle.

2.1 CIM Modeling Frameworks

As the different types of CIM systems are the bases of future companies, the exact description in all of their life-cycle phases has vital importance. There are different expressions for defining the complex modeling of dynamic, stochastic, large-scale, complex systems like CIM. Reference Architecture, Reference Model, Modeling Framework are everyday elements of the CIM modeling terminology.

The widest range of models are covered by modeling frameworks that provide structures which clarify the links among the different types of models that are representing the knowledge about a system on different abstraction levels, from different aspects and in different phases of its life-cycle. The modeling framework also contains the description of methodologies, processes that make possible the stepwise transformations between the different models.

There are different approaches to CIM (and CIM components) description with frameworks. The best known approaches are the IMPACS modeling framework [3], the CIM-OSA Architectural Framework [4], the Purdue Enterprise Reference Architecture [5], the ICAM project 1105 (Conceptual Design for Integrated Manufacturing), CAM-I DPPM

(Discrete Part Manufacturing Model) and the NBS AMRF [6]. In [7] an introduction and a detailed comparison of these frameworks is given.

As the methodology and the reference model developed at our laboratory is based on the CIM-OSA Architectural Framework, we give a very short introduction into it. The CIM-OSA, an Open System Architecture for CIM, is being developed by the AMICE ESPRIT Project (No.688 and 5288) to support enterprise-wide modeling of a particular enterprise from requirement definition to implementation description. The CIM-OSA also supports the execution of the CIM system model in order to control the everyday enterprise operations, and to provide physical support to full integration of CIM functions, components and information.

The CIM-OSA Modeling Framework is probably the most complex, and this complexity offers many possibilities. Its approach is more theoretical, needs more work to handle it efficiently. This framework (known also as CIM-OSA cube) provides a three dimensional structure with three orthogonal axes (Fig.1). The CIM-OSA framework covers closely our development directions (flexible design/redesign of FMSs) offering the possibility to fit all types of models we are dealing with into a general architecture.

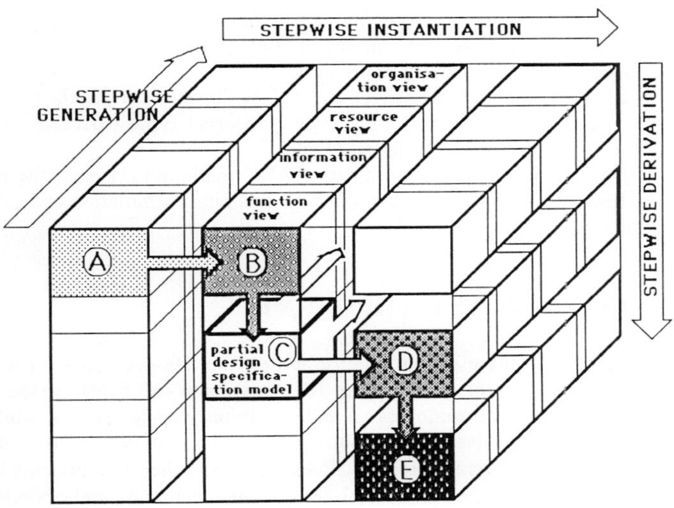

Figure 1.The place of the CIMRL Reference Model in the CIM-OSA Cube

2.2 Application of Reference Models for Design

This chapter shows the steps leading from the reference models to the implementation models for a CIM system based on the CIM-OSA. The basic goal of a reference model on the Partial Design Modeling Level is to give a set of models describing a type of enterprise. From this model the Particular Design Specification Model can be built up with instantiation reduced with the requirements gained from the Particular Requirements Definition Models. This model refers to a specific enterprise with all details needed for its design.

This is the point in system design, where more design versions can be generated, what is very important from the aspect of future reconfigurations. A well-designed system has an extendible life-cycle. "Well-designed" means in this case that the system is flexible enough to produce different kinds of products within the predefined limits with slight modifications, redesigns, if necessary and possible.

The Particular Implementation Description Model derived from the Particular Design Specification Model contains all components of the Information and Manufacturing Technology for the enterprise. Based on this implementation model the physical system can be realized.

2.3 Modeling Methods, Techniques and Tools

A set of methodologies, techniques and tools have been developed for the design and analysis of complex systems from different aspects. These methods have various strengths and weaknesses relating to the different phases in the life-cycle of a system. Usually the modeling fields of an enterprise are divided into information, decision and physical system modeling [6], [8]. In physical system modeling the IDEF0, Petri-Nets, GRAFCET, Simulation, GEMMA systems/methodologies are applied.

According to the conventional classifications of the abstraction levels there are the conceptual, structural and realization modeling levels. Different types of modeling methods are appropriate in the same field on the different levels.

We started with functional modeling using the SADT methodology and the realization phase has been reached with a step by step approach. In this phase simulation and different AI programming techniques have been applied.

3. DESIGN OF CIM AND ITS COMPONENTS

The effect of the design phase on the quality and on the life-cycle cost of a system is significant. It is stated by different authors (e.g. [9]) that about the 80 % of the life-cycle costs of a system is determined at the conceptual and preliminary design stages, while the real costs of these design phases are less then about 20 % of the cumulative costs during the system's life-cycle. An important remark is that CIM systems and their components have to be handled as product innovation as their successful design, implementation and operation can be guaranteed only in this way [10].

The life-cycle of a CIM system can be divided into the System Requirements Specification (e.g.Behavior Analysis), System Design (e.g.Implemented Component Selection), System Build and Release (e.g.Verify Implementation), System Operation phases. (Those sub-phases we are dealing with are in the brackets.)

The complex products, or large-scale systems that can satisfy the overall productivity and quality demands, can be rationally designed by the proper application of the concurrent engineering concept.

3.1 Concurrent Engineering Approach to CIM Components Design

The life-cycle engineering approach to product development (PD) goes beyond the life-cycle of the product itself. By its definition [11] "Concurrent Engineering (CE) is a systematic approach to the integrated concurrent design of products and their related processes, including manufacture and support. This approach is intended to cause the developers, from the outset to consider all elements of the product life-cycle from conception through disposal, including quality, cost, scheduling and user requirements."

CE approach assists to consider the product throughput time reduction as a main objective, beside meeting the customer's quality requirements and delivery dates [2]. The CE technology requires a parallel, interactive, and cooperative team approach to the product and process design in opposition to the traditional design practice, which is sequential, iterative, and distributed.

3.1.1 Phases of the Product Life-Cycle

The life-cycle of a product can be divided into the definition, development (analysis, design, planning) manufacturing, marketing and application phases.During the birth of a new product there are three coordinated life-cycles going on at the same time: conceptual design considering manufacturing; manufacturing process design including production planning, equipment selection, plant layout, process planning, etc.; the development of the product service and maintenance system (marketing, distribution, etc.).

In order to make the design activities better structured, the second life-cycle (manufacturing process design) has been divided into two parts, namely into process and facility design. The first one contains production planning, process planning etc., and the second one means equipment selection, the design of plant layout etc.

3.1.2 Concurrent Engineering for Facility Design

Usually production facilities are used "as is", and CE approach to product design is simplified only to simultaneous product and manufacturing process design. This simplified approach is also called as design for manufacture (DFM).

For small and medium lot sizes (or in the case of different disturbances) existing facilities should be 're-used' after some modifications or re-organizations. This need for reconfiguration/redesign is valid in an extended degree for the one-of-a-kind production (OKP) systems.

The basic reasons and types of reconfiguration are: machine break down, increasing or decreasing need of production volume of the same product, changing the product, introduction of new product, connection/disconnection of resources.

To help in these reconfigurations, or for a more thorough basic configuration that takes into consideration the different layout versions, a broader product range, etc. the CE concept can be effective, it decreases investment and operation costs by integrating manufacturing process and facility design.

3.2 Goals of the CIMRL Reference Model

A simplified and at the same time extended CIM reference model has been developed in our laboratory based on the CIM- OSA architecture. We intend to couple the design of the product and the manufacturing system that produces the product. At this stage of the research we focus only on that phase of the product design (technology process planning) which directly effects the manufacturing system design. The sequence, the parallelism, the loops and the interactions among the different design steps, phases have been defined with the analysis.

The basic goal of a reference system on the Partial Design Modeling Level is to give a set of systems describing a type of enterprise. In our case reference system(s) include the discrete part manufacturing for machine industry.

The main goal was to develop a reference system that gives also the possibility for the design of different production system types (job shop, batch, mass and OKP) by stressing the stronger connection between the information and material flow even in the first stage of the design [12]. The additional goal of our project was to create a reference system that is appropriate to handle the simultaneous product and production system (manufacturing and assembly) design (also in special cases like OKP) by connecting the product and production system life-cycles.

As the first step of our project we developed the reference system with the above described characteristics for FMS and FMC design/redesign. We have developed a prototype knowledge based system for FMC design based on this reference system [13]. In this paper only those models of discrete manufacturing will be discussed that usually are applied in the technical design, installation and operation phases.

4. KNOWLEDGE BASED HYBRID SYSTEMS

Hybrid systems are the results of integration. In the following the "hybrid" word refers to systems that are structurally integrated and consist of conventional (algorithmical), or a conventional simulation program and programs developed by using any kind of artificial intelligent technology. Three basic types of hybrid systems can be distinguished from the point of view of the applied tools and the integration level. In this case there are systems that have been developed

- by using hybrid languages,
- by using hybrid shells/tools, or
- separately and connected to each other later.

The goals and reasons of building hybrid systems are as follows:

- to solve complex, difficult problems,
- separately neither symbolic nor numerical computing can successfully solve the problem,
- the different systems can compensate each other's disadvantageous characteristic,
- to increase efficiency,
- using existing conventional systems.
- to produce more user-friendly systems.

4.1 Hybrid languages

The direct tools of hybrid system development are the hybrid languages. Some well known hybrid languages are the T- PROLOG, TC-PROLOG, CS-PROLOG [14], [15], the simulation LISP [16], and the CAYANE [17].

4.2 Hybrid tools

The hybrid tools offer a higher level, user friendly environment to develop hybrid systems. Today these tools are the most popular ones because highly flexible, effective systems can be built in a relatively short time with them.

These tools are usually knowledge based shells that have an embedded simulation part and also have good graphical possibilities and different language interfaces.Only to list the best known ones of the field; ART, SimKit, Simulation Craft, Nexpert Object, ALL-EX, ALL-EX PLUS, G2.

4.3 Hybrid, coupled systems

A system can be called a hybrid one if it has been composed from a minimum of two systems, one of which uses conventional techniques while the other one is based on artificial intelligence techniques. They can be logically distinguished in some extent after the have been coupled, and they work together in some way to produce an acceptable solution (optimal, or suboptimal). From structural aspect according to [18] there are two ways for connecting systems:

- by coupling two systems only on the "surface", a shallow coupled system can be created, which means that they (or one of them) use only the results produced by the other system. In this case the systems are "black boxes" for each other.
- in the other case the two systems are in a deeper connection, they have common part(s). This type of the connected systems is called as deep coupled hybrid system.

The examples handle always only the connection of two systems, but of course three or even more systems (belonging to different system-types) can be connected, creating a very complex, multiple coupled hybrid system. In the case of blackboard architectures more, different types of "knowledge sources" can be connected creating really complicated hybrid systems.

The connection of expert systems (ES) and simulation always had a big importance. This special group of hybrid systems can be divided into four main categories [15],[19],[20]:

1. Embedded
 a. Simulation model embedded in an ES
 In this case the ES can use the simulation model in decision making during the problem solving process.
 b. Simulation model contains an ES
 In this form the simulation model can get help from the ES in the case of simulating e.g. uncertain processes.
2. Parallel
 a. The simulation can access the results of the decisions made by the ES.
 b. With this architecture
 -the ES can control the simulation model,
 -the ES can be tested on a simulation model,
 -instead of the simulation a real system can be controlled by the ES (real-time process control).
3. Cooperative
 The ES contains some knowledge about the simulation model (how it can be modified) as well as about the field of application. This form of combination is very useful for users non experienced in simulation, because the ES helps them building or modify the model.

4. Intelligent front-end
   The ES is an intelligent front-end for the user, generates the necessary instructions for him, interprets and explains the results from the simulation package.

Of course there are other combinations of hybrid systems only the basic types have been listed above.

5. HYBRID SYSTEM DEVELOPMENTS FOR CIM IN CIMRL

In the last years some prototype hybrid systems have been developed in the CIM Research Laboratory (CIMRL). Three different, but corresponding rapid prototype systems were written before we started the development of the recent one.

There were some additional experiments to test languages, shells,etc but only the most important stages in the field of hybrid systems will be introduced in the following. Finally a comparison, evaluation of these activities also will be made.

5.1 Hybrid system for configuration and simulation

The first version [13] was written in CS-PROLOG language [21] running on a microVAX(VMS) and a networked PC(MS-DOS) was used with EGA graphic for animation. CS stands for communicating sequential. This is a special Prolog language where the simulation of complex systems could be implemented in the form of parallel running processes. This was a straightforward simulation, however the language made some decisions possible during simulation run suggesting for the developers to open the system towards scheduling.

This system was connected to our first experimental expert system written in the rule-based ES shell ALL-EX, to assist in system configuration and reconfiguration [22]. The ALL-EX shell is written in the CS-PROLOG language and is running on PC under DOS. The simplified architecture of the system can be seen in Fig. 2/a.

Later on emphasis was given to simulation and the development of an industrially applicable configuration/ reconfiguration expert system is going to be started soon.

5.2 Hybrid system for simulation, scheduling and evaluation

In the next version the PC based CS-PROLOG simulation was already combined with expert systems implemented in the frame-based ALL-EX PLUS expert system shell. This way questions during simulation run could be answered by the parallel running scheduler advisor which made it possible to have our first scheduler. Another expert system was added to the simulator which ran after the simulation to evaluate results. The system architecture is introduced in Figure 2/b.

5.3 Hybrid system for simulation and input data preparation

The third version [23] was the same as the previous one except of an additional parallel running expert system (quality control advisor) and some accordingly designed changes in inputs and outputs. To make I/O easier and more user-friendly an additional system was implemented in ZORTECH-C++. The system runs on a PC(MS-DOS)-PC(OS/2) network. An overview of system structure is given in Figure 2/c.

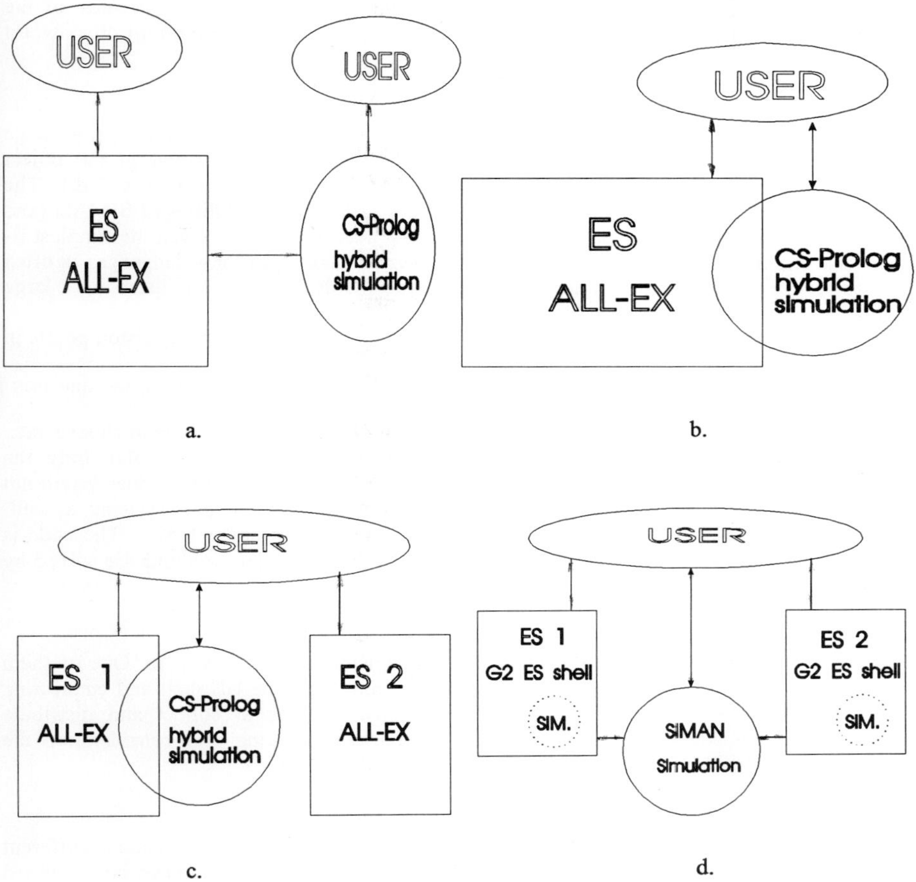

Figure 2. Different realized hybrid architectures

5.4 Hybrid system for simulation, scheduling and quality test

The recently developed system has the main objectives as defined in [13]. Knowledge based simulation and scheduling should be solved including expert evaluation and some quality control functions as well. Additional expert systems help in the reconfiguration of the system. These systems take into consideration the results of the simulation in two extents: the quality of the manufactured product and the throughput of the system. The hybrid system (framework) as a whole balances the quality and the quantity of the production.

Even a preparation advisor is planned to be added to the system, but because of the unexpected complexity of the data preparation problem the development of this expert subsystem was postponed into the future.

5.4.1 Structure of the system

The system is a so called hybrid system as a 'classical' simulation-animation package (SIMAN/Cinema [24]) is used together with expert systems implemented in the object oriented, real-time expert system shell G2 [25] (Architecture see in Figure 2/d.). The application of expert systems is explained by the very restricted possibilities of SIMAN (and all other traditional simulation packages/ languages) in decision making. Even the simplest if-then- else structure which is generally used for decision making in these languages is often hard to realize. However it is rather difficult to write the decision rules if there are a large number of conditions to satisfy at the same time.

If a manufacturing schedule is created with simulation, there are some decision points in the process of the simulation. For example:

-if there are more than one workpieces in a parallel buffer (temporary storage), the question is which workpiece can leave the buffer first, or

-more than one machine tool is suitable for a certain operation, which machine to choose, etc.

The application of a 'real-time' shell is not necessary if our goal is to solve 'only' the above tasks. The main reason of using G2 is that in the future we plan to use the system not only for simulation, but for real-time control of manufacturing and assembly systems, as well.

Recently the system is implemented on a SUN Sparcstation under UNIX. The code is written using the possibilities of G2 and of SIMAN/Cinema. Special problems are solved by routines written in C.

5.4.2 The main problems of software implementation

There were two types of problems during implementation of the system. One of them was the implementation of different expert systems using the same G2 shell and connecting them to the SIMAN/Cinema system. The other problem was to collect and distribute knowledge among knowledge bases of the advisor systems taking into consideration the restricted and used decision power of CINEMA.

5.5 Lessons learned from the different experiments

While developing the prototype systems a lot of experiences were cumulated. Different knowledge representation forms, simulation languages, graphical editors have been applied under different operation systems.

There are lessons have been learned like it is not worthwhile to connect an expert system running under OS/2 with a simulator using MS-DOS.

It is worth to use special simulation packages for manufacturing system analysis because of the built in possibilities.

The animation can help in simulation model validation besides making easier the evaluation of the simulation run.

6. CONCLUSIONS

A short summary of CIM modeling methodologies and the application of concurrent engineering approach for CIM design was given in the paper. The short overview of knowledge based hybrid systems introduced the description of these type of systems developed in the CIM laboratory.

It can be stated based on our experiences that knowledge based hybrid systems are proper tools for analyzing, evaluating complex, stochastic systems like CIM components.

Coupling simulation with expert systems in different ways can increase the efficiency, the speed of simulation experiments. The first experiences with the G2 expert system shell and the SIMAN/Cinema simulation package are promisable. On the SUN Unix environment there are additional possibilities to extend the prototype system with quality analysis, etc.

The result of the project will be a flexible, knowledge based environment for CIM system design that is based on international standards. Within this environment the products and the production components of CIM (automated manufacturing and assembly systems) can be designed in a concurrent way.

7. ACKNOWLEDGMENTS

This research has been supported by the National Committee of Technical Development (OMFB-grant No.91-97-08-0217) and the Hungarian National Science Foundation (OTKA-grant No.T 4122) research grants. Special thanks to our colleagues for their work in implementing the referenced systems.

8. REFERENCES

1 M.P. Groover, Automation, Production Systems, and CIM, Prentice-Hall, Int. Editions, London, 1987.

2 G.J. Olling, CIM Status and Direction in the U.S.A., CAPE '91, Bordeaux, France, 10-12 Sept., 1991, Elsevier Publ., 1991, pp 33-40.

3 G. Doumeingts, B. Vallespir, M. Zanettin, D.Chen, GIM - GRAI Integrated Methodology, A methodology for designing CIM systems, University Bordeaux I, LAP/GRAI, Version 1.0, May 1992,

4 CIM-OSA AD 1.0, Architecture Description, ESPRIT Consortium AMICE, 1991.

5 T.J. Williams, The Purdue Enterprise Reference Architecture, Report number 154, Purdue Laboratory for Applied Industrial Control, December 1991.

6 G. Doumeingts, Modelling Techniques for CIM, Workshop on Computer Integrated Manufacturing Open Systems Architecture, Brussels, 7-8 March 1990, Workshop Proceeding.

7 Architectures for Integrating Manufacturing Activities and Enterprises, Technical Report of IFAC/IFIP Task Force, Editor T.J.Williams, March 1993.p515.

8 K. Mertins, W. Süssenguth, R. Jochem, Integrated Information Modelling for CIM: An Object-Oriented Method for Integrated Enterprise Modelling, CAPE '91, Bordeaux, France, 10-12 Sept., 1991, Elsevier Publ., 1991,pp 315-323.

9 J.L. Nevins, D.E. Whitney, Concurrent Design of Products and Processes, McGrow-Hill, New York, 1989.
10 J. Ranta, The Impact of Electronics and Information Technology on the Future Trends and Applications of CIM Technologies, 2nd IIASA Ann.WS. on CIM: Future Trends and Impacts, January, 1989, pp 130-186.
11 K.E. Winner,et.al., The Role of Concurrent Engineering in Weapons System Aquisition, Report R-338, Institute for Defense Analisys, Alexandria, VA., 1988.
12 G. Kovács, I.Mezgár, E.Szelke, M.Grint, One-of-a-Kind Production- A Concurrent Engineering Approach, IFIP WG 5.7 Working Conf. "New Approaches towards 'One-of-a-Kind' Production", 12-14 Nov., 1991 Bremen, Preprint Eds.:B.E. Hirsch, K.D.Thoben, pp.115-128.
13 G.L. Kovács, I.Mezgár, S.Kopácsi, Concurrent Design of Automated Manufacturing Systems - Using Knowledge Processing Technology, Preprint of The First CIRP Workshop of the Intelligent Manufacturing Systems, Seminars on learning in IMS, Budapest, Hungary, 6-8 March, 1991, pp.205-222.
14 I. Futó, et.al., System simulation and cooperative problem solving on a PROLOG basis, in Implementation of PROLOG,Ed.:J.A.Campbell,Ellis Horwood,1984.
15 I. Futó, AI and Simulation on PROLOG Basis; Int. Symp. on AI, Expert Systems and Languages in Modelling and Simulation, 2-4 June, 1987, Barcelona, Spain, Vol.I, pp. 1-6.
16 R.M. Sergio, et.al.,Towards a knowledge based network simulation environment, in Winter Comp. Sim. Conf., 1985.
17 S.Ruiz-Maier, J.Talavage,, A hybrid paradigm for modeling of complex systems, SIMULATION, 48:4, 135-141 (Apr.1987).
18 C.T. Kitzmiller, J.S.Kowalik, Symbolic and Numerical Computing in Knowlwdge-Based Systems,in Coupling Symb. and Num.Comp. in ESs.,Workshop,Bellevue,Washington, 27-29 Aug.1985, Elsevier S.P., 1986, pp 3-17.
19 A. Lehmann, Taxonomy and application of expert systems in simulation, Int.Symp. on AI, Expert Systems and Languages in Modelling and Simulation, 2-4 June,1987, Barcelona,Spain, Vol.II.,pp 1-6.
20 R. O'Keefe, Simulation and expert systems - A taxonomy and some examples, Simulation 46:1 (Jan. 1986), pp 10- 16.
21 CS-PROLOG Version 3.2-Reference Manual, MULTILOGIC COMPUTING Ltd., Budapest,Hungary,1989.
22 ALL-EX Reference Guide, Multilogic Computing Ltd.,1991.
23 S.Kopácsi, N.T.Koussoulas, D.Gavalcová, Knowledge Based Modelling, Simulation and Evaluation of Flexible Manufacturing Systems, paper to be presented at the KNOWHSEM '93 Working Conference.
24 C.D.Pegden, R.E. Shannon, R.P.Sadowski, Introduction toSimulation Using SIMAN, McGraw-Hill Book Company, New York, 1990.
25 G2 Reference Manual, Version 3.0, Gensym Corporation, 1992.

KNOWLEDGE BASED PRODUCT DESIGN

Knowledge Based Hybrid Systems (B-11)
I. Mezgár and P. Bertók (Editors)
Elsevier Science Publishers B.V. (North–Holland)

Computer support in the (re)design of mechanical products

A new approach in feature based design, focusing on the link with CAPP

O.W. Salomons, J.H. Kappert, F. van Slooten, F.J.A.M. van Houten, H.J.J. Kals

University of Twente, Mechanical Engineering Department, Laboratory of Production & Design Engineering, P.O. box 217, 7500 AE Enschede, The Netherlands

Abstract

Concepts are proposed to improve the computer support in the (re)design of mechanical products. These concepts are combined with a proposal for an improved link between CAD and CAPP systems. The concepts resulted from a methodic design approach aiming at the design of a re-design support system. The concepts incorporate a mixed top-down/bottom-up design strategy. The multiple views problem arising from the use of design and process planning features is dealt with differently when compared with previously published methods. To solve the multiple views problem it is proposed not only to consider the different form features at component level but to also use abstract features and assembly relations in order to yield manufacturing features. Prototype implementations are under development to evaluate the concepts.

Key word codes: I.1.2.1, I.3.5., D.1.5.
Key words: Artificial intelligence; applications and expert systems, Computational geometry & object modelling, object oriented programming.

1. INTRODUCTION

In mechanical engineering design a high percentage of all the design tasks can be considered as re-design tasks. In re-design a great deal of the principles, the structure, the components as well as the relations between these components are already "known". Nevertheless, a lot of time is spent on detailing "known" structures, components and relations. Because of this and the time limits inherent to design practice, designers often overlook manufacturing and assembly aspects. Routine design tasks are therefore attractive for computer support and automation. Improved support of routine tasks will give designers the opportunity to spend more time on the interesting and significant parts of the design. However, true re-design support has not been achieved yet. This also holds for the functional link between CAD and CAPP.

This paper presents concepts in order to improve both the computer support during re-design tasks as well as the functional link with CAPP. The functional link with

CAPP will be shown to be dependent on the required support of the re-design tasks. The concepts have been inferred from a methodic design approach aiming at the design of a re-design support system. This paper focuses on the first phases of the methodic design approach (problem analysis and functional design). A future paper will discuss the last phases of the methodic design approach (configuration and detail design) in more detail.

The paper is organized as follows. Section 2 reports on the background of re-design support. The methodology that has been followed, is explained in section 3. Section 4 details on the problem analysis. Section 5 discusses the functional design of a re-design support system. Section 6 roughly presents the 'embodiment design' which is proposed to support the concepts presented. In section 7 detail design and implementation are discussed briefly. Section 8 deals with results and conclusions.

2. BACKGROUND

The background of re-design support will be outlined according to the following subjects: feature based design (2.1) and assembly modelling (2.2). Refer to [1] for a more detailed review of these subjects. These subjects fall within the broader category of engineering design, which is reviewed in [2]. In [2] three different views on the design process are discerned: descriptive, prescriptive and computer views. Having a descriptive view on design means studying design practice in order to derive requirements for a.o. computer support of design. Having a prescriptive view towards design means that the focus is on a prescribed sequence of activities to be carried out to arrive at a good design. A computer view on design means that the focus is on computer support and automation of the design process. Although the term computer view has generally been accepted, the term information view would be more suitable as it stresses the importance of information analysis and structuring of information as a prerequisite for computer support or automation of design.

2.1 Feature based design

Some progress in the sense of better computer support in (re)design has been achieved by feature based design. Although research in feature based design is rapidly progressing, true design support has not been reached yet [1]. Also an adequate functional link between feature based CAD and CAPP has not been established [1]. The literature on feature based design sofar shows a predominant computer view (or rather an information view) towards the design process. In feature based design the focus has been on the design of single components by using predefined form features [1]. A lot of feature definitions did appear in features literature; for an overview refer to [1]. In this paper we will adopt the following definition of a form feature: "*A generic shape with some engineering meaning.*" (definition by L. Wingard [3]).

Form features often served as geometric building blocks with no other engineering meaning than that of an easy to use piece of geometry for the documentation of the shape of a single component.

In features literature the multiple views problem has been identified. For an example of this refer to the work by Hummel et al. [4]. Hummel et al asked a design engineer to model a single component from a detailed drawing of this component with a commercial feature based CAD system. A process planner was asked to identify the manufacturing features of the same component. Hummel et al observed that the features used by the design engineer were different from those used by the process planner. Hence multiple views would exist between design form features and manufacturing form features (and other application form features). Solutions to the multiple views problem as identified by Hummel et al have been presented in the features literature. The idea behind these solutions is generally the same; translating a component's design form features into manufacturing or other application form features. Depending on the technique that has been applied, these solutions have been given different names. Examples are e.g. feature mapping [5], feature conversion [6] and feature relaxation [7]. However, the situation of the designer as studied by Hummel et al probably does not correspond with actual design practice.

In design practice designers design assemblies focusing first on functioning and assembly aspects. The components that constitute the assemblies are detailed later, taking into account the previously determined assembly and functioning aspects as well as manufacturing aspects. Sometimes the designer does the detailing of the components himself but often this work is done in drawing offices. Usually drawings or sketches together with some written instructions are the medium of communication between design department and drawing office. Therefore design practice generally has a top-down trend.

When CAD systems made their entrance in companies, most often they were used for detailing single components first and building assemblies from these components later; they supported a bottom-up approach. The commercial feature based design systems that have been introduced sofar also supported this bottom-up kind of approach. Therefore these systems do not support the general top-down trend of design practice enough. The fact that these systems are most often used in drawing offices only is illustrative for this. Thus, the design features that Hummel et al identified were just generic shape elements related to the modeller used and were suitable only to document the final geometry of the component. The design features as identified by Hummel et al do not have a functional meaning or design intent; they could have been used by the traditional drawing office, but hardly by actual designers.

The ideas on design practice as described above, were confirmed by descriptive research as described in [8]. A designer was asked to re-design a small assembly and a process planner was asked to make the process plan for each individual component of the resulting design. Only paper and pencil were used. During the design process that was studied, vague notions of geometry developed towards concrete notions of geometry. Therefore it seemed that both form features and abstract features were important in re-design practice. Abstract features can be defined as: "*Entities that cannot be evaluated or physically realized until all variables have been specified or derived from the model.*" (definition by J. Shah [9]). From [8] it appeared that often features resulted from assembly relations. Attempts have also been made in [8] to identify the multiple views on design form features and manufacturing form features as reported by Hummel et al. In [8] multiple views between some design form features and some manufacturing form features could be identified. However, they were not as different that they could not be combined if necessary [8]. From the descriptive studies as described in [8], it also became apparent that no straightforward design strategy is followed in practice. Similar results were reported in [10] for creative design.

2.2 Assembly modelling

Assembly modelling has begun to become a topic since the mid-eighties. One of the first publications in this area was by Lee et al [11]. Ko et al defined some face-face relations by which bottom-up assembly design was made possible [12]. At present some commercial systems already incorporate this possibility to model assemblies. These systems can model assemblies by putting components together that have been detailed out completely by relating faces of one component to faces of another component. These relations are the relations that a completed design will have and which can be used when applying a bottom-up approach to assembly design.

Assembly modelling as part of a more or less top-down design strategy has been described in [13] and [14]. Kimura et al developed a concept called context which is very similar to what will be called a relation in the remainder of this paper [13]. Kimura et al define a context as a combination of a function and a set of shapes (form features). In [14] Gui presents concepts to support design in a top-down fashion using object orientation which - in a way - is similar in spirit to this paper. The same spirit is also found in papers from Zamaniam et al [15], Sodhi et al [16], Roy et al [17], Nielsen et al [18] and Finger et al [19], [20].

3. METHODOLOGY

This paper mainly applies an information view (or computer view) towards re-design support. However, having an information view means that one finally aims at designing software for computer systems. In this case the aim is at designing a re-design support system. Doing so, a methodic design approach, according to van den Kroonenberg [21], has been applied. Therefore the design process of the re-design support system has been divided into the following phases: problem analysis, functional design, configuration design and detail design. As software is to be designed and not some mechanical system, the methodic design approach was followed very loosely and combined with a methodology from object oriented software design [22]. This methodology combines the best characteristics of other object oriented analysis and design methods such as the methods by Booch [23] and Coad & Yourdon [24].

Because of the application dependence of features (and other objects), case studies were applied to two real world problem domains. The domain of the first case study is the (re)design of gearbox assemblies, while the second case study is the (re)design of frame assemblies for aeroplanes. The remainder of this paper focuses on the problem analysis and the functional design phases of the methodic design approach. The configuration and detail design phases of the design of the re-design support system will briefly be discussed in this paper but will be discussed at length in other papers.

4. PROBLEM ANALYSIS

The problem is to design a re-design support system. The system should fulfil the needs/requirements of the users who have to work with it and the companies that introduce it in their organizations. From the background of re-design support (section 2) a general philosophy on re-design support has been developed. From this philosophy general functional needs and requirements for re-design support can be inferred. The general philosophy and the perceived needs and requirements are summarized in section 4.1. Section 4.2 deals with application specific functional requirements that were inferred from the case studies and the object oriented analyses.

4.1 General philosophy and functional requirements of re-design support

A re-design support system should enable the assimilation of existing designs. In doing this it will generally be required to know more about the design than just its geometry (as in today's CAD systems). It is therefore necessary that the designer's intentions are recorded during the design process for later use. Thus, the re-design support system would need some mechanism for design history recording and playback. Recording a design history should distract the designer from his work as little as possible. Except for providing for the assimilation of existing designs, it seems useful to provide the designer with templates of assemblies that can be detailed further.

A lot of tasks in re-design can be automated while a lot of other tasks need a high degree of user interaction. Therefore a re-design support system should sometimes be system-driven and sometimes be user-driven; in this sense the system should be hybrid. The question when it should be user or system driven depends on the application. The system should therefore be configurable depending on the application. Probably there will be different users of the re-design system: the user who configures the system to some application (the system manager) and the user who is actually supported by the system in doing re-design work (the end-user). Most probably these different users will put different requirements on the system.

End-users would like to work with entities that have meaning to them: assemblies, components and features that are relevant in a certain application. They also would like to remain as free as possible; to be able to work top-down when they like and bottom-up when they like. When applying a more or less top-down design style, features often result from relations with other components. Therefore a re-design support system must provide for assembly modelling. Component modelling functionality is also required when the trend is more or less bottom-up. Moreover, component and assembly modelling will often be closely related; assembly relations (functions) will give rise to constraints to be imposed on the components. Therefore constraints should be modeled and managed in close relation to assembly and component modelling. The re-design support system must complement for the cognitive limitations of end-users as much as possible. For example, the limited memory capacity of human end-users could be complemented by the external memory facilities of a re-design support system.

System managers need tools to customize the re-design system for a certain application. They define features, components and assemblies. They also determine the freedom that is given to the end-users; they determine when designers are allowed to work top-down or bottom-up. System managers can also define prescribed sequences of actions to be carried out. It is therefore very important that system managers have in depth knowledge of the design process and the products for which they are configuring the re-design support system.

When designing assemblies, often standard components have to be chosen and ordered; a re-design support system must also support this functionality. Usually vendors of standards components offer catalogues in computer readable format. It would therefore be preferable to use existing catalogues instead of putting a lot of effort in creating one's own catalogues. The re-design support system could be linked with inventory control systems in order to inform the designer which parts or materials are available. The automatic generation of ordering lists is another possibility connected with the catalog selection functionality. Selection from standards tables and other catalog information should also be allowed for. Calculations also would have to be provided for.

As in top-down (re)design processes abstract notions of geometry usually develop into detailed notions of geometry, abstract or incomplete geometry (abstract features) can be useful. Form features remain necessary for bottom-up modelling of components.

The kind of company which introduces the re-design support system also is of influence to the general requirements imposed on the system. For example, the question whether the company designs and produces by itself or whether it only designs and orders its products from manufacturing companies will put different requirements on the re-design support system. The kind of applications which should be built in or linked with are dependent on the kind of company.

4.2 Application specific requirements; cases and object oriented analyses

The case studies confirmed many of the general functional requirements that were mentioned in the previous paragraph. For instance, making selections from catalogues, standards or company internal technical handbooks. A great need for feature mapping has not been identified because the multiple views problem between design form features and manufacturing form features was seen as minor in both case studies.

A need for the use of assembly relations however, could clearly be identified. What will be called a relation in this paper can be considered as an assembly relation. A relation is defined as follows: *a relation relates two or more objects*. Objects in this case are assemblies, components and features (abstract features and form features). Relations can exist at multiple levels of abstraction and can be decomposed, finally resulting in complete component descriptions in terms of form features and low level relations between these components (face-face relations). It must be noted that when reference is made to relations, feature-feature relations within the same component are excluded. These relations between features within a part have been researched a lot and are closely related to the way in which features are represented. For an overview of research on feature representations refer to e.g. [1]. An example for the need of relations is taken from the gearbox case. Different possibilities can exist for the connection relation between a shaft and a gearwheel, when designing gearboxes. A weld connection or a key connection could be possibilities that are frequently used within a company to realize the shaft-gearwheel connection relation. Choosing for a particular key-connection could finally result in the automatic generation of a key component and keyway features in both shaft and gearwheel.

Within many companies it will be required that a re-design support system links with existing software. For example, existing CAD systems or databases etc. In the aeroplane case study for instance there is the requirement to be able to link with the CATIA™ CAD system.

5. FUNCTIONAL DESIGN

The main function of a re-design support system is shown in figure 1 as a black box together with its in- and outputs. The main function could be subdivided into the following sub-functions:

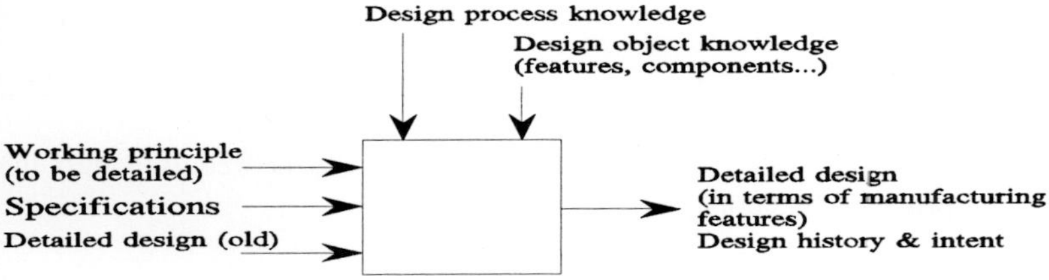

Figure 1. Main function of a re-design support system shown as a black box, together with its inputs and outputs.

- end-user interface.
- modelling (of assemblies and components, including catalog selection, recording of design histories etc.).
- kernel modelling (geometry engine).
- applications (mapping to the applications and applications themselves).
- interfacing with other systems.
- database interface and database (for handling and storing data on e.g. product models, catalogues, features, relations etc.).
- system manager user interface.

In figure 2, based on the above functions, a system architecture for a re-design support system is shown. In the following each of the subfunctions that was determined will be discussed and sometimes even be decomposed again.

5.1 User interface for the end-user

This functional module should allow the designer to work top-down or bottom-up whenever this is allowed and/or required. For example, if a designer does not want to choose a standard connection between a gearwheel and a shaft, company policy could be to not let him allow to model his own alternative.

Sketch input would be desirable, at least if results from descriptive studies of "paper and pencil" design processes like [8] and [10] can be extrapolated for computer aided

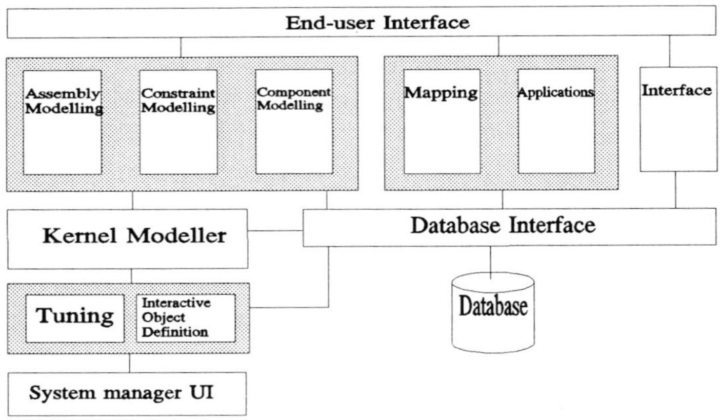

Figure 2. Architecture of the re-design support system.

(re)design support systems. However, more research seems to be required into the need for and the provided functionality of a sketching module.

5.2 Modelling (of assemblies and components)

In most present CAD systems assembly modelling functionality is not present. If it is present, it only allows one to relate components which have fully been detailed out previously in terms of solids by means of low level face-face relations. This represents a bottom-up design style, which does not completely correspond with design practice in which top-down styles are also applied (as was observed in [8], [10]).

The object oriented case studies indicated that assembly relations and abstract features could help in supporting a top-down design style when modelling assemblies. Form features remain necessary for the bottom-up style design of single components. Constraints can be derived from functions (top-down) or can be imposed on single components by judgement of the designer (bottom-up). The modelling function can therefore be subdivided into assembly modelling (1), constraint modelling/management (2), component modelling (3) and auxiliary modelling (4).

- 1. Assembly modelling --- Assembly modelling could be done top-down by making use of relations and abstract features as described previously. Bottom-up assembly design is possible by using low level face-face relations. Detailing relations or objects means that constraints such as tolerances will appear as well as form features. This gives rise to the "grey glue" in figure 2 which closely connects assembly modelling with constraint modelling/management (2) and component modelling using features (3).

- 2. Constraint modelling/management --- Constraints such as tolerances and dimensions could be inferred both from detailing abstract relations or objects (top-down) or just from modelling single components first, defining tolerances on them and putting them in the assembly later (bottom-up). Constraint management is thus related to both the assembly and component modelling functional units.

- 3. Component modelling --- Components can be modeled by both using abstract features or by using form features; top-down and bottom-up design style. Components are usually part of an assembly and thus there is a strong link with (1). Also constraints can be imposed on single components which explains the link with (2).

- 4. Auxiliary modelling functionality --- The "grey glue" in figure 2 reflects the close connection between the above described functional units. Services that could be provided for are e.g. recording and playing back design histories.

5.3 Kernel modeller (geometry engine)

The kernel modelling part is closely connected to the previously described modelling part. The functionality of the kernel modeller mainly consists of providing geometry and solids modelling functions. This functionality can be provided by present commercial packages like - for example - the ACIS™ kernel modeller or the ICAD™/Parasolid™ modeller.

5.4 Mapping and applications

In figure 2 "mapping" is shown together with "applications". By "applications", applications are meant that are built in within the re-design support system. Manufacturability evaluation could be an example of a possible built-in application. However, other applications than manufacturability evaluation such as design evaluation, assembly evaluation and analysis could be possible as well. The mapping function will be addressed in more detail in section 5.8.

5.5 Interface

The re-design support system should be able to generate output data, to read in input data and to exchange data with other (CAX) systems. Thus, interfacing with applications outside the re-design support system should be provided for. Mapping could be required again to prepare for this interfacing. In figure 2 an interface is shown which could be of any type e.g. STEP, IGES, system specific etc. The use of a system specific interface should be prevented when possible.

5.6 Database interface and database

A lot of information needs to be accessed by the different functional modules; e.g. standard component catalogues, predefined features, old designs, design histories etc. For doing this a database interface and a database are required.

5.7 User interface for the system manager

Components, features, relations, mapping rules etc. are all domain dependent and a re-design support system should allow customization per application domain. This customization is the task of the system manager. In customizing an application the system manager has a need for some easy to use tools as well as an own user interface. In figure 2 tuning and interactive object definition are shown as some of the main tasks to be fulfilled.

5.8 The link with CAPP

One of the possible (usually external) applications following design is CAPP. In the following sections proposals will be made to realize the link with CAPP systems, depending on the design style that a component was created with; i.e. by a top-down or a bottom-up style.

- *The link with CAPP for components resulting from a top-down design style* --- Relations besides objects could serve for achieving the CAPP link for components resulting from a top-down design style. Rather abstract objects and relations will develop towards more concrete objects and relations during the design process. This gives one the possibility to store and revise previous decisions such that one could speak of a design history capture and play back. A design history can be used advantageously not only within re-design but also within process planning. If problems arise in the process planning phase of a certain component its history could be played back to find the rationale behind the trouble causing features or tolerances. As many features are due to assembly relations, it seems important to also include their history and intent. Alternative assembly relations can give rise to new features which might cause less problems with regard to manufacturability. Components resulting from this kind of top-down design style, in the end of the design process, will be described in terms of manufacturing form features. Thus the need for transforming design form features into manufacturing form features will have vanished. Final detailing or modifying might even be done automatically. Certain design decisions could even be delegated to a manufacturing evaluation module or CAPP system.

- *The link with CAPP for components resulting from a bottom-up design style* --- For those parts of the re-design task for which there is no computer support by predefined relations and components, the designer could work in a bottom-up fashion to obtain the components needed. Components can be modeled bottom-up using features in roughly two ways; by using generic geometric building blocks (the design form features as described by Hummel et al in [4]) or by using manufacturing form features. In the latter case, i.e. when manufacturing form features are used in design, there is no need for feature mapping. However, in the case of using generic geometric building blocks or design form features, the design form features have to be translated to manufacturing form features in some way. This could be done by using feature recognition or by "mapping" techniques similar to those presented in [5], [6] and [7].

6. CONFIGURATION DESIGN

How each of the previously determined functions could be fulfilled and which alternatives have been chosen will be discussed at length in a future paper. For some functions, some possibilities for fulfilling them have been already indicated in the preceding sections. One of the alternatives that has been considered is that of embedding the ICAD™ system in a generally applicable re-design support system. The use of existing CAD systems for specific re-design applications has also been investigated. In the following we will focus on the data structure for the assembly modelling functionality and end-user interface of a generally applicable re-design support system.

7. DETAIL DESIGN

The data structure of the assembly modelling module consists of objects and relations. Objects in the end will be physical objects, while relations in the end give rise to form features and low level face-face relations. Both relations and objects are modelled as objects in an object oriented approach. Both objects and relations can and should be classified. The exact classification will depend on the application

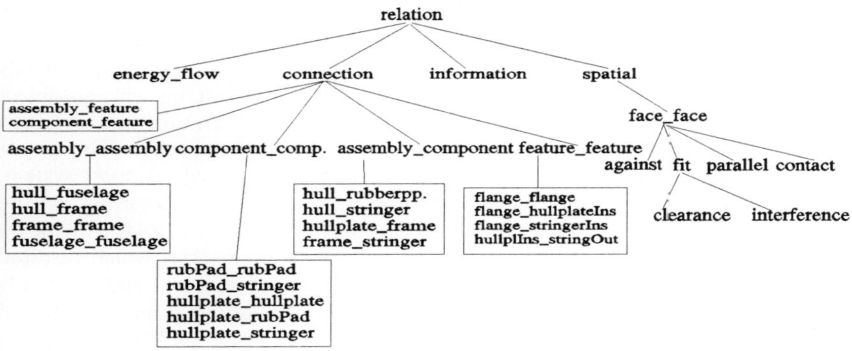

Figure 3. Relation classification hierarchy for aeroplane frame assembly design.

domain. Relations are - just like features - domain dependent. Private communication with prof. D. Ullman resulted in the preliminary classification of relations into spatial relations, connect relations, energy flow relations and information relations. In figure

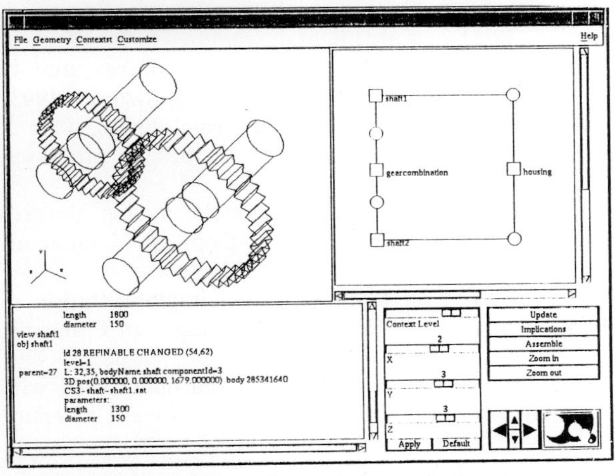

Figure 4. End-user interface of the re-design support system for gearbox re-design.

3, the final relation classification hierarchy of the aeroplane frame domain is shown; both a problem independent and a problem dependent part can be identified.

How the end-user interface presently looks like is shown in figure 4. However, the end-user interface as shown in figure 4 is only a preliminary one; more research seems required for the development of proper end-user interfaces for CAD systems. This holds especially for the question of sketching. In a future paper we hope to elaborate on these developments.

8. RESULTS AND CONCLUSIONS

A methodic design approach towards the design of a re-design support system has been presented. Methodic approaches from mechanical engineering design and object oriented software design have been combined. Results from literature and previous research have been taken into account in this methodic design approach. This paper only presents a first 'iteration' of the process of designing a re-design support system. Applying a methodic design approach seems useful as future design iterations can be taken into account, as it gives one a guideline during the design and as it provides a means for documenting and discussing the design. The main concepts which resulted from the methodic design approach are those of abstract feature and assembly relations. The use of abstract features and (assembly) relations seems to be of importance in improving the functionality of feature based design and in overcoming the multiple views problem.

Acknowledgements

This research is supported by the Technology Foundation (STW) and Fokker Aircraft B.V.

References

1 Salomons O.W., Houten F.J.A.M. van, Kals H.J.J., Review of research in feature based design, submitted to the Journal of Manufacturing Systems, 1993.

2 Finger S., Dixon J.R., A review of research in engineering design, Part I: descriptive, prescriptive and computer based models of design processes, Research in engineering design, vol.1, 1989, 51-67.

3 Wingard L., Introducing form features in product models, a step towards cadcam with engineering terminology, Licentiate Thesis, Dept. of Manufacturing systems, Royal Institute of Technology, Stockholm, 1991.

4 Hummel K.E., Brown C.W., The role of features in the implementation of concurrent product and process design, ASME Winter Meeting, San Francisco, 1989, 1-8.

5 Bhatnagar A.S., Implementation of feature mapping and reasoning shell with application to group technology coding, MSc Thesis, Arizona State University, August 1988.

6 Rosen D.W., Dixon J.R., Dong X., A methodology for conversions of feature based representations, Int. Conf. on Design Theory and Methodology, DTM '91, ed. Stauffer, DE Vol.31, 1991, 45-51.
7 Mäntylä M., Opas J., Puhakka J., Generative process planning of prismatic parts by feature relaxation, in: Advances of design automation 1989, presented at the ASME Design Technical Conferences 15th design automation conference, Montreal, Quebec, Canada, 1989.
8 Salomons O.W., Slooten F. van, Houten F.J.A.M. van, Kals H.J.J., Observations from redesign and process planning practice, a study of human practice aimed at improving feature based CAD/CAPP, proceedings of the twelfth International Conference on CAD/CAM, Computer Graphics and Computer Aided Technologies (MICAD '93), Vol.1, Hermes, Paris, February 9-12, 1993.
9 Shah J.J., Conceptual development of form features and form feature modelers, Research in engineering design, Vol.2, 1991, 93-108.
10 Ullman D.G., Dietterich T.G., Stauffer L., A model of the mechanical design process based on empirical data, Artificial Intelligence for engineering, design and manufacturing, vol.2, 1, 1988, 35-52.
11 Lee K., Gossard D.C., A hierarchical structure for representing assemblies, part 1, cad, vol.17, no.1, 1985.
12 Ko H., Lee K., Automatic assembling procedure generation from mating conditions, cad, vol. 19, no.1, 1987, 3-10.
13 Kimura F., Suzuki H., Tanaka I., A pattern directed design system for machine assembly, Annals of the CIRP, vol. 40, 1, 1991, 127-130.
14 Gui J.K., Object-oriented assembly and assembly design process modeling, journal of Engineering Design, vol.2, no.2, 1991.
15 Zamaniam M.K., Fenves S.J., Thewalt C.R., A feature based approach to structural design, Engineering with computers, 7, 1991,
16 Sodhi R., Turner J.U., Representing tolerance and assembly information in a feature based design environment, Advances in Design Automation, DE-Vol.32-1, ASME, 1991, 101-106.
17 Roy U., Banerjee P., Liu C.R., Design of an automated assembly environment, cad, vol.21, no.9, 1989, 561-569.
18 Nielsen E.H., Dixon J.R., Zinsmeister G.E., Capturing and using designer intent in a design with features system, Proceedings Design Theory and Methodology, DE-Vol.31, ASME 1991, 95-102.
19 Hoover S.P., Rinderle J.R., Finger S., Models and abstractions in design, Int. Conf. on Engineering Design, ICED '91, Zurich, August 27-29,1991.
20 Finger S., Safier S.A., Representing and recognizing features in mechanical designs, 2nd. Int. Conf. on Design Theory and Methodology, DTM '90, 1990.
21 Kroonenberg H.H. van den, Design methodology as a condition for computer aided design, VDI-Berichte 565, VDI Verlag, Düsseldorf, 1985.
22 Aksit M., Bergmans L., Object oriented analysis, design and implementation, course notes, University of Twente, 1991.
23 Booch G., Object oriented design with applications, Benjamin Cummings Publishing Company, 1991.
24 Coad P., Yourdon E., Object-oriented analysis, 2nd ed., Yourdon press, 1991.

Knowledge Based Hybrid Systems (B-11)
I. Mezgár and P. Bertók (Editors)
Elsevier Science Publishers B.V. (North–Holland)

Concurrent engineering with the PRODES system

I. Horváth[a], P. Kulcsár[a] and L. Horváth[b]

[a]Institute of Machine Design, Technical University of Budapest, Műegyetem rkp 3, H-1111 Budapest, Hungary

[b]Department of Manufacturing, Bánky Donáth Polytechnic, Népszínház u. 1, H-1081 Budapest, Hungary

Abstract

This paper describes a preliminary report on a project to develop the core of a computer-aided concurrent engineering system, called PRODES. First an overview is given on the different approaches to feature-based design, which has been further developed to result in the shape design and modelling methodology of the PRODES system. This methodology, detailed in the second part, is based on the abstract object concept. Abstract objects have varying structural topology and shape geometry. They can be qualitatively characterised by and selected based on their functions. In the third part of the paper, the architecture and the different operating states of the PRODES system are described. Finally, the semi-generative methodology of manufacturing process planning with the PRODES system is described.

Keyword Codes: J.6; I.3.5; I.2.1
Keywords: Computer-Aided Engineering, Object Modelling, Expert Systems Application

1. INTRODUCTION

Traditional product development is a sequential, distributed and iterative process. To satisfy the requirements of up-to-date product development and manufacturing, however, we need parallel, integrated and straightforward processing [1]. The approach that makes these possible, is concurrent (or simultaneous) engineering. In a concurrent engineering methodology, different development and manufacturing activities exert dynamic effects on each other, therefore, productivity and quality can be increased significantly. Implementation of concurrent engineering (CE) requires: (i) integrating the product development facilities of the company to form a flexible, reconfigurable and open system; (ii) application of creative group techniques early in the conceptual engineering phase in order to use the collective knowledge for solving engineering problems, and (iii) using sophisticated computer-aided concurrent engineering systems supporting interactive and parallel knowledge processing of the different disciplines involved.

A competitive company must apply relevant software packages which, in themselves, are able to implement the CE methodology. Traditional CAD systems do not represent the most appropriate solutions in this respect. Up to now, more emphasis has been placed on the teamwork aspects of CE than on the development of specific software systems. As a matter of fact, concurrence has been attacked from the point of view of the methodising and synchronisation of sub-processes rather than of parallel generation and evaluation of solutions by an integrated system. Among the used approaches, we should mention the methodologies of design for functioning, design for manufacturing, design for assembly, design for quality and design for testability. Concurrent engineering systems are expected to address this problem as well. Software development for CE raises at least three questions:

(i) Where are the limits of making conceptual, structural, geometric and technological design activities parallel?
(ii) How can the product and the process be modelled to facilitate the integration of the design activities?
(iii) How can the design knowledge and data be captured, formalised and represented in a design systems?

The primary challenge facing computer-aided concurrent engineering systems is the integration of the information generated and used by the different disciplines (e.g. functional design, geometry design, engineering analysis, manufacturing planning and so on) into a shared product model, or a meta-model [2].

With the financial support of the Technology Development Committee of Hungary, research and system development was initiated at the beginning of 1991 at the BTU, Institute of Machine Design, to outline and program a core system with a design methodology, which facilitates concurrent engineering. To provide the multi-faceted methodology, the concept of feature-based design was adopted and modified to satisfy the requirements. At the launching of the project, four specific requirements were defined for this system: (i) it should increase the productivity and consciousness through the use of the pre-programmed libraries of design and manufacturing features; (ii) a product modelling methodology should result from providing a unified object oriented representation scheme for the functional, design and manufacturing features and for the design sub processes; (iii) it should help engineers of different professions think in accepted technical and engineering terms (concepts) rather than in neutral geometric primitives and drawing entities, (iv) it should behave as a consulting expert system by providing tools for the capturing, representation and processing of qualitative functional, morphological and production information in a unified manner.

Recent CAE systems have made significant progress in the simultaneous execution of design activities, since the data stored in their database can be accessed directly by engineers using different modules of the systems concurrently. But these systems do not give answers to the questions of qualitative design knowledge processing nor do they produce more than one solution to the problems, respectively. CE processes require advanced product modelling and an information management environment. The hardware architecture has to make the distributed handling of the product information possible in heterogeneous networks of mainframes, servers, workstations, personal computers and sensor processors. With the advent of solid modelling, it became possible to handle the engineering quantities, structure topology and, partly, the manufacturing aspects of the mechanical products. Progression has been made in the formal specification of feature-based design technology. In this respect the Integrated Product Information Model specified by the Standard for the Exchange of Product Definition Data (STEP) is the most significant [3].

The feature concept is a very important in realisation of product models, since they represent the tools for the integration of geometrical and manufacturing knowledge [4]. A feature is a distinctive or characteristic region of a part (workpiece) defining a geometric shape which is specific from the point of view of manufacturing [5]. Features, by definition, are always technological process oriented. Therefore they are often called technological features. Technological features includes keyways, holes, ribs, bosses, undercuts and so on. They are the bridge between geometry description and process planning. Certain features are in conformity with functional parts of the workpiece or assembly [6]. They can be called design or constructive features [7]. Design features reflect the realisation that geometric models represent the design in greater detail than is useful for the experts. The features must also include the quantitative data and the qualitative attributes. The quantitative data describes, for instance, the geometry, topology, material specification and manufacturing parameters. The qualitative data are used to describe the functionality and behaviour of the object. Those features that describe the physical phenomenon and operating functions are functional primitives.

There are several methodologies to handle features in product models. The most significant ones are: (i) feature specification; (ii) feature recognition; and (iii) design with features. Feature specification is the simplest way of using features. If on a geometric model of a part a certain combination of the geometric primitives appears, it can be considered a feature. Having identified this group of geometric primitives, process planning systems are able to compare it to the elements of a predefined feature library. A more interesting case is the automated extraction of manufacturing features from a previously defined geometric model, which is called feature recognition [8]. If features are recognised, the design can be evaluated for manufacturability and the process plan can be recomposed from the specific manufacturing operations. Feature recognition may be implemented based on 2D, 2.5D and 3D models with different representations [9]. By the types of machining features recognisable by the system, the rotational and the prismatic part feature extraction approaches can be separated [10]. A design with features approach assumes a database of design features that can be combined to form the geometric model and to provide the information for manufacturing. Since features are higher level entities than the primary geometric primitives, their use is more practical for the designers and results in higher performance. However there are still many unanswered questions and certain limitations have already been revealed. Design with features makes the design process more object oriented. For systems using the design with features concept, the most important system module is the feature catalogue that should be filled by the user according to the demands of the application. Design with features can again be based either on the CSG or B-rep representation schemes. Features should be taxanomized by group technology. Unfortunately, no theory and methodology exist to decompose an object into manageable (and functional) features and then to recompose them into a product. Most of the research work on feature-based CAD addresses a portion of the features required for a particular area.

2. ARCHITECTURE OF THE PRODES SYSTEM

The PRODES system is developed on an ACERFRAME 3000 MP50 four-processor, workstation level microcomputer with 32 MBytes RAM. For the time being, the software package runs under the MS WINDOWS 3.1 operating system [11]. For storing the executable files, the feature libraries and the data bases, a 560 MBytes hard-disk storage is needed. The system is programmed in C++ object oriented programming language. The architecture of the

PRODES system is different to both the integrated CAD systems and knowledge based expert system. The system is modularised and, with respect to its software components, can be separated into two parts; the system shell and the operative software pool. Their main modules and the control flow among them can be seen on Figure 1. This specific architecture is required by the design methodology that can be implemented in the PRODES environment. The system

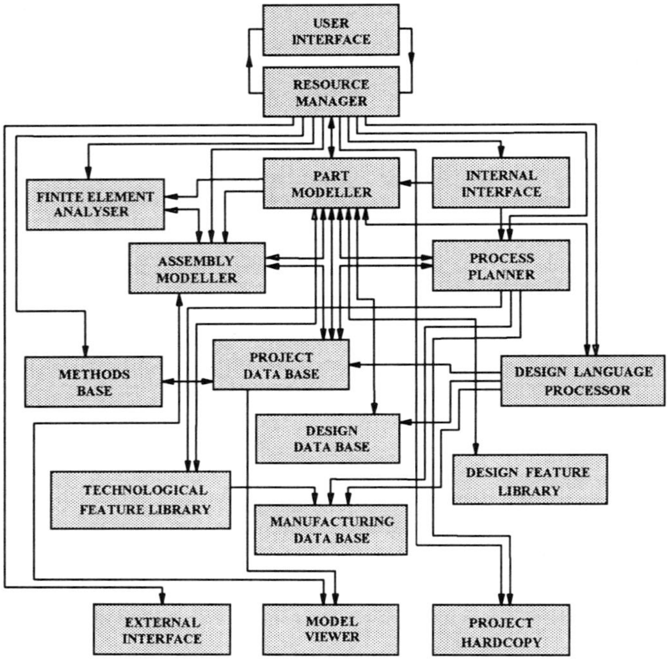

Figure 1. The modules and architecture of the PRODES system

shell comprises of the user interface and the system manager. The graphical user interface allows the selection of the needed operating states of the system, commanding the resource manager and maintaining graphical and numerical data communication with the system. It uses nearly all of the pre-programmed program building resources and system programming tools of the WINDOWS. The PRODES system is a knowledge-based system. The knowledge of the system is made of the sets of the design and technological features, the facts describing the manufacturing environment, the design data base, the object oriented configurer and, finally, the numerical analysis methods based on simple mechanical or more sophisticated finite element models [12].

The resource manager is one of the most important modules. It controls the operation of the system by invoking and linking the software needed in the various operating states. Note that the PRODES system has seven main operating states. These are: (i) resource development; (ii) conceptual design and configuration; (iii) part modelling and viewing; (iv) assembly definition and management; (v) methods application and finite element analysis; (vi) manufacturing process planning; (vii) data communication and documentation. The resource manager sets the system parameters that belong to the required state, initialises the system tools, opens the needed libraries and databases and controls the operation.

A privileged module of the PRODES system is the part modeller. It provides the tools for generating the shape models based on the combination and evaluation of abstract and concrete feature objects, respectively. The operation of this module compares with that of a traditional CSG modeler. The geometric primitives, however, are design and technological features with specific internal representation. The part modeller provides tools for deriving object instances based on these features. The manufacturing database of the PRODES system contains both knowledge for automatic decision support procedures and facts for the manufacturing possibilities available in the manufacturing environment (factory) being considered [13]. The files of factory data are for the manufacturing departments, machine tools, cutting tools, fixtures and their interconnections. There are files for state changes of primary shape surfaces and technologic features and process structure patterns for primary shapes [14].

3. METHODOLOGY OF CONCURRENT ENGINEERING BASED ON THE ABSTRACT OBJECT CONCEPT

The design with features approach passes the problem of mapping the product conception onto a manufacturable and operable geometry to the designer. When this methodology is applied, designers should have an original and proved conception for the object to be designed. All implementation of design with feature methodology suppose that the preliminary design and the embodiment have already been done and the geometric/structural parameters of the object are known. On the other hand, we may say, that the above mentioned methodologies presume the existence of one or more formal models of the object. These formal models can be considered design prototypes, which can be generated either by the modification of a previously existing object or by synthesising that of constitutional entities. The use of systems involved with the implementation of the design with features methodology is the most effective if the design prototype is thoroughly worked out. During designing, it is the designer who selects the feature entities based on their functionality and combines them into products providing the needed performance. In case of the PRODES system, the designer may obtain support from the system to conceptualise the prototype and to elaborate the functional domain of the object.

Being a quasi expert system, the PRODES system is application domain oriented. The intended field of application of the system is belt conveyers for transporting bulk and piece material. The PRODES system is designed to support the initial phases of engineering design, that is, conceptual design, shape design and analysis and process planning. Its functional modules were selected and arranged to open up new vistas in the implementation of the concurrence. The architecture of PRODES strives towards getting rid of the problems of sharing parametric solid models among individual systems for conceptual design, shape design, numeric analysis and manufacturing planning. Solving this problem is regarded as the most important question in the advancement of computer-aided engineering.

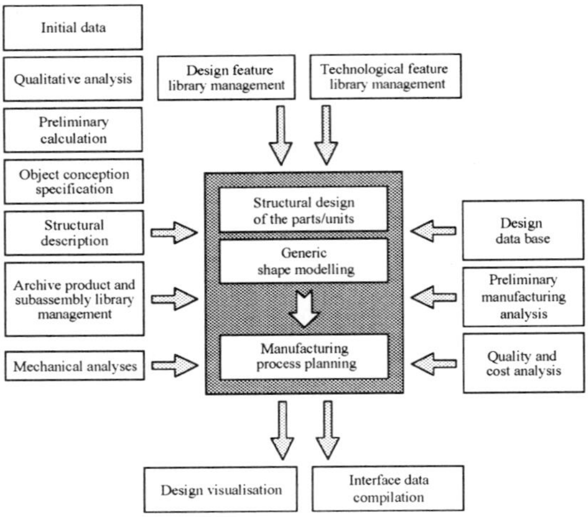

Figure 2. Process model of the PRODES system

Process model of the concurrent engineering, carried out by/with the PRODES system, can be seen on Figure 2. In every single operating state of the system, certain set of functional modules are active or can be activated. For instance, when the system is prepared for object-dependent application, first the design language processor is called through the user interface and the resource manager, respectively. By use of the high-level PRODES-L language, the system can be controlled, entities for the design and technologic feature libraries can be generated and mapped onto data structures. Furthermore, files for the design and manufacturing database can be inputted. The design language processor has a built-in compiler that interprets the lexical structures of the input source file, executes syntax checking and generates internal data structures or a runable command sequence file. Visual checking and correction of the shapes of the design and technologic features requires the use of the part modeller.

The PRODES system is for generating design prototypes of belt conveyers by concurrent engineering rather than for producing drawings for those objects. Therefore, the options for drawings and dimensioning are not included in this system. The facility to transfer model data to commercialised drawing system, however, provided through the built-in IGES converter. This external interface makes, the otherwise integrated, architecture open and extendible in the future. For the time being, an interactive interface is provided between the project database and the manufacturing process planner, to interrogate the project database and specify the auxiliary geometric and non-geometric data of the modelled part.

When an object designing session is commenced, the resource manager activates either the preliminary designer software in the methods base or the part modeller. From the part modeller the design and technologic feature libraries, the design and manufacturing databases can directly be accessed, as required. The part modeller gives a simple wireframe representation of the objects. If a wireframe representation with hidden line elimination or surface shading is needed, the model viewer is to be activated. If not a single part, but an assembly of parts is to be designed, the user should change for the assembly modelling operation. This is also the case, if an original specification and coding of an assembly is to be done. Having the object definitions for all of the parts, their shape design by the part modeller can be started. Again, the assembly modeller provides a simple wireframe representation, which can be converted into a hidden line removed or shaded solid representation by the model viewer. For managing and applying analysis programs, the PRODES system provides the method base. In fact, this is a library of specific analysis programs that can be executed in the PRODES environment by the interrogation of the project database. For deeper analysis of the behaviour of the objects, a finite element analyser is included with direct interfacing to the part modeller and the project database.

The preliminary calculation is oriented to determine the global geometric (e.g. conveyer belt length and width, required material cross section, support arrangement) and operating parameters (e.g. belt speed, power requirement, live loads). The preliminary designing software contains the object conceptions (notations for sub assemblies, units and parts) which are to be used to built up the conveyer belt with the required functionality. From the known possible configurations, the software specifies those that fully or mostly cover the requirements. This is comprehensively called object conception specification. In addition, the software gives the list of the needed objects. Based on this, the designer may give a final specification for the required libraries. Embodiment starts with the component design made by the designers. Taking into consideration the required functionality and structural connections, the designer is to conceptualise the shape model of the part. Then the best fitting design feature is to be selected from the relevant library. Newly designed parts can be generated based on the abstract objects.

When the shape design sub process has been finished and the appropriate type of material has been chosen, mechanical analyses can be made to check for the behavioural adequacy. Moreover, preliminary manufacturability analyses can be performed to expose, for example, parts with dimensions that do not match to the available machine tools, or unnecessarily fine tolerances and surface roughnesses, and also to forecast the probable machine tool loading. Based on the calculated engineering quantities, limited quality and cost analysis can also be implemented.

Having completed the shape design and analysis, the designer may pass over to manufacturing process planning. The data base of the PRODES process planning module contains models for manufacturing these building blocks of components. This makes the realisation of the concurrence of design and planning possible. In the PRODES system, functions are available for the examination of manufacturing possibilities of a primary shape or a technological feature. For this reason the primary shape and technological feature checker system components offer the possibility of examining an object alone. This includes answering questions such as: Are there manufacturing methods for the object? Are there machine tools, cutting tools, which can be used to realise the manufacturing? The manufacturing of individual objects are interdependent so that these functions serve only examination and are not available for planning

4. GEOMETRIC AND ASSEMBLY MODELLING WITH THE PRODES SYSTEM

The PRODES system has a unique geometric and assembly modelling approach. This modelling methodology has it's roots in the design with features approach. In fact, it is a further developed version of this approach, which will provide significant increase in the efficiency of product development. Modelling entities of PRODES are called objects. These objects are fully parametric and, therefore, can be reshaped in compliance with the user requirements. Their treatment is comparable to that of traditional CSG primitives. The objects are mapped onto an extended B-rep data structure, which has been enabled to represent non-manifold geometries.

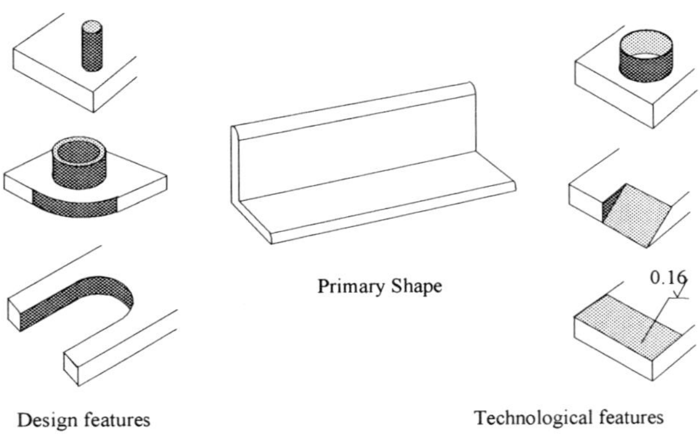

Figure 3. An example for design and technological features

Objects can represent different features. In the present form of PRODES, designing, FE analysis and technological features are defined (Figure 3). Based on their geometric and topological contents, the objects can be of three types: abstract objects, concrete objects and instance objects. They have been specified by considering a large number of previously developed designs (Figure 4). The abstract design feature (ADF) objects are specific shape models with varying structural topology. They comprise, in general, one primary shape carrying solids and one or more secondary shapes. With the secondary shapes, the designer can modify the primary shape in order to provide the needed functionality. All ADF objects have parametric topology and geometry. The abstract technological feature (ATF) objects are in connection with one of abstract design feature objects. They describe the manufacturing aspects of the ADF objects. The ATF objects are parametrized from the point of view of the machining operations. Because of their parametrization, ADF and ATF objects provide large flexibility in designing parts. It should be noted that the PRODES-L language makes possible to define new objects.

By selecting the relevant secondary shapes and merging them with the primary shapes, the designer specifies objects with fixed topology. These objects are called concrete objects. They are not fixed geometrically, that is, different dimension values can be specified to the geometric parameters. When both topology and geometry have been fixed, we speak about instance objects. In fact, these are the objects, which can directly be mapped onto an extended B-rep data model.

To facilitate assembly management, the PRODES system allows the specification of parameters for the relationship of one or more objects. These parameters, called global, may define relative positions, resulting dimensions, gaps, measure of overlap and so on. The designer can express the global parameters by the parameters of the abstract objects. It is possible to specify constraints on these parameters.

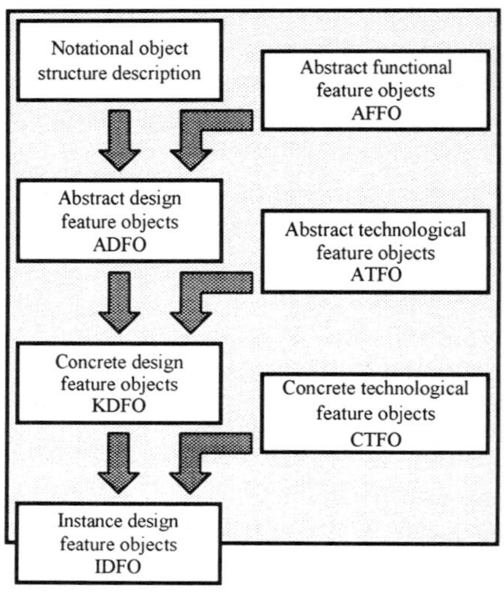

Figure 4. Modelling with functional, design and technological features

5. MANUFACTURING PROCESS PLANNING WITH THE PRODES SYSTEM

The purpose of process planning in the PRODES system is twofold. There are functions which the designer can use to evaluate the manufacturability of machine parts or the form features placed on them, taking into account the available data of the manufacturing environment. Once a prototype of the component has been specified, the planning of its manufacturing process takes place. The process planning module of the PRODES system manages the development of the process plan model and uses the workpiece model from the part modeller (Figure 5).

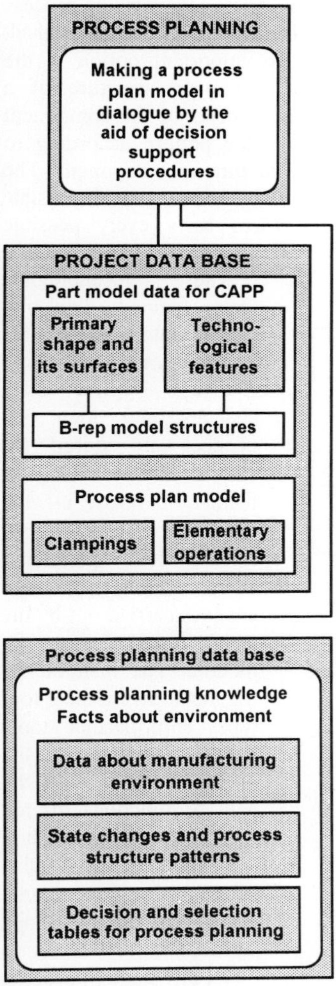

Figure 5. Process planning with the PRODES system

The elements and element groups of the process plan are attached to the part's construction elements which were used by the designer at the modelling of the part. The process planner selects process plan structure for the primary shape, while the process plan sub processes are oriented to the secondary shapes (technological features). The comprehensive choice of parts, which can emerge at the planning practice of PRODES, required special considerations with respect to the process planning strategy. In the PRODES system, a special implementation of the semi-generative process planning approach offers the possibility of both manual and automatic process plan modelling. The planning process consists of a sequence of steps. There are built-in procedures for support decision making at practically every step. These procedures offer various solutions based on the knowledge in the process planning data base.

When the designer develops the model of a part, he first selects a primary shape, then secondary shapes, and places them on the primary shape. The process planner develops the model of a process plan in a similar way. He selects a process structure pattern for the primary shape of the workpiece, then selects manufacturing sub process for each of the secondary shapes, and finally builds these into the process plan structure according to instructions placed in the pattern.

The process structure pattern is stored in the process planning data base and applied for a primary shape with well-defined geometrical and technological properties. There is a choice of patterns, and new patterns can also be elaborated. The process plan pattern contains manufacturing state changes, each of which is dependent on a primary or a secondary shape surface. As was mentioned earlier, the secondary shapes are technological features, the choice of which is available for the designer.

The process planning data base contains manufacturing information for each technological feature.

The feature state changes selected for the technological features are placed into the process structure pattern, where the state changes are grouped into clampings. Basically, in this stage of planning the process plan consists of clampings, each of which contain given state changes of given primary shape surfaces and technological features in a given sequence. The remainder of the planning proceeds from this state change sequence, and involves the selection of manufacturing methods, machine tool types and machine tools, the generation of elementary operations as well as the selection and arrangement of the cutting tools.

Figure 6 shows a schematic diagram of process planning sub modules. The planning process starts with a conceptual stage. The conception sub module assists at the specification of blank and the required non-machining operations. However the most important option is the definition of clampings. The process planner can determine the main structure of a manufacturing process by grouping certain state changes of given objects of the component into the same clamping. The primary shape sub module selects the pattern according to component characteristics and basically adapts it to the real structure of component. The process structure pattern is a state change sequence of an abstract object on which all possible surfaces of the primary shape exist and each of these surfaces carry every possible technological feature.

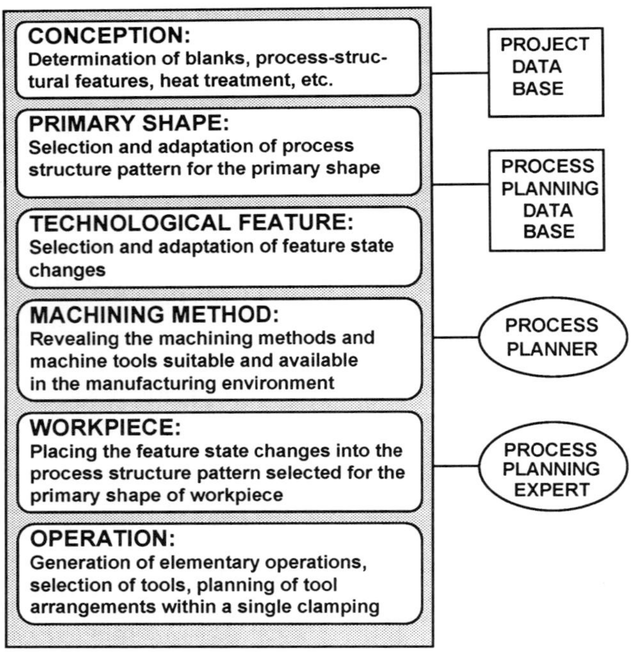

Figure 6. Main sub modules of the process planning module

The technological feature gains its final machined state in the course of one or more state changes. Typical state changes and their sequences are stored for each of the technological feature defined in the PRODES system. The selection and parameter value adding of state changes carried out by the technological feature sub module. The manufacturing methods, machine tool types and machine tools can be selected by the process planner with the help of functions offered by the machining method sub module. This also stands for the state changes of primary shape surfaces.

At this stage of process planning machining state changes are available for the primary shape and the technological features. For the surfaces of the primary shapes the state changes have been grouped into clampings according to the process structure pattern selected. The state changes of technological features have been placed in sequences. The following step of the process planning, assisted by the workpiece sub module, is the placement of state changes selected for the technological features in the process, according to the information gained from the process structure pattern. If clampings have been defined by the appropriate function of the conception sub module, and these deviate from those defined in the process plan pattern, a special function of the workpiece module rearranges the state changes according to the clampings defined in the conception.

The operation sub module assists the detailed planning of the process within a clamping. First, the elementary operations are generated according to instructions placed in the process planning data base, then cutting tools are chosen for the elementary operations and co-ordinated within the clampings assigned to the same machine tool. The decision support functions of process planning gain specific knowledge about the primary environment from the so called technological tables. These tables can be defined by attaching solution parameters to specific sets of conditional parameters. A record in these tables contains information similar to a production rule.

6. CONCLUSIONS

Concurrent engineering (CE) is a design philosophy which is aimed at taking into consideration all of the factors that influence the products during their life-cycle. As a methodology, it requires company-wide communication among the different experts and systems that are integrating their activities by providing unified product modelling facilities. It has been proved that by considering the influencing factors as early as possible, the overall design will be improved. CE requires the simultaneous execution of the product design and manufacturing process planning activities. The most crucial question of CE is the sharing and collecting of information. Therefore, CE assumes handling a generic product model and specific methodologies to support design activities. For this purpose we propose the abstract object based design methodology that applies design knowledge processing, and can be implemented in a quasi expert system environment. It promotes parallel generation of design prototype variants, which is essential in increasing both the productivity of design and the technical merit of the product.

The proposed modelling methodology provides a parametric object representation and is based on the systematised combination of feature entities. Generating a concrete object means selecting the required modifying secondary shapes of the given set of the abstract design object and combining them with the primary shape. In this generic shape modelling, design and manufacturing information is merged through design and technological features. This approach requires significant pre-programming effort, however, it increases the efficiency of the modelling significantly. It is also more user friendly, since objects can be selected by meaningful notations. By the use of the shape model, the PRODES system supports the development of process plans. Typical process plan structures are stored for the primary shapes in the form of process structure patterns. The system selects the appropriate pattern according to component characteristics and adapts it to the real structure of component. The process structure pattern is a state change sequence of an abstract object on which all possible surfaces of the primary shape exist and each of these surfaces carry every possible technological feature.

7. ACKNOWLEDGEMENT

The authors gratefully acknowledge the support provided by the Technology Development Committee of Hungary under grant OMFB-GSZI 92–97–47–0326.

8. REFERENCES

1 I. Horváth, Intelligent CAD: Solved Problems and Open Issues; in Proceedings of CAMP'92, ed. by Hencsey, G., Renner, G., Budapest (1992) pp. 46-53.

2 W. Seiler, Technische modellierungs- und Kommunikationsverfahren für das Konzipieren und Gestalten auf der Basis der Modell-Integration, VDI Fortschritt-Berichte, Reihe 10, Nr. 19, VDI-Verlag, Düsseldorf (1985).

3 J. J. Shah, A. Mathew, Experimental investigation of the STEP Form-Feature Information Model; Computer Aided Design, Vol. 23, No. 5, (1991) pp. 282-296.

4 J. M. Pinilla, S. Finger, F. B. Prinz, Shape Feature Description and Recognition Using an Augmented Topology Graph Grammar; in NST Engineering Design Research Conference, University of Massachusetts, Amherst (1989) pp. 285-300.

5 J. J. Cunningham, J. R. Dixon, Designing with Features; in Computers in Engineering, ed. by Tipnis, V. A., Patton, E. M., ASME, San Francisco (1988) pp. 237-242.

6 I. Horváth, Selection and Combination of Part Segments by Function, in Proceedings of the 1990 International Conference on Engineering Design, Volume 1, ed. by: Hubka, V., Kostelic, A., Heurista-JUDEKO, Zagreb (1990) pp. 91-98.

7 I. Horváth, A Segment-Based Approach to Systematic Design Synthesis; in Artificial Intelligence in Design '91, ed. by J. Gero, Butterworth-Heinemann, Ltd., Oxford (1991) pp. 703-720.

8 S. Finger, S. A. Safier, Representing and Recognizing Features in Mechanical Designs; Proceedings of the Second International Conference on Design Theory and Methodology, Chicago, September (1990).

9 S. Joshi, T. C. Chang, Graph-Based Heuristics for Recognition of Machined Features from 3D Solid Model; Computer Aided Design, Vol. 20, No. 2 (1988) pp. 58-66.

10 Y. C. Lee, K. S. Fu, Machine Understanding of CSG: Extraction and Unification of Manufacturing Features; Computer Graphics and Application, (1987) pp. 20-32.

11 K. Dorozsmai, I. Horváth, P. Kulcsár, L. Horváth, Conceptual and Development Plan of the PRODES System, Volume 3: Design and technological features, PRODES-L design language, Design and technological databases; Research and Development Report No. 270.004/92, TUB Institute for Machine Design, Budapest (1992) (in Hungarian).

12 I. Horváth, P. Kulcsár, Conceptual and Development Plan of the PRODES System, Volume 2: Geometric and Structural modelling; Research and Development Report No. 270.004/92, TUB Institute for Machine Design, Budapest (1992) (in Hungarian).

13 L. Horváth, A Methodology for Computer Aided Planning of Manufacturing Process Based on Feature Oriented Workpiece Modelling; in Proceedings of CAMP'92, ed. by Hencsey, G., Renner, G., Budapest (1992) pp. 54-61.

14 K. Dorozsmai, L. Horváth, I. Horváth, Conceptual and Development Plan of the PRODES System, Volume 4: Methodbase, Technological process planning; Research and Development Report No. 270.004/92, TUB Institute for Machine Design, Budapest (1992) (in Hungarian).

COUPLING LEARNING WITH OTHER PARADIGMS

Knowledge Based Hybrid Systems (B-11)
I. Mezgár and P. Bertók (Editors)
Elsevier Science Publishers B.V. (North–Holland)

Exploiting Domain Knowledge, Neural Networks and Genetic Algorithms to Harvest Traffic Simulation Results

Mark Foy and Carl Uhrik

Laboratorio di Ingegneria Informatica, Dipartimento di Informatica e Studi Aziendali, Università degli Studi di Trento, 38050 Mesiano-TN ITALY
e-mail: mfoy@itnsun4.cineca.it and carl@itnsun4.cineca.it

Abstract

This paper discusses the development of an incremental hybrid simulation-learning system that can be applied to problems with (1) many features, (2) high cardinality features, and (3) many classification categories. First, this hybrid system integrates a simulation module and a previously developed learning algorithm called a genetic algorithm (GA). Secondly, this hybrid system is extended to include another previously developed learning algorithm call a neural network (NN). Lastly, the initial NN is constructed with domain knowledge, to facilitate learning in large space environments. The theory behind adding domain knowledge is that by *starting* the NN closer to what is believed to be the optimal/converged configuration, the final system will more accurately generalize unseen examples and be generally more robust.

Specifically, this investigation has its testing grounded in the domain of automobile traffic signal control because this domain offers many features (road network conditions), many classification categories (traffic signal timing strategies), and available domain knowledge that can be integrated into the construction of a NN.

This paper first presents the positive results of combining a GA and a simulation, and applying this to the problem of determining traffic signal timings. In this system, the GA has been able to converge on what a traffic engineer would regard as optimal or near-optimal solutions. Next, this framework is expanded to integrate a NN, and preliminary results of this integration are given. These early results indicate difficulties in training the currently described NN. We observed that the present domain knowledge, used in our system to help build the initial NN, did not have a large positive effect on the training of the NN under the conditions discussed in this paper.

Keyword Codes: I.2.1; I.2.6; I.6.3
Keywords: Artificial Intelligence, Applications and Expert Systems; Artificial Intelligence, Learning; Simulation and Modeling, Applications

1. INTRODUCTION

Automobile traffic signal control is a large, non-Boolean, erratic, multi-modal (i.e., two or more optimal solutions may exist for a specific traffic situation), non-linear, gray, and hard-to-predict domain. Due to these characteristics, many traditional control strategies are not sufficient to handle the task of configuring traffic signals to allow for the efficient flow of cars. Specifically, since closed-form analytical solutions for calculating signal timing configurations are not available (i.e., they are infeasible), simulation is often used in traffic engineering systems. Additionally, because the traffic signal control domain is large (i.e., there are many possible ways in which to configure a network of traffic signals), it is desirable to have an intelligent system specify the timings for traffic signals.

This paper first presents a traffic simulation implemented within a genetic algorithm (GA), an intelligent search algorithm that searches by manipulating populations of structures (i.e., binary strings representing data structures which symbolize possible solutions to a problem) into new populations using operators patterned after natural genetic operators [Holland, 1975]. A GA offers a good way of searching the space of possible traffic signal timing strategies because it minimizes the solution space points it needs to evaluate by intelligently searching this space. The GA implemented in this paper uses the simulation module repeatedly to evaluate different signal timing strategies, eventually choosing a near-optimal signal timing strategy for the given traffic conditions.

Even though the GA searches intelligently, it still needs to evaluate many points in the solution space. The time to perform one complete GA run with multiple simulations can be prohibitive, therefore, although it is not necessary, it is desirable to implement a system which could learn how the GA "acts". This system should take the same input as the GA, and predict how the GA would respond to this input. A neural network (NN) has this ability because it learns by iterating over a given set of "examples" (i.e., input-output or condition-action pairs). A NN can also predict output for examples it has not seen by extrapolating and interpolating over the examples which it has "seen" (i.e., the examples that were used to train the NN).

Furthermore, by integrating a NN into this hybrid GA/simulation system, traffic engineers could add partial or incomplete domain knowledge. Domain knowledge is desirable because it could make the NN easier to train. Moreover, it would seem that the integration of domain knowledge could not hinder that NN because the NN can use approximate domain knowledge and then modify the constraints of this domain knowledge as it sees fit. Therefore, this strategy is mostly designed to give the NN a "head start" on learning. The initial idea of employing domain knowledge was taken from concepts presented in [Towell et al., 1990], [Shavlik and Towell, 1989], and [Katz, 1989].

This paper first outlines the reasons for implementing this system in the traffic domain. Next, a simple example that specifically addresses traffic signal control is described. This example will be used throughout this paper. This will lead to a discussion of the overall framework and a specific description of the systems contained in this framework. First, the utilization of a GA with a traffic simulation will be discussed, followed by the results of applying this GA/simulation hybrid system to the simple traffic control situation. Second, the motivation for applying a NN will be presented, followed by a discussion of how domain knowledge was included in this framework. This will lead to an analysis of this integration, examining some of the difficulties of using NNs and domain knowledge together. Specifically, explanations to the difficulties encountered during testing will be suggested, as well as proposing fixes to these difficulties .

2. WHY A TRAFFIC DOMAIN?

The state of traffic control today is in need of an intuitive, robust system to continually optimize traffic signal timings. Intelligent computer traffic control systems are needed to dynamically handle changing traffic conditions.

In standard pre-timed controllers, traffic signal timings are fixed at what is determined to be the most effective timing strategy. Timing determination involves either extensive analysis of traffic data or after short observations of traffic trends, making it fairly time consuming task. Due to the time constraints, timing determinations are done infrequently, making the pre-timed control method a static model. Therefore, with this method, signal cycle times and offset times are calculated once, for current conditions, and are then set into the individual traffic signals for an extended period of time (i.e., months). The signal cycle timings do not change with demand. This static characteristic is a clear disadvantage, motivating the development of more dynamic methods.

The hybrid framework described here is a more dynamic method and its development is encouraged by the lack of traffic control systems that do intelligent, dynamic control of traffic signals.

3. THE TRAFFIC SIGNAL CONTROL DOMAIN

The problem addressed in this study entails finding a near-optimal traffic signal timing configuration at all intersections of a street network given the current intersection characteristics. The current characteristics consist of: (1) the current number of cars at each lane of each intersection and (2) the external arrival volumes. The system should produce a near-optimal timing configuration for north/south green phase and east/west green phase for the current conditions for all the intersections in the network. First, the inputs and outputs of the hybrid system described in this paper will be defined.

There are many ways in which traffic control can be viewed. To facilitate understanding, a typical traffic situation is constructed that is both manageable and comprehensible. The street network has 4 intersections shaped in a square/circle configuration, each intersection adjacently connected to two other intersections by perpendicular roadways (see Figure 1). This lets us define traffic control as a function of two input vectors:

(input.1) $= [n_{111} \cdots n_{ijk} \cdots n_{443}]$ (1)

where n_{ijk} = number of cars on lane k of approach j of intersection i.

For the example in this paper, i=1 to 4, j=1 to 4, k=1 to 3, and total length is 48.

(input.2) $= [v_{11} \cdots v_{ij} \cdots v_{44}]$ (2)

where v_{ij} = external arrival volume on approach j of intersection i.

For the example in this paper, i=1 to 4 , j=1 to 4 , and total length is 16 (note, in this example, there are only 8 locations where cars can enter the network, therefore the other 8 approaches have an external arrival volume of zero and were removed).

where the result of evaluating these two vectors through our system is three output vectors:

(output.1) = [tgt] (3)

where tgt = total green time given to each intersection for one full cycle.

The same total green time is used for all intersections in the current system, but their is no reason this can not be changed.

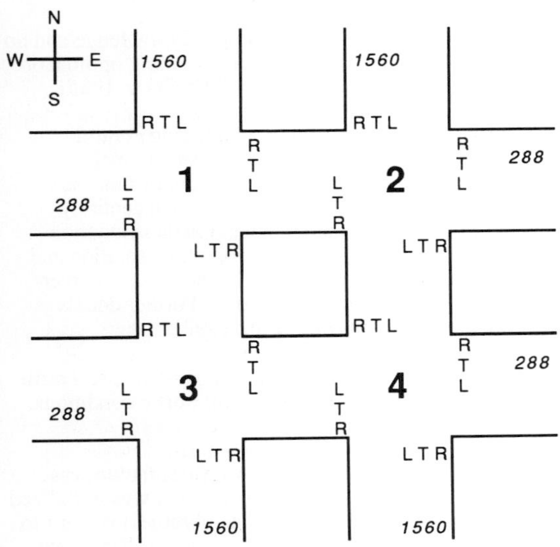

Figure 1 - Street Network Configuration - intersections numbered from 1 to 4 - L=turning Left R=turning Right T=going straight Through - numbers appearing at external components are the external arrival volumes in cars per hour used for the example GA run described later (output.2) $= [d_1 \;\; . \;\; . \;\; . \;\; d_i \;\; . \;\; . \;\; . \;\; d_4]$ (4)

where d_i = direction in which the first green phase will allow traffic to flow at intersection i, either north and south or east and west.

For the example in this paper, i=1 to 4.

(output.3) $= [nsgt_1 \;\; . \;\; . \;\; . \;\; nsgt_i \;\; . \;\; . \;\; . \;\; nsgt_4]$ (5)

where $nsgt_i$ = proportion of total green time (output.1) that will be allocated for the north/south green phase at intersection i.

For the example in this paper, i=1 to 4.

This view, considered in a typical setting, results in many high cardinality feature variables and many classification categories. For this example, the feature space size (i.e., the number of possible distinct inputs) is approximately 10^{92}. Likewise, the classification space size (i.e., the number of possible distinct outputs) is approximately 10^7. This demonstrates that this domain, and certainly many more domains, could benefit from frameworks that operate efficiently in large space problems.

Due to these large spaces, many algorithms have been ruled out because they do not have the capability to handle this type of problem. It is difficult to envision ID3 [Quinlan, 1983] (and many other induction algorithms) classifying 10^{92} different instances into more than 10^7 categories, though it may be possible. In our framework a GA is implemented first so that our system could take advantage of GAs' global, parallel search characteristics and their black box evaluation strategy. Secondly, NN are considered because they may have the right types of characteristics to attack large space problems and because they could be used to supplement the GA.

4. OVERALL FRAMEWORK

4.1. Genetic Algorithm (GA)

GAs are useful because they can operate in environments with no domain knowledge and no care about the size of the instance space. They also can be very effective at finding optimal or near-optimal solutions to dynamic real-world problems [Goldberg, 1989] [PICGA, 1985] [PICGA, 1987] [PICGA, 1989].

Our initial framework first employs a GA to find "near-optimal" signal timing control strategies for "typical" traffic situations. The GA bases its selection of a "near-optimal" solution on the "total average wait time per car", trying to minimize the amount of time cars have to wait. The GA is given the two input vectors defined above. These inputs entirely describe one traffic situation. The GA is initiated, and by using the given traffic simulation module to evaluate points in the solution space, it converges on a "near-optimal" solution and outputs this solution. This "near-optimal" solution consists of timings for north/south green phase and east/west green phase for all intersections in the traffic network. Further details regarding the structure of this "Traffic GA" can be found in [Foy et al., 1992].

4.2. GA Results

The results of applying a GA to this traffic control problem have been positive. The Traffic GA produced very reasonable traffic signal timing plans for many different traffic conditions. One typical Traffic GA run is described below.

4.2.1. Traffic Environment

To illustrate the performance of the GA in this framework, a simple traffic situation was constructed. First, the (input.1) vector (i.e., the number of cars at all locations), was initialized with typical numbers. Next, the (input.2) vector (i.e., the external arrival volumes) was set to the values corresponding to the volumes shown in Table 1 (and also in Figure 1). The north and south approaches were given 5 to 6 times as much traffic as east and west directions.

4.2.2. GA Performance

To obtain a stable, unbiased, average result, five GA runs were executed. Each run was executed for 60 GA generations with a GA population of 50. This means that for each run, the traffic simulation was executed 3000 times. At each generation, the average fitness of the

Table 1 - Street Network Configuration

Approaching	Arrival Volume (cars/hour)
from North to Intersection 1	1560
from West to Intersection 1	288
from North to Intersection 2	1560
from East to Intersection 2	288
from South to Intersection 3	1560
from West to Intersection 3	288
from South to Intersection 4	1560
from East to Intersection 4	288

generation was calculated and the member in the population with the best fitness value (i.e., the shortest wait time) was identified. Refer to Figure 2 for a graphical illustration of how the GA starts with bad solutions (i.e., solutions that produce on-average high wait times) and locates good solutions (i.e., solutions that produce on-average short wait times). To produce Figure 2, the average fitness of each GA generation (from generation 0 to 60) from each of the 5 Traffic GA runs were averaged together. This produced the "Average Wait Time of Population" curve. The "Member With Minimum Wait Time" curve was produced in the same manner by averaging the fitness of the best-of-generation members for each of the 5 Traffic GA runs.

Figure 2 shows that the population seems to converge to the optimum or near-optimum member by the 20th or 30th generation. Therefore, it is possible to terminate the GA after 20 generations, instead of after 60 generations, and still obtain a near-optimal solution. This reduced generation scenario would reduce the number of simulations from 3000 to 1000. The graph also shows that typical minimum wait time values were around 40 seconds for these traffic conditions. After the last generation, which in this case was the 60th generation, the member with the maximum fitness (minimum wait time) can be selected as the best signal timing configuration and called the solution from the Traffic GA.

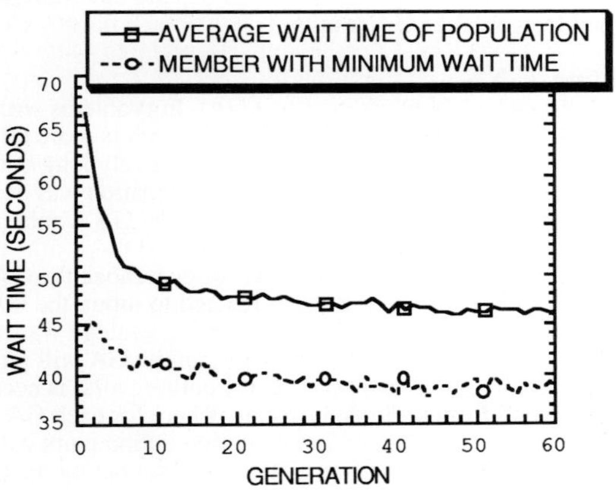

Figure 2 - The Average of 5 Traffic GA Runs - showing generation average (average wait time) and best-of-generation (minimum wait time) results

Table 2 - The GA's Maximum Fitness (Minimum Wait Time) Timing Strategy

Intersection Number	First Direction	Green Time (sec)	Second Direction	Green Time (sec)	Total Cycle Time (sec)[a]
1	E/W	12	N/S	42	60
2	N/S	36	E/W	18	60
3	N/S	42	E/W	12	60
4	E/W	9	N/S	45	60

[a] Total Cycle Times have two yellow phases - of 3 seconds each - added, in addition to the two green times.

4.2.3. GA Output

A typical maximal fitness member, actually found by one of the GA runs executed on the traffic environment described in Table 1, is shown in Table 2. Note that for all intersections, the green phase time for the north/south (N/S) directions was considerably longer than for the east/west (E/W) directions. Observe that total cycle times are equal because this constraint was put on the GA. Additionally, it should be pointed out that the Traffic GA selected a total cycle time of 60 by itself, this number is not programmed into the GA. Also recognize that the GA found a strategy that utilized very similar green phase times for the north/south directions and also for the east/west directions. We expect this behavior because similar green phase times often allow the best flow of traffic because cars can move through the network with fewer stops if there is some type of synchronized cycle time [Reljic, 1988]. Next, notice that the GA could have given green phase times up to 114 seconds, but only went as high as 45 seconds. This is because the GA was searching for a strategy that would *minimize* wait time, and if it were to allocate more green phase time to north and south directions, the wait times for cars coming from the east and west would increase too dramatically to make this a beneficial action. Lastly, we were impressed with this solution because it provides two 33 second green bands, one for northbound traffic through intersections 1 and 3, and another for southbound traffic through intersections 2 and 4. These long north/south green bands are desirable because they allow smoother traffic flow on these major north/south arterials. Overall, the hybrid system with the GA and a simulation module has performed very well. Next, we will expand this framework to integrate a NN which will be designed to generalize GA results.

4.3. Neural Network (NN)

4.3.1. Motivation for Integrating a NN

There is one primary motivation for using a NN in addition to a GA. In problems with large hypothesis (or classification) spaces like traffic control, even tough GA search is more efficient than traditional search methods, the run time of the GA can become prohibitively long due to the simulation module. If every time an optimization of a specific traffic situation was desired, the GA had to be rerun, then execution times would become too long and the GA would not be practical. A NN has the potential to addresses this issue.

To reduce repetitive computations, a NN could be used to "learn" or generalize the GA's behavior (i.e., how it reacts to input). If a NN knew how the GA reacted to input, the GA would not have to be executed every time a new near-optimal signal timing strategy was needed for a new traffic situation. To "teach" a NN how to "predict" what output the GA will produce given certain input, a training set of condition-action pairs (i.e., input-output pairs) is needed. These condition-action pairs can be collected from multiple GA runs, where for each GA run, different inputs (i.e., traffic conditions) are given. A set of GA condition-action pairs will be used as the training set for a NN. Figure 3 illustrates the framework for augmenting the GA with a NN.

4.3.2. Format of NN Input and Output

The format of the input and the output from the GA is simple and is easily manipulated to work as training instances for the NN. Combining the input to the GA with the corresponding

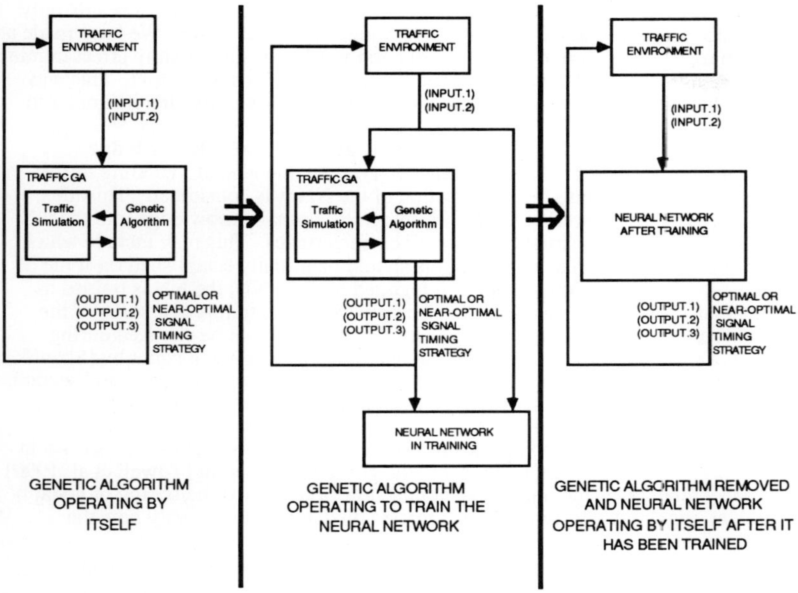

Figure 3 - Framework for Combining the GA with a NN

output, a condition-action pair is formed with the following format:
+ the input, or the traffic situation (taken directly from (input.1) and (input.2))

x1=cars turning *left* from the *north* side of intersection *1*
x2=cars going *straight* from the *north* side of intersection *1*
.
x47=cars going *straight* from the *west* side of intersection *4*
x48=cars turning *right* from the *west* side of intersection *4*
x49=external arrival volume from the *north* to intersection *1*
x50=external arrival volume from the *west* to intersection *1*
x51=external arrival volume from the *north* to intersection 2
.
x56=external arrival volume from the *east* to intersection *4*

+ the output, or the optimal signal timing strategy (taken directly from (output.1), (output.2), and (output.3))

y1=the total green time given to each intersection for one full cycle
y2=the direction in which the first green phase will allow traffic to flow at intersection *1*
y3=the direction in which the first green phase will allow traffic to flow at intersection *2*
y4=the direction in which the first green phase will allow traffic to flow at intersection *3*
y5=the direction in which the first green phase will allow traffic to flow at intersection *4*
y6=the proportion of total green time that will be allocated for the north/south green phase at intersection *1*
.
y9=the proportion of total green time that will be allocated for the north/south green phase at intersection *4*.

The GA input features were actually combined or reduced for input to the NN. The NN only receives 12 combined features that represent the 56 features described above. We have made an assumption (which may not be correct) that this reduction would not significantly affect training because the combined features are believed to have the same effects on the output. This was done to reduce the number of distinct inputs into the NN in the hopes of making it simpler to train.

4.3.3. Establishing the Structure of the NN Using Domain Knowledge

To help the NN more easily learn how to predict GA output, the idea of including traffic engineering domain knowledge into the construction of the NN was considered. This idea, which was first discussed in [Towell et al., 1990], integrates domain knowledge into the NN by converting if-then clauses into structural network characteristics. This may include which nodes should be connected as well as what weights should be initially established on some of those connections. After domain knowledge is integrated into the NN, the NN is trained as usual, adjusting connection weights as needed. The only exception to this is if some of the connection weights are constrained such that they are fixed and can not be changed during training. Two other similar approaches exist in the open literature. First, a paper by [Shavlik and Towell, 1989] that presents an approach very similar to the one adopted here, and second, a paper by [Katz, 1989] that focuses on improving learning time rather than improving classification.

In this paper, traffic engineering domain knowledge is used to construct the NN shown in Figure 4. This configuration was formed by using the guidelines given in [Towell et al. 1990] regarding the connecting nodes, the initializing connection weights, and the setting of nodal biases. Refer to the original paper for more information on determining exactly how the domain knowledge is used.

For this paper, the domain knowledge rules come in a form similar to the following:

1) IF [[the average number of cars coming from the north is large] AND [the average number of cars coming from the south is large]] THEN [there are many cars coming from the north/south]

2) IF [there are many cars coming from the north/south] THEN [[the first direction to travel at inter 1 should be north/south] AND [the proportion of total green phase time given to the north/south green phase at inter 1 should be long]]

These vague, general rules then construct the following connections in the NN (see Figure 4):

from rule 1) [node 0 to node 12 with weight near 3.0; node 2 to node 12 with weight near 3.0]

from rule 2) [node 12 to node 23 with weight near 3.0; node 12 to node 24 with weight near 3.0]

Additionally, after considering these rules, the biases are established based on information in [Towell et al. 1990]. The rest of the weights and biases are established similarly, and the network in Figure 4 is produced.

These rules are of course only approximately correct, so it is important that the system be allowed to and able to adjust most NN weights to modify incorrect rules/theories.

Once the structure of the NN has been established using the domain knowledge, the NN is ready to be trained using the condition-action pairs obtained from multiple GA runs. The next section will discuss the initial results observed from using a NN in this hybrid system.

4.4. NN Testing

Here we present the NN training and testing scenario. Some initial results are given, followed by a discussion of why some of the results are counter intuitive.

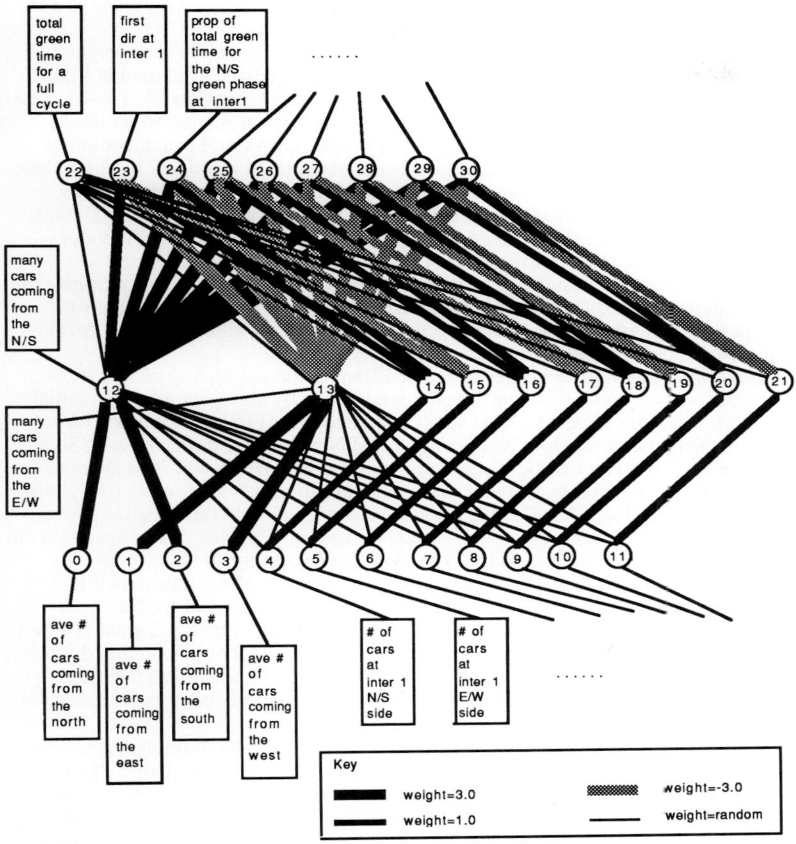

Figure 4 - The NN Constructed with Domain Knowledge - input nodes (at the bottom) represent the traffic situation and output nodes (at the top) represent the optimal signal timing strategy

4.4.1. NN Training Results

Four different implementations (or NN configurations) were constructed and used to test the usefulness of blending domain knowledge into a NN. The commonalties between configurations include: (1) all have 31 nodes, (2) all have 12 inputs and 9 outputs, and (3) all use back propagation for learning after initialization (whether the initialization is random or controlled by domain knowledge). The four network configurations follow.

(Net Config 1)=(No domain knowledge used) A network with complete interconnection between the input and hidden layers and between the hidden and output layers. Initialized with all random connection weights. No nodal biases.

(Net Config 2)=(Domain knowledge used to establish structure of NN) A network with reduced interconnection as shown in Figure 4. Initialized with all random connection weights except the weights on the connections between nodes 4 through 11 to nodes 14 through 21. These special connections were fixed at 1.0. Random biases on nodes 12 and 13 and all output nodes.

(Net Config 3)=(Domain knowledge used to establish structure of NN and weights) A network identical to Figure 4. Connection weights initialized as shown in Figure 4. All weights could change during back propagation except 0 to 12, 1 to 13, 2 to 12, and 3 to 13, which were fixed to values within 0.01 of 3.0 based on the information gained from [Towell et al. 1990]. The other weights that could not change were between nodes 4 through 11 connected to 14 through 21, which were fixed at 1.0. All weights between the hidden and output layers were preset to within 0.01 of 3.0 (or -3.0 for negated antecedents), but were allowed to change during training. Biases were fixed (without being allowed to change) at values determined by the scheme discussed in [Towell et al. 1990].

(Net Config 4)=(Domain knowledge used to establish structure of NN and weights, but connectivity reduced even more) Same as Net Config 3 (similar to Figure 4) except connectivity is reduced even further by eliminating all connections between nodes 4 through 11 and nodes 12 and 13.

The training set consisted of 22 examples (i.e., condition-action pairs), and all networks were trained 10 separate times with different random seeds (thereby changing all of the initial connection weights). Training targeted a total sum of squared error (TSS) (a common error measuring value for NNs) of 16.0. Both Net Config 1 and Net Config 2 achieved a TSS value of 16.0, but Net Config 3 and Net Config 4 could not achieve this error level. The average number of training epochs required by the four network configurations were respectively: 46, 158, >20,000, and >20,000. Both Net Config 3 and Net Config 4 could only realize average TSS values of 19.88 and 25.41, respectively, after 20,000 training epochs. These results seem to show that networks prefer total connectivity to restricted connectivity.

Three test sets were then used to check the generalization abilities of the four network configurations. Each test set contained 10 examples. Test Set 1 consisted of traffic configurations that were very similar to those used in the training set. Test Set 2 encompassed fairly random traffic situations, most of which are uncommon, and Test Set 3 contained typical, more conventional traffic situations. All examples within the test sets, as well as the training set, were generated by the GA. The average results obtained from these test sets are shown in Table 3.

This table shows that as the test sets become further removed from the training set, errors increase. Likewise, as the networks become more restricted by domain knowledge, errors generally increase, except for the case "Test Set 3 with Net Config 4" in which general traffic domain knowledge is applied to typical traffic situations. Some of these results are unexpected, and in the next section an analysis of some of these counter intuitive results, as well possible explanations for them, will be given.

4.4.2. Analysis of NN Results

The results of applying a NN, trained with sparse data, to generalize how a GA reacts to input has not been as successful (i.e., not as accurate) as using a GA by itself. Although the results above are partially disappointing, some information can be gained from examining them. First, Net Config 4 seems to handle typical traffic situations (like the type included in Test Set 3) better than the others. This is one of the few cases in which a domain knowledge structured

Table 3 - NN Testing Results
Total Sum of Squared Error (TSS) Values

Test Set	Net Config 1	Net Config 2	Net Config 3	Net Config 4
1	5.91	6.40	9.50	11.20
2	10.39	9.32	12.31	12.65
3	18.09	21.95	18.03	13.33
All Test Sets	11.46	12.56	13.28	12.39

network was able to out perform the random network with random weights (Net Config 1). The other instance where a knowledge derived network outperformed Net Config 1 was with Net Config 2 processing Test Set 2. Net Config 2 was the best at classifying situations similar to the cases in the training set (examples in Test Set 2). All other comparisons between Net Config 1 and the domain knowledge constructed networks indicate that Net Config 1 was equivalent, or often times better than the other networks. The question is: why did domain knowledge help in only a few cases?

4.4.2.1. Was the Instance Space Too Large and Erratic?

Generally, NNs can learn mappings regardless if they are linearly separable or not, but this space may be too large and erratic to be reasonably *learned*. Expecting a NN to learn a mapping of 10^{92} instances to 10^{7} classification with only 31 nodes may be asking too much. Additionally, most instances do not have just one optimal classification to which they can be associated. When the GA is given a traffic situation for which an optimal control strategy is need, it will produce one, of maybe many, control strategy that would work well on the situation. Therefore solutions are not distinctly right or wrong, giving this domain many gray areas. This whole question of training instances leads to the next question.

4.4.2.2. Were Too Few Training Instances Used?

There is good reason to believe that too few training examples were used. First, the fraction of all possible instances that appeared in the training set was very small (only 22 instances in the training set). This very small set is generally not considered enough to train a 31 node NN; the network is not likely to develop a good set of weights with these limited examples. Even if the network is trained for thousands of epochs, this would not help because the training set is not diverse enough and all the network can do is overfit the examples. Overfited networks cannot generalize to produce reasonable results on unseen instances. The most desirable action is to increase the number (and diversity) of examples in the training set. Some experiments have been run with a training set consisting of about 50 examples, and improved performance has been observed. This should certainly be one of the first directions for future work, to insure that the hybrid system is properly tested. Overall, it may be asking too much for a network to develop a robust set of weights when so few instances are available for training.

4.4.2.3. Was the Domain Knowledge Incorrect?

The domain knowledge used to construct the NN was taken from [Foy et al., 1992], from the results of the GA run presented above, and from intuitive notions about traffic characteristics. Although it is very likely that this domain knowledge was only approximately correct, we believed the NN could modify it as needed.

One may think that the correctness of this knowledge could be checked by exploring the NN weights after training, but this idea should be approached with caution. Some trained networks were examined as to their final weights, and it was found that sometimes very little change had taken place in the preset weights. On the other hand, other similar cases showed large preset weight changes. This is a tricky area because weight changes depend on the weights as a set, not individually. Different random starting weights (exclusive of the preset weights) affect all weight changes. Therefore, changes are hard to generalize across all networks because the random weights are generally so different. Interpreting connection weights is mentioned as a future research area in [Towell et al., 1990].

Overall, we believe the domain knowledge was approximate, but not incorrect.

4.4.2.4. Was Knowledge Used in the Wrong Way?

It is possible that knowledge was used in the wrong way to establish NN structure. Maybe it would have been better to use the knowledge to develop a more elaborate *class* hierarchy of nodes with additional intermediate nodes associated with intermediate concepts (e.g., "north/south flow is greater than east/west for entire traffic network" or "intersection 1 is at oversaturation"). The current structure may have been incongruent with the best learning conditions, thereby resulting in sub-optimal learning.

5. CONCLUSIONS

This paper presented a hybrid system that was designed to utilize a simulation module and handle large space problems. A need for systems that approach larger more real-world problems was identified, and this research is a step in that direction. The genetic algorithm (GA) proved that it was able to identify "near-optimal" signal timing strategies, but the idea of integrating a neural network (NN) and domain knowledge needs to be further explored.

It is important to realize the benefits of using a GA and the potential benefits of integrating a NN. The GA especially has proved to be useful for attacking the problem of determining traffic signal timings even though it does require significant computational effort. It can be postulated that because some success has been reported, other problems, which involve large feature spaces and large classification spaces, could especially gain from using a GA, and also may gain from implementing a similar hybrid system.

On the other hand, our preliminary tests which used domain knowledge to structure NNs did not seem to consistently improve NN learning or classification. The work done in this area is only preliminary, and further study is needed to determine the absolute value of using a NN in this framework.

We have learned that integrating knowledge may not be as easy as it appears, but we have outlined different ideas which could help improve the performance of the NN in the future.

6. REFERENCES

[Foy et al., 1992] Foy, M., Benekohal, R.F., and Goldberg, D.E. "Signal Timing Determination Using Genetic Algorithms" to be published in The Transportation Research Record, in 1993.

[Goldberg, 1989] Goldberg, D.E. Genetic Algorithms in Search, Optimization, and Machine Learning. Reading, MA: Addison-Wesley, 1989.

[Holland, 1975] Holland, J.H. Adaptation in Natural and Artificial Systems. Ann Arbor, MI: University of Michigan Press, 1975.

[PICGA, 1985] Proceedings of an International Conference on Genetic Algorithms and Their Applications. Carnegie-Mellon University, July 24-26, 1985. Edited by John J. Grefenstette. U.S. Navy Center for Applied Research in Artificial Intelligence, 1985.

[PICGA, 1987] Proceedings of the Second International Conference on Genetic Algorithms. Massachusetts Instituted of Technology, July 28-31, 1987. Edited by John J. Grefenstette. L. Erlbaum Associates, 1987.

[PICGA, 1989] Proceedings of the Third International Conference on Genetic Algorithms. George Mason University, June 4-7, 1989. Edited by J. David Schaffer. Morgan Kaufmann Publishers, Inc., 1989.

[Katz, 1989] Katz, B. "EBL and SBL: A Neural Network Synthesis" in Proceeding of the Eleventh Conference of Cognitive Science School, 683-689, 1989.

[Quinlan, 1983] Quinlan, J. R. "Learning Efficient Classification Procedures and their Application to Chess End Games" in Ryszard Michalski (Ed.) Machine Learning: An Artificial Approach, Tioga, 1983.

[Reljic, 1988] Reljic, S. "TRAFSIG: A Computer Program for Signal Settings at an Isolated, Under- or Oversaturated, Fixed-Time Controlled Intersection" in Traffic Engineering and Control, Vol. 29, No. 11, Nov. 1988, 562-566, 1988.

[Shavlik and Towell, 1989] Shavlik, J.W. and Towell, G.G. "An Approach to Combining Explanation-based and Neural Learning Algorithms" in Connection Science, Vol. 1, No. 3, 231-251, 1989.

[Towell et al., 1990] Towell, G.G., Shavlik, J.W., and Noordewier, M.O. "Refinement of Approximate Domain Theories by Knowledge-Based Neural Networks" in AAAI-90, 861-866, 1990.

Knowledge Based Hybrid Systems (B-11)
I. Mezgár and P. Bertók (Editors)
Elsevier Science Publishers B.V. (North–Holland)

POSSIBILITIES OF APPLICATION OF MACHINE LEARNING IN MANUFACTURING PROCESS PLANNING

László Horváth - Imre J. Rudas

Bánki Donát Polytechnic, Népszinház u. 8. Budapest H-1081 Hungary

Abstract

Manufacturing process planning forms an important link between CAD and CAM and essentially determines the organization, economy and productivity of the production. In this paper the knowledge acquisition problem of CAPP systems is investigated. First CAPP as a problem area is described, then an approach to knowledge based CAPP is presented. Based on the investigations on knowledge requirements for CAPP the knowledge is sorted out to four levels, and the appropriate learning methods are discussed.

Keyword Codes: J.6; I.2.4; I.2.6
Keywords: Computer-Aided Engineering, Knowledge Representation, Learning

Introduction

The overwhelming majority of CAPP systems applied in present process planning practice is not able to support the process planner at his decision making. The advent of first generation expert systems has not brought considerable change in this field despite of the comprehensive research and development effort. The problems are originated from the shortage of appropriate planning methods and the difficulties of the elaboration of knowledge bases in practice. In the future the demand for better utilization of flexible manufacturing systems will be emphasized in consequence of the requirements for shorter product cycles [9], [10]. The decision support features of CAPP systems lead to quick process plan generation taking into account the momentary conditions of the manufacturing environment.

In this paper the knowledge acquisition problem of CAPP systems is investigated. First CAPP as a problem area is described, then an approach to knowledge based CAPP is presented, and the knowledge requirements of CAPP are discussed, finally the knowledge is sorted out to levels and the appropriate learning levels are discussed.

1. CAPP as a Problem Area

Manufacturing process planning forms an important link between CAD and CAM. It essentially determines the organization, economy and productivity of the production. For that reason numerous research and system development projects have been initiated during the last two decades practically without complete successes regarding the applications for real industrial environments.

At the earlier eighties the first generation of expert systems offered the promise of solution. The researchers tried implement rule-frame based hybrid expert systems for CAPP. Most of these CAPP expert systems utilized backward chaining techniques [1],[4,],[9]. The reasoning process starts from the finished part and seeks for process element which results in the finished state. Omitting the influence taken by that process element on the workpiece, a less-machined workpiece is defined. Then a new process element is seeked which results in this less-machined workpiece state. The reasoning process is continued until the less machined workpiece is the blank. The incomplete models did not make the elaboration of practically applicable decision making procedures possible. At the same time the application of these expert systems require knowledge base, which are too complicated, special and hardly understandable for company personnel.

The aforementioned CAPP systems suffer from simplified problem solving features, unstructured knowledge base, lack of learning capabilities and inadequate process planner and expert interfaces.

A variety of CAPP methods have been presented in Hungary over the past twenty five years. On the basis of earlier research on conventional CAPP systems [12] the development of the GLEDA process planning system at late eighties resulted comprehensive knowledge storage and retrieval possibilities and improved user interface [15]. The experiences with the GLEDA system has been utilized at our recent study of CAPP decision support and learning procedures,as well as knowledge bases taking into account the appropriate methodical tools of present day AI.

2. An Approach to CAPP

The process planning approach presented in this paper has been elaborated for knowledge based CAPP systems [11],[13]. It is described schematically in Fig. 1.

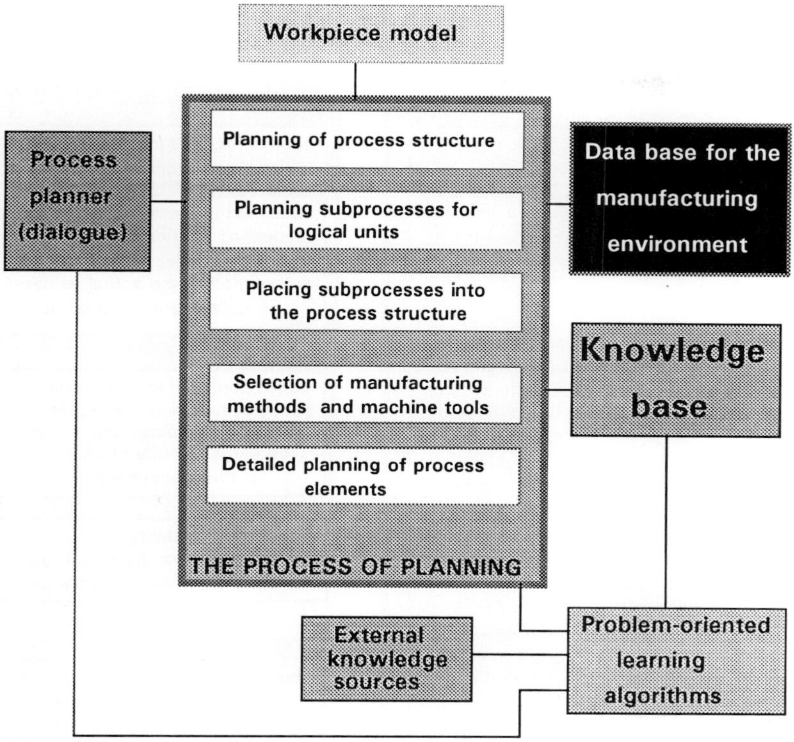

Fig. 1.

The planning is based on workpiece model which contains general data of the workpiece and data of logical units. The logical unit comprise workpiece surfaces which have to be examined as an unit in the course of process planning. It defined specially for the CAPP. The main steps of process of planning (left hand side of Fig. 2.) is outlined below.

- *Determination of the structure of the process plan.* The main structural units of the process plan are groups of manufacturing state changes. Within a group, the machine tool and the clamping position are unchanged. These groups have to be composed on the basis of typical process structures and process structural features. These groups are referred as clampings. Manufacturing state changes should be defined for each logical unit types and descripted by attributes of the logical unit representing its input and output state. The CAPP system can generate a process plan structure according to rules and structure patterns.

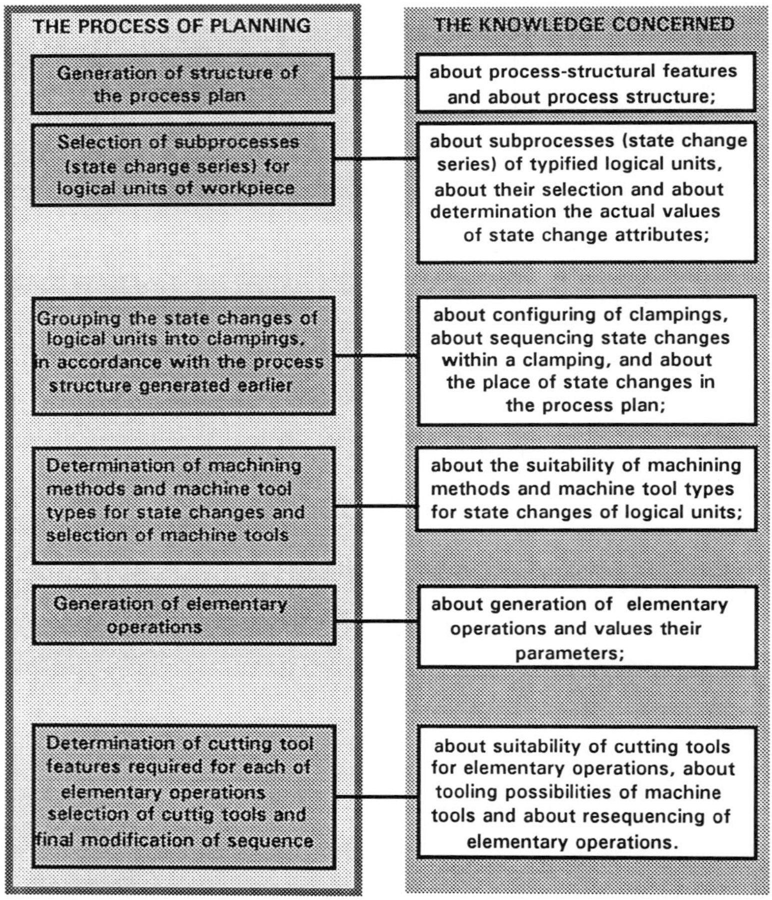

Fig. 2.

- *Selection of geometrical state changes for logical elements.* The manufacturing of a logical unit requires one or more state changes in predefined order. The state changes are selected according to values of logical unit attributes which determine the manufacturing task. The attributes of logical units obtain values for the input and the output state.

- *Placing the manufacturing state changes into the process plan structure.* The state changes are grouped into predefined clampings. The sequence within a clamping is determined based on precedence rules.
- *Selection of manufacturing methods and machine tools.* The state change can be accomplished by one of the suitable manufacturing methods. The manufacturing method can be available on certain machine tool types. The suitability of a machine tool can be judged on the basis of facts stored in data base of manufacturing environment. The machine tool selection can suggest drawing clampings together.
- *Detailed planning of machinings grouped in a clamping.* At this point the process plan consists of state changes selected for the logical units of the workpiece. The final, detailed process plan contains elementary operations which have to be generated from the state changes, depending on machine tool characteristics. For example the elementary operations for conventional machine tools differ from the one which is intended to be an input for NC program generation. There are rules for generation of elementary operations and for determination of machining dimensions. The elementary operation can be realized using one of the suitable tool types. The tool selection is influenced by their availabilty on the machine tool choosed earlier.

3. Knowledge Requirements for CAPP

The aim of the manufacturing process planning of workpieces is to determine a process plan which is not in contradicion with manufacturing knowledge both applied in general and in the planning environment. Moreover the process planning should be based on real data of the manufacturing environment available. The workpiece, the process plan and the manufacturing environment (factory) can be represented by object structures (fig. 3).

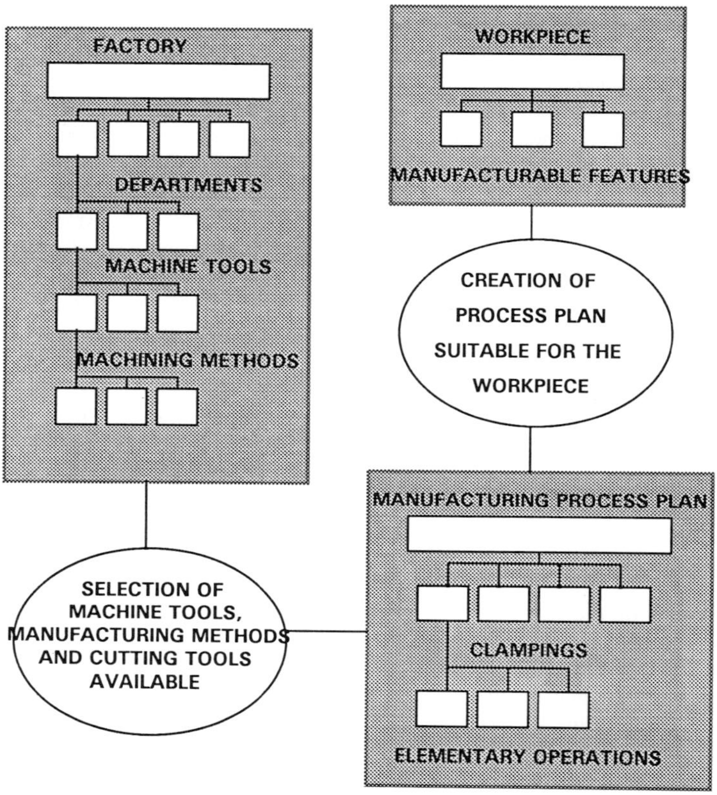

Fig. 3.

The knowledge required for the CAPP (right hand side of Fig. 2) can be divided into three groups (fig. 4):

- Knowledge which generally applicable for the domain of process planning. For example roughing a surface preceeds smoothing that surface always. Turning is applicable for machining surfaces which are axisymmetric.
- Knowledge which applicable in process planning environments where the process planners accept it. Examples: One can achieve a dimensional tolerance of IT7 by the application of a special fixture at turning. The main bores of a prismatic workpiece having a diameter larger than 40 mm should be machined by boring rod.

- Knowledge which is valid for only the given manufacturing environment. This is originated mainly from facts of manufacturing resources available.

The source of knowledge can be acquired from books, product literature, interwievs, process plans, rules of thumbs and facts of the manufacturing and process planning environment.

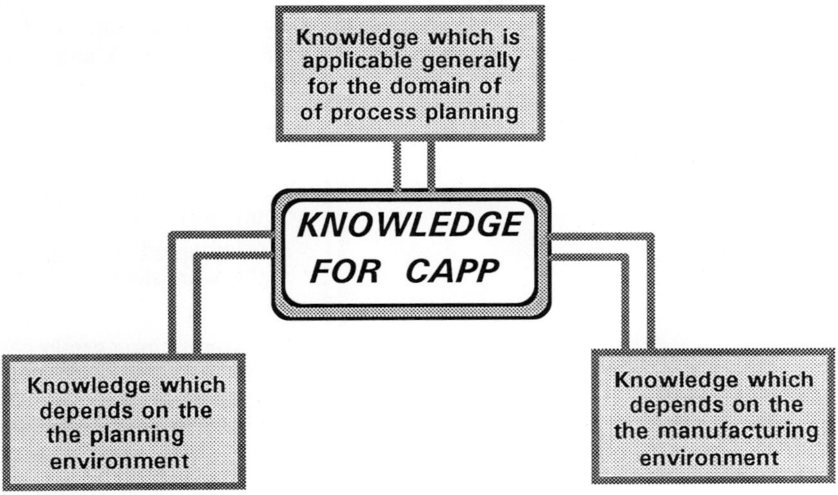

Fig 4.

The planning process requires knowledge for definition and attribute objects, for interrelation between attributes of various objects and for adding values to attributes. Knowledge has to be acquired into the system for determination of process structure, identification of logical units on the workpiece, selection of manufacturing state changes for the logical units, sequencing of manufacturing state changes, selection of manufacturing methods for manufacturing state changes, selection of machine tools, generation of elementary operations and selection of cutting tools.

The structure of the manufacturing process is the result of grouping of process elements into clampings. The related knowledge can be acquired in form of heuristic rules or typical process plan structures. It can be determined primarily on the basis of general attributes of workpiece (fig 5).

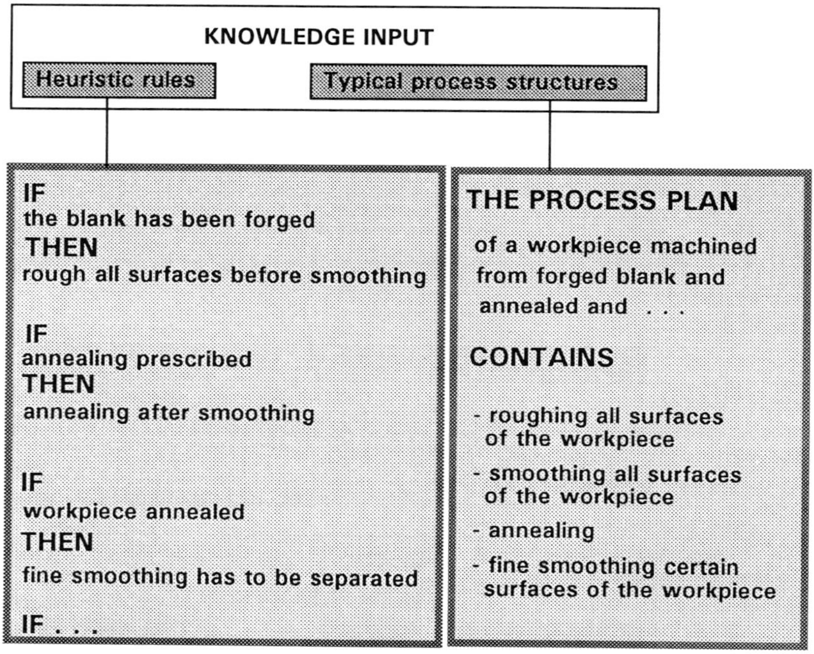

Fig. 5

The logical unit of the workpiece comprises surfaces which are machined at the same machining process or for which the process planner selects the same manufacturing subprocess. Knowledge is required for the definition of logical units and manufacturing subprocesses for them as well as for choosing the right subprocess for a given object (Fig. 6). The knowledge can be acquired in the form of heuristic rules or typical subprocesses.

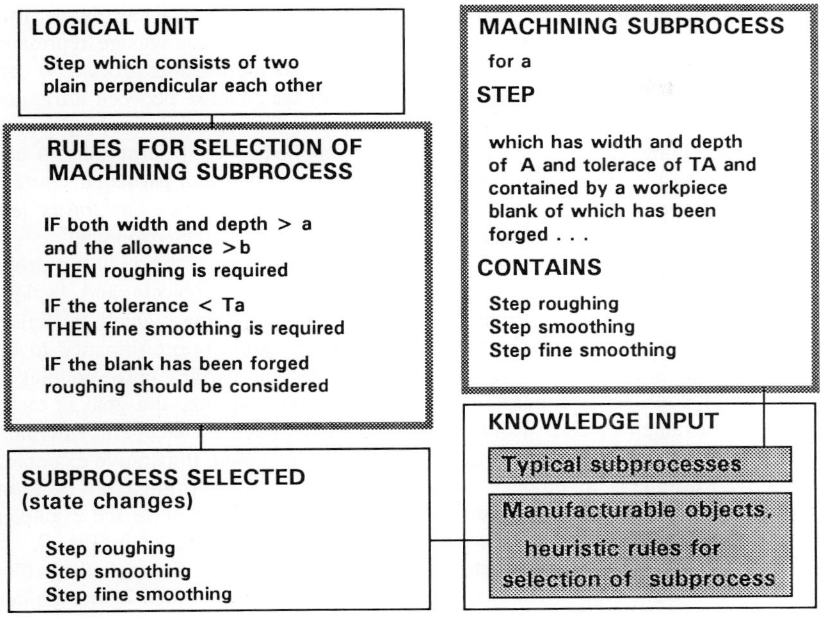

Fig 6

The sequencing of the manufacturing state changes selected for logical units in this case means only placing them into the process plan structure determined earlier. The knowledge can be acquired in the form of heuristic rules.

The taxonomy and and attribute of the objects are also important parts of the knowledge. Typical object models and structures can be acquired. At the same time information can be obtained from the practical experience for the selection of object attributes which should be taken into account at a given step of the process planning. There are typical and less typical relationships and dependences between the attributes of various objects.

4. Levels of Knowledge and Levels of Learning

According to the discussion of the previous chapter the knowledge will be sorted out to four levels. The *first level of knowledge* consists of the knowledge about objects and background knowledge. Objects are described by their definitions and attributes, logical connections of attribute-value pairs, relation between individual attributes using a description language. Background knowledge expressed by a knowledge representation scheme - such as rules, frames and semantic nets - assists the learning process. It contains additional information about the attribute values, on the relation between attributes and structural knowledge about the objective concepts. The *second level of knowledge* consists of the knowledge of concepts, which are supplied by their definitions. A concept is represented by a common description of objects that belong to a particular cluster. The *third level of knowledge* consists of the relation among concepts. The *fourth level of knowledge* consists of the knowledge about adaptation to external environment.

Corresponding to the four levels of knowledge four levels of learning are proposed. The first level of learning is acquiring knowledge about objects and background knowledge. This learning is realized by advice-taking learning, which is defined as follows; Learning by adding information that has been carefully preformatted to fit into appropriate slots in the system's memory [14]. The second learning corresponding to the knowledge of concepts is concept learning. In concept learning, the goal is to induce from a set of examples a classification procedure for the general concept they define.

This type of learning is represented by the following simple illustrative example. The aim of the learning is to produce expert system - like rules from a set of examples described by means of attribute values. Suppose that the description of the examples, by means of attributes X, Y, Z, together with their classification are given by Fig. 7.

There are many sophisticated learning algorithms which can be applied to solve the problem. For the illustration the simple AQ -algorithm is used which learns the concept by means of the given positive and negative examples. Based on the algorithm, the most general description of the concept can be obtained as follows.

$$\left[(v(x)=T)\wedge(v(y)=C)\right]\vee\left[(v(x)=T)\wedge(v(y)=F)\wedge(v(z)=S)\right]\vee$$
$$\vee\left[(v(x)=TM)\wedge(v(y)=F)\wedge(v(z)=S)\right]\vee\left[(v(x)=TM)\wedge(v(y)=C)\wedge(v(z)\neq E)\right]$$

The third learning corresponding to the the relation among concepts is organizing knowledge. The fourth learning is adaptation to external environment mostly done by advice-taking learning.

Attributes	Hole (U)	Manufacturing method (V)	Machine tool (W)
Attribute values	In the axis of the rotational part (T) Perpendicular to the axis of the rotational part (TM)	Drilling with twist drill (C) Boring with boring rod (F)	Lathe (E) Turning center (EK) Radial drilling machine (S)

Positive examples:

	U	V	W
P1	T	C	E
P2	T	C	EK
P3	T	C	S
P4	T	F	S
P5	TM	C	EK
P6	TM	C	S
P7	TM	F	S

Negative examples:

	U	V	W
N1	T	F	E
N2	T	F	EK
N3	TM	C	E
N4	TM	F	E
N5	TM	F	EK

Decision tree:

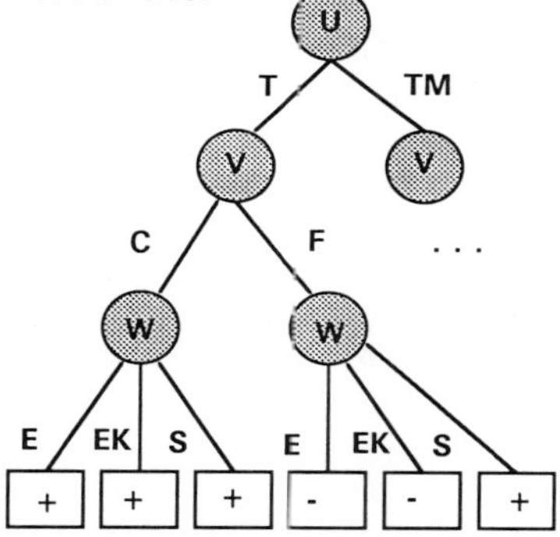

Fig. 7.

5. Conclusion

A general approach to application possibilities of machine learning in Manufacturing Process Planning is given. After the discussion of CAPP as a problem area a novel approach to such systems is presented. Based on the detailed discussion of knowledge requirement for CAPP, the knowledge and the appropriate learning methods are investigated.

REFERENCES

1. Ham, I. - Stephen, C.: Computer - Aided Process Planning: The Present and the Future, Annals of the CIRP, No. 2, 1988 p:591-601
2. Iwata,K. - Sugimura, N.:An Integrated CAPP System with "Know hows" on Machining Accuracies of Parts, Transaction of the ASME, Maj, 1987, p:128-133
3. Eskiciough, H. - Davies, B. J.: An Interactive Process Planning System for Prismatic Parts (ICAPP), Annals of the CIRP, No. 1, 1983, p:365-370
4. Erve, A. H. - Kals, H. J. J.: XPLANE, a Generative Computer Aided Process Planning System for Part Manufacturing, Annals of the CIRP, No. 1, 1986,. p:325 - 329
5. Züst, R. - Taiber, J. : Knowledge-based Process Planning System for Prismatic Workpieces in a CAD/CAM Environment, Annals of the CIRP, No. 1, 1990, p:493 - 496
6. Neelamkavil, J.: Artifical Intelligence Applications in Production Process Planning, Engineering Digest, 15. November, 1985, p:28-31
7. Miller, R. K. : Artifical Intelligence: A new tool for manufacturing, Manufacturing Engineering, April, 1985, p:56-62
8. Eversheim, W. - Marczinski, G. - Cremer, R.: Structured Modelling of Manufacturing Processes as NC-Data Preparation, Annals of the CIRP, No. 1, 1991. p:429 - 432
9. Davies, B, J. - Darbyshire,I. L.: The Use of Expert Systems in Process-Planning, Annals of the CIRP, No. 1, 1984. p:303 - 306
10. Maloney, L. D.: Ways to Quicken Your Product Cycle, Design News, 24, February, 1992 p:74-77
11. L. Horváth: Knowledge-based CAPP system for machining, CAMP '91 conference, Budapest, 1991, p: 113-125
12. M. Horváth: Computer aided manufacturing planning, new methods and applications, 3 emes JST Production Automatisee Conference, Toulouse, 1981.
13. L. Horváth: A methodology for Computer Aided Planning of Manufacturing Process Based on Feature Oriented Workpiece Model, CAMP '92 conference, Budapest 1992, p: 54-61
14. L. Uhr: Multi-Computer Architectures for Artificial Intelligence. John Wiley & Sons, Inc. New York, 1987.
15. L Horváth: The GLEDA computer-aided manufacturing process planning system; Intertechno '90 Conference, Budapest, 1990, p: 157-168.

PROBLEMS OF MANUFACTURING AUTOMATION

Knowledge Based Hybrid Systems (B-11)
I. Mezgár and P. Bertók (Editors)
Elsevier Science Publishers B.V. (North–Holland)

Graph Theoretic-Based Decision Support for Facilities Layout

L.R. Foulds

Department of Management Systems, University of Waikato, Private Bag, Hamilton 3105, New Zealand

Abstract

We describe various graph theoretic heuristics for facilities layout. These heuristics underpin a new facilities layout decision support system called LayoutManager, which is described elsewhere in these proceedings.

Keyword Codes: G.2.2; I.2.1; J.6
Keywords: Graph Theory; Applications and Expert Systems; Computer-aided Engineering

1. INTRODUCTION

Facilities layout, an important research topic in physical system design, has received much recent attention from operations researchers and others. This is due, in part, to increased global competition in manufacturing which has spurred renewed efforts to reduce manufacturing costs. Efficient physical layout is critical in the ability to achieve and maintain manufacturing competitiveness. Indeed, Tompkins and White (1984) reinforced this point by observing that up to half of the operating costs in a manufacturing system are related to materials handling and layout. These concepts are interrelated, as improved layout design often brings about reductions in materials handling, transportation, congestion, and work-in-process. The resulting benefits in productivity have been summarized by Sule (1988). Recent advances in the field of facilities layout have been reported by Heragu (1992) who surveyed recent models and techniques in a special issue of the **European Journal of Operational Research**, which was dedicated to facilities layout and edited by him. It also contains the following articles. Chajed et al. (1992) concentrated on the flow network component. Askin and Mitwasi (1992) integrated layout with process selection and capacity planning. Van Camp et al. (1992) presented a nonlinear optimization approach. Herague and Alfa (1992) and Kouvelis et al. (1992) experimented with simulated annealing algorithms. Kaku and Rachamodugu (1992) and Ram and Vesivanadham (1992) considered the layout of flexible manufacturing systems. Sarin et al. (1992) presented a multiattribute decision-theoretic approach. Leung (1992) described a graph-theoretic heuristic for designing loop-layout manufacturing systems. Rosenblatt and Golony devised a distance assignment approach and Urban (1992) and Balakrishnan, Jacobs and Ventakaramanan (1992) considered dynamic problems.

We concentrate on layout design with the aim of producing a scale plan of the physical system to be designed. This is a specialized, but important, concern within the more general endeavour of manufacturing systems design. We are concerned with the physical arrangement of a given number of facilities, of arbitrary shape, within a given configuration. Classical approaches to this exercise, as described by Foulds (1983), have adopted a single optimality criterion based on either qualitative relationships between facilities or quantitative transportation costs, without the possibility of user interaction or experimentation. In contrast, Harmonosky and Tether (1992) developed a multi-factor plant layout methodology.

2. A GRAPH THEORETIC APPROACH TO FACILITIES LAYOUT

We first discuss the layout problem to be addressed here and then describe various graph theoretic heuristics which accommodate various objective functions for it. Consider a given set of n **facilities**, denoted by *1,2,...,n,* where the first is an **exterior facility**, included to represent the relationships between the other (interior) facilities and their environment. The interior facilities each has a given area and can assume an arbitrary shape. They are to be laid out in a rectangle R, of given dimensions, with the perimeter of each facility comprising straight line segments, each of which is parallel to one of the sides of R. The exterior facility, occupies the rest of the plane outside R. The sum of the areas of the interior facilities (from now on we drop the word **interior**) equals the area of R. We wish to identify a tessellation of R into subregions, corresponding to the facilities drawn to scale, which satisfies an optimality criterion based on some performance measure of the layout.

The tessellation of R is commonly called a **block plan**. Most of the usual layout objective functions depend, at least in part, upon which pairs of facilities (including the exterior facility) are actually adjacent in the block plan. The adjacency structure of any block plan can be modelled in terms of graph theory by what is called an **adjacency graph**. (For undefined graph theoretic notation and terminology see Foulds (1991)). An adjacency graph can be constructed for any block plan by associating a vertex of the graph with each facility and joining two vertices by an edge whenever their corresponding facilities are adjacent.

Thus the adjacency structures of promising block plans can be identified by searching among the adjacency graphs. This search can take place among the spanning planar subgraphs of a certain parent graph. The parent graph has n vertices, one representing each facility, and an edge joining two of its vertices whenever the adjacency of the corresponding two facilities is feasible. Each edge is commonly assigned weights to reflect the importance that its end point facilities are adjacent or somewhat near if not actually adjacent. In some instances, the adjacency of certain pairs of facilities is compulsory and for other pairs it is forbidden. The feasible adjacency subgraphs are invariably planar because the layout takes place within part of the plane, namely R. Once an adjacency graph, which is suitable according to the adopted criteria, has been identified, its dual can be constructed in order to create the block plan. Rinsma et al. (1990) have provided a method to create a block plan from any given adjacency graph. We outline a similar approach to this question of block plan construction in the next section.

Unfortunately the identification of an optimal spanning planar subgraph of a complete edge-weight graph is NP-hard (Giffin (1984)). Problems with relatively small n can be solved optimally by the algorithm of Foulds and Robinson (1976). However difficulties in using the method with increasing n has reinforced recent attempts to devise graph theoretic heuristics for the problem. The utility of the application of graph theory in facilities layout has been evaluated by Hammouche and Webster (1985) and by Hassan and Hogg (1987).

We now discuss various objective functions and graph theoretic heuristics which adopt them.

3. THE DELTAHEDRON HEURISTIC

One measure of performance we can use to construct a planar graph is to assign a closeness rating, as specified in a relationship chart, as weights of the respective edges of the graph. We can then find the planar graph which maximizes the sum of the weighted edges. Three graph theoretic heuristics for this problem were compared experimentally by Foulds et al. (1985). The best of the three, the **Deltahedron heuristic**, is described next.

The Deltahedron heuristic reduces the complexity of finding the maximum weighted planar subgraph by using the concept of a **combinatorial deltahedron**. The construction properties of combinatorial deltahedra have been discussed by Foulds and Robinson (1979). A combinatorial deltahedron is a maximal planar graph and, as such, all of its faces are triangles. A triangle represents a set of three facilities which are mutually adjacent. The deltahedron heuristic is a construction procedure which adds successive vertices to a deltahedron to create a planar graph containing all facilities. The positioning of each vertex in the deltahedron is chosen in an attempt to maximise the sum of weighted edges present.

The process begins by defining an insertion order for the facilities. The insertion order should be chosen to improve the possibility of finding the optimal solution. An example of a useful ordering is to order facilities according to their column sums in the relationship chart. The column sum for a facility is the sum of its closeness ratings with all other facilities. Ordering in this fashion is likely to provide an improvement in the deltahedron score over a randomly ordered solution as facilities inserted early are usually adjacent to more facilities. the closeness ratings of facilities with high column sums are likely to benefit the deltahedron the most if they are inserted early.

An alternative method of choosing the insertion order is to consider the block plan that will be generated from the deltahedron. The method used to construct a block plan may produce a more workable arrangement of facilities if the larger facilities are inserted first. Because the final block plan is an important objective of the user, LayoutManager inserts facilities in order of non-increasing size. Ties are settled using a lexicographic order.

Having decided on an insertion order, the Deltahedron heuristic constructs a complete graph on four vertices from the first three facilities plus the exterior facility. The process used to construct the block plan requires the initial graph to include the exterior facility vertex. This graph is termed a tetrahedron and specifies four triangles in which other vertices may be inserted. The remaining vertices are inserted one at a time, in the specified order. Each vertex is inserted in the existing triangle which provides the largest increase in the deltahedron score. A vertex will be inserted adjacent to exactly three existing vertices by three connecting edges. The next vertex can then be inserted into one of the three newly created triangles or any other existing triangle. Insertion continues until all vertices have been inserted. The resulting graph is guaranteed to be planar and there is a routine in LayoutManager which will construct a block plan from this.

We now illustrate the method by using it to solve the following numerical example. This example will also be used for each of the methods to be discussed later. Consider the following closeness rating matrix for a seven-facility layout problem.

0	51	49	61	93	111	120
51	0	97	60	125	36	39
49	97	0	131	70	141	103
61	60	131	0	85	89	139
93	125	70	85	0	53	59
111	36	141	89	53	0	41
120	39	103	139	59	41	0

The areas of each facility are:

Facility 2 : 400,

Facility 3 : 256,

Facility 4 : 676,

Facility 5 : 484,

Facility 6 : 144, and

Facility 7 : 900.

Facility 1 is assumed to be the exterior facility.

Using the areas of each facility to determine the insertion order, the initial tetrahedron comprises of vertices: 1, 5, 4, 7 and the remaining vertices will be inserted in the order: 2, 3, 6. The scores associated with inserting vertex 2 in the triangles of the tetrahedron are:

(1,5,4) : $\omega_{21} + \omega_{25} + \omega_{24} = 236$

(1,5,7) : 215,

(1,4,7) : 150, and

(5,4,7) : 224,

where ω_{ij} is the closeness rating between facility *i* and facility *j*. Hence we place vertex 2 in (1,5,4). The resulting deltahedron now comprises triangles (1,5,7), (1,4,7), (5,4,7), (1,5,2), (1,4,2), (4,5,2), and we evaluate the benefits of inserting the next vertex (vertex 3), in each of these. We choose triangle (5,4,7) with a benefit of 304. Similarly, vertex 6 is inserted into triangle (5,4,3) completing the construction. The deltahedron solution score is 1383. Further details of the Deltahedron heuristic have been given by Foulds and Robinson (1978).

3.1 AN N-BOUNDARY MODEL

The previous model of the facilities layout model measured only the performance of a solution based on physically adjacent facilities. It is more realistic to consider the effect of nearly-adjacent facilities as well. In the N-Boundary model we wish to maximize the sum of benefits between all pairs of facilities, where the benefit of any pair of facilities is dependent on the least number of boundaries that must be crossed in order to travel between them.

Boundaries in a block plan are equivalent to edges in the adjacency graph. The least number of boundaries between any two facilities can be determined by finding the length of the shortest path between corresponding vertices of the graph. Grunbaum (1967) has shown that for given n facilities there are at most $[(n\text{-}2/3] + 1$ boundaries that must be crossed in travelling between any pair. If the closeness rating for each pair of facilities is given for each possible number of boundary crossings we can use the basic deltahedron heuristic described to find a solution.

3.1.1 THE N-BOUNDARY-DELTAHEDRON HEURISTIC

When inserting vertices into the deltahedron, a method for calculating the length of the shortest path from each newly inserted vertex to every other vertex in the deltahedron is required. Because of the nature of the deltahedron we can make some simplifications to a shortest path algorithm proposed by Dantzig (1967), allowing us to calculate shortest paths efficiently.

When adding a vertex, v, to the deltahedron, the shortest path between existing pairs of vertices is unaltered, so we need only find the lengths of the shortest path from each vertex to the newly inserted vertex. As a vertex is connected to exactly three vertices, x, y, and z, the shortest path to all other vertices must pass through one of these. For every other vertex, v, the length of the shortest path from u to v will be one more than the minimal shortest path length from v to x, y, or z.

To construct the deltahedron, the vertex is temporarily inserted into each candidate triangle and the shortest paths and resulting location is calculated. The triangle insertion which provides the best contribution to the deltahedron score is chosen and the deltahedron and shortest paths are updated to reflect this insertion.

To illustrate the method using the previous example, we require relationship matrices for each possible distance between facilities. For seven facilities, we require $[(7\text{-}2)/3] = 2$ relationship tables. Arbitrarily, we set the entries in the table for a distance of two to be 0.55 of the given table. The order of insertion of facilities is the same as that used in the deltahedron heuristic, as is the initial tetrahedron. The benefit of this initial tetrahedron is

$$\omega^1{}_{15} + \omega^1{}_{14} + \omega^1{}_{17} + \omega^1{}_{54} + \omega^1{}_{57} + \omega^1{}_{47} = 557$$

where $\omega^d{}_{ij}$ represents the benefit of placing facilities i and j at a shortest path distance of d.

The next facility to be inserted is facility 2. If we consider each triangle, in turn, vertex 2 will be at distance 1 from the vertex of the triangle and at distance 2 from the remaining vertex. For example, the benefit of inserting vertex 2 in triangle (1,5,4) is

$$\omega^1{}_{21} + \omega^1{}_{25} + \omega^1{}_{24} + \omega^1{}_{27} = 257.45$$

The benefit of inserting vertex 2 in the other triangles are:

(1,5,7) : 248,

(1,4,7) : 218.75, and

(5,4,7) : 252.05

Hence, we insert vertex 2 in triangle (1,5,4). Continuing, we insert vertex 3 in triangle (5,4,7) and vertex 6 in triangle (5,4,3). The N-boundary solution benefit is 1585.15. Further details of the N-Boundary-Deltahedron heuristic have been given by Giffin and Foulds (1987).

3.2 THE MINIMUM TRANSPORTATION PROBLEM

Although the N-Boundary model has considered the contribution of both adjacent and near-adjacent facilities to the performance of the layout, the model does not take into account the actual physical distance that must be travelled between facilities. The minimum transportation model considers the cost of transportation between adjacent and non-adjacent facilities, measured as a function of distance. It is assumed that there is given a chart whose entries represent the cost per unit distance travelled and the volume of transportation involved. The objective of the facilities layout problem, in these terms, is then to minimize the sum of the product of traffic volume and the distance travelled between all pairs of facilities. A graph theoretic heuristic is again used to find a solution to this problem.

3.2.1 THE SUPER DELTAHEDRON HEURISTIC

The Super Deltahedron heuristic uses the basic construction technique of the deltahedron heuristic to attempt to minimize the transportation cost. An insertion order is defined for the facilities and the first three facilities and the exterior facility are used to construct an initial tetrahedron. The remaining facilities are added to the tetrahedron in the defined order by finding the triangle of the current deltahedron in which the vertex contributes the minimum cost.

To determine the best triangle, the distance from the facility when inserted into each candidate triangle to every other facility is required. Because the exact arrangement and shape of facilities is not determined until after a suitable planar adjacency graph is found, the actual distance travelled is undetermined, so an estimate of the distance travelled between any pair of facilities is used. Assuming each facility will be square in shape in the final block plan layout, and using the shortest path between a given pair of vertices, the estimated distance between facilities p and q is taken to be

$$\frac{1}{2}\sqrt{a_p} + \sum_i \sqrt{a_{v_i}} + \frac{1}{2}\sqrt{a_q}$$

where a_i is the area of facility i and v_i are the facilities on the shortest path from facility p to facility q, excluding p and q.

Using this distance estimate, the cost is calculated by multiplying these distances by the appropriate closeness rating. The facility vertex is then inserted into the triangle of lowest cost by connecting it to the three vertices in the triangle.

To illustrate the Super Deltahedron heuristic we can use the seven facility problem already specified. The heuristic requires an area for the exterior facility, facility 1. When considering the distance from a facility to the exterior region, we will assign facility 1 a zero area. Specifying a zero area implies that the calculated distance is the distance from a facility to the layout boundary. However, when calculating the shortest path between other facilities, a zero area for the exterior region is incorrect as it implies facilities on opposite sides of a layout can be reached with zero cost by moving through the exterior region. In fact, the shortest path, based on rectilinear distance, between any two facilities in a block plan layout cannot contain the exterior region. Paths containing the exterior facility are therefore excluded from consideration as an intermediate vertex.

Again, we use the same insertion order of facilities yielding a solution score for the initial tetrahedron of

$$11\omega_{15} + 13\omega_{14} + 15\omega_{17} + 24\omega_{54} + 26\omega_{57} + 28\omega_{47} = 11082$$

and a rectilinear shortest path matrix of:

0	11	13	15
11	0	24	26
13	24	0	28
15	26	28	0

To insert the next vertex, vertex 2, we calculate distances and benefits for each candidate triangle. For example, when inserting vertex 2 into triangle (1,5,4), the distances, d_{2j}, to each vertex j is:

$$d_{21} = \frac{1}{2}(\sqrt{a_2} + \sqrt{a_1}) = 10,$$
$$d_{25} = \frac{1}{2}(\sqrt{a_2} + \sqrt{a_5}) = 21,$$
$$d_{24} = \frac{1}{2}(\sqrt{a_2} + \sqrt{a_4}) = 23,$$
$$d_{27} = \min(d_{21} + d_{17}, d_{25} + d_{57}, d_{24} + d_{47}) = 47.$$

This yields a solution contribution of

$10\omega_{21} + 21\omega_{25} + 23\omega_{24} + 47\omega_{27} = 6328,$

which when compared to the scores of the other triangles is the best choice for inserting vertex 2.

The shortest path matrix for vertices 1,5,4,7 and 2 is then updated to:

0	11	13	15	10
11	0	24	26	21
13	24	0	28	23
15	26	28	0	47
10	21	23	47	0

In a similar manner, for vertex 3, we choose triangle (5,4,7) with a contribution of 11800 and for vertex 6 we choose triangle (5,4,3) with a contribution of 10559 for an overall solution score of 38789. Further details of the Super Deltahedron heuristic have been given by Foulds and Giffin (1985).

3.3 DELTAHEDRON IMPROVEMENT HEURISTICS

3.3.1 VERTEX RELOCATION

Given a final deltahedron with all vertices inserted, vertex relocation can be applied to improve the deltahedron score. When a vertex is added to the deltahedron it is made adjacent to exactly three other vertices. The triangle of vertices into which the vertex is inserted was chosen on the basis that it increased the deltahedron score by the most from all triangles currently present.

In retrospect however, the insertion of the vertex into the chosen triangle may not have been the best possible insertion, as insertions of other vertices may have created triangles in which the vertex may be inserted to provide a larger score. The previous decision to insert a particular vertex can be reconsidered, if in the final deltahedron, no other vertex has been inserted adjacent to the facility. That is, provided no other vertex is dependent on the existence of the triangles created by the insertion the vertex may be inserted elsewhere. If we examine all vertices in the final deltahedron that have exactly three adjacent edges we can attempt to improve the deltahedron score through vertex relocation. Relocating these vertices has no effect on the placement of other vertices or their contribution to the final deltahedron score. By temporarily removing these vertices they can be reinserted, using the appropriate objective function, into the triangle that provides the best improvement. The strategy considers all such vertices, finding the vertex and insertion which gives the best improvement.

3.3.2 EDGE REPLACEMENT

Consider an edge of the deltahedron and the two associated vertices. Because of the way vertices were inserted into triangles of a planar graph, these two vertices will have in common another pair of vertices to which they are both adjacent. These four vertices then make up two triangles with the considered edge as a base. The edge replacement operation involves the removal of the existing edge and its replacement with the opposite edge that connects the other two vertices.

In performing this operation, the resulting graph remains planar but the score will need to be recalculated. For the deltahedron heuristic, the effect of a single edge replacement is restricted to the removing of the benefit of the current edge and replacing it with the benefit of the opposite edge. For the Super Deltahedron and N-Boundary-Deltahedron heuristics, where shortest paths are used, the affect of replacing an edge requires a complete recalculation of the score of all vertices. The edge replacement strategy used in LayoutManager finds the best single edge replacement based on the improvement in the score. It is possible that the opposite edge with which we wish to replace the current edge may already exist. The ramifications of this possibility are discussed by Foulds and Robinson (1979). In this situation, the current edge is not considered for replacement. Also, to maintain a deltahedron structure, edges of the initial tetrahedron can be replaced only if the replacement edge will create an alternative tetrahedron which still contains the vertex, corresponding to the exterior facility.

We can use the deltahedron produced by the deltahedron heuristic in the previous example to illustrate the vertex relocation and edge replacement improvement heuristics. Beginning with vertex relocation, there are two candidate vertices that can be relocated: vertex 2 and vertex 6. Vertex 6 will not improve the solution score as it was the last vertex inserted and as such has already been inserted into the best triangle. It is possible that the benefit of inserting vertex 2 into a triangle created by the insertion of the vertices 6 and 7, which were inserted later and did not exist at the time vertex 2 was inserted, may provide a better benefit. By calculating the benefit for each triangle, it is found that inserting vertex 2 in triangle (2,6,7) gives a benefit of 258 which is an improvement over its current benefit of 236. We can then relocate vertex 2 to triangle (2,6,7) to create a deltahedron with a higher score, whose solution score is now 1405.

Using the deltahedron improved by the vertex relocation operation above, we can attempt to further improve its solution score using edge replacement. For edge replacement we consider all edges of the deltahedron whose associated opposite edge does not already exist. We must also exclude edge replacement operations which destroy the initial tetrahedron structure. For the given example the only edges we need consider are (2,7), (6,7), and (2,6) whose opposite edges are (3,5), (3,5), and (4,5) respectively. Using the Deltahedron heuristic, the improvement of replacing these edges are:

$\omega_{35} - \omega_{27} = 7$

$\omega_{35} - \omega_{67} = -84,$ and

$\omega_{45} - \omega_{26} = -31.$

Hence we replace edge (2,7) with edge (3,5) and improve the solution score to 1412.

Both vertex relocation and edge replacement can be applied repeatedly until no further improvement is possible.

The above improvement strategies make individual improvements to the deltahedron in isolation, and may therefore overlook other possible improvements. If we consider a sequence of changes whose net benefit to the deltahedron score may be positive but which contains a vertex relocation or edge replacement step that actually decreases the score, the above strategies will not attempt to make the improvement. Without doing a computationally expensive search of all possible vertex relocation and edge replacement combinations, situations such as this cannot be found. The author is currently exploring the possibility of other strategies, including simulated annealing and tabu search. Further graph theoretic methods for facilities layout have been described by Leung (1992) and by Boswell (1992). We now go on to describe a decision support system which implements the above ideas.

4. LAYOUTMANAGER

LayoutManager is a user-friendly menu-driven decision support system for facilities layout which can accommodate any of the objective functions mentioned in the previous section. It employs the graph heuristics previously described and implements routines to output a colour-coded, labelled block plan. It is written in Pascal, within the Microsoft Windows environment, for an IBM-compatible personal computer. A simple interactive computer program for designing a layout with the objective of minimizing transportation costs only has been reported by Wascher and Chamoni (1987).

The user provides LayoutManager with a particular problem to solve by loading a standard ASCII text file that provides all the information necessary to specify the problem. Input files may be created and edited with any suitable text editor or they can be created and edited with the built-in editor. The input file must specify the objective function to be adopted and the number of facilities to be considered, including the exterior facility. There is a soft limit imposed by the program of 40 facilities. This limit may be increased through recompilation if necessary, provided the additional computer memory requirements can be met.

For the objective functions of Sections 1.1 and 1.2, relationship charts of closeness ratings between every pair of facilities is required. For the N-Boundary heuristic, relationships charts are required for each distance from 1 to $[(n - 2)/3] + 1$ where n is the number of facilities.

For any given pair of facilities the closeness rating can be specified as a floating point number allowing the user to define exactly the input parameters; or when the measure is less precise and only a rough estimate of closeness is available the user can specify the closeness rating using a restricted set of values represented by the letters A,E,I,O,U and X.

The letters A,E,I,O,U,X are representative of actual floating point values in decreasing order where A represents the highest value and closeness rating and X is the lowest. The user may specify, in the input file or interactively, the equivalent floating point value for each letter. Default values of 64.0, 32.0, 16.0, 4.0, 1.0, 0.0 and -1024.0 will be assigned to A,E,I,O,U,X respectively if no values are specified. The large default negative value of the letter X is designed to be used for facility pairs whose adjacency is infeasible. The letter representation is loaded in by LayoutManager and is internally converted to the equivalent floating point value, but is still viewed as letter values.

To generate a block plan and for use in the Super Deltahedron heuristic, we require the area of each facility. Areas are expressed as positive floating point values. A unique name must be assigned to each facility so that it can be identified.

An optional property of facility colour can also be assigned to each facility. The facility colour is used to distinguish the facilities in the block plan and as a key to the facility names. Each facility colour is specified as a background colour code, a foreground colour code and a pattern index. The values for the background and foreground colour are limited to the capability of the output display device. The pattern index specifies one of the following patterns: no pattern (background colour only), horizontal lines, vertical lines, left and right diagonal lines, crosshatching and diagonal crosshatching. If no facility colour is specified Layout Manager will assign a unique colour combination.

LayoutManager is described in further detail by Foulds (1993), elsewhere in these proceedings.

5. SUMMARY AND CONCLUSION

We have described a number of graph theoretic heuristics for facilities layout. They accommodate various objectives, such as the maximization of adjacency scores and the minimization of transportation costs. The heuristics make it possible to adopt some flexibility when choosing an objective in facilities planning. They are used to underpin a new decision support system for facilities planning called LayoutManager which is fully described elsewhere in these proceedings.

REFERENCES

Askin, R.G. and M.G. Mitwasi. 1992. Integrating facility layout with process selection and capacity planning. *EJOR* **57**, 162-173.

Balakrishnan, J., F.R. Jacobs and M.A. Ventakaramanan. 1992. Solutions for the constrained dynamic facility layout problem. *EJOR* **57**, 280-286.

Boswell, S.G. 1992. TESSA - A new greedy heuristic for facilities layout planning, IJPR, **30**, 1957-1968.

Chhajed, D., B. Montreuil and T.J. Lowe. 1992 Flow network design for manufacturing systems layout. *EJOR* **57**, 145-161.

Dantzig, G.B. 1967. All shortest routes in a graph. In *Theory of Graphs*. Proceedings of an international symposium held in Rome, 1966. Gordon and Breach, New York.

Foulds, L.R. 1991. *Graph Theory Applications*. Springer-Verlag, New York.

Foulds, L.R. and D.F.Robinson. 1976. A strategy for solving the plant layout problem. *Oper. Res. Quart.* **28**, 845-855.

Foulds, L.R. and D.F. Robinson. 1978. Graph theoretic heuristics for the plant layout problem. *IJPR*. **16**, 27-37.

Foulds, L.R., Gibbons, P.B., and J.W. Giffin. 1985. Graph theoretic heuristics for the facilities layout problem: an experimental comparison. *Opns. Res.* **33**, 1091-1106.

Foulds, L.R., and D.F. Robinson. 1979. Construction properties of combinatorial deltahedra. *Disc. Appl. Math.* **1**, 75-87.

Foulds, L.R. and J.W. Giffin. 1985. A graph theoretic heuristic for minimizing total transportation cost in facilities layout *IJPR* **23**, 1247-1257.

Foulds, L.R. 1993. LayoutManager : A decision support system for facilities planning. These proceedings.

Giffin, J.W. 1984. Graph theoretic techniques for facilities layout. Ph.D. Dissertation. Department of Economics, University of Canterbury, Christchurch, New Zealand.

Giffin, J.W., and L.R. Foulds. 1987. Facilities layout generalized model solved by n-boundary shortest path heuristics. *EJOR* **28**, 382-391.

Grunbaum, B. 1967. *Convex Polytopes.* John Wiley, New York.

Hammouche, A., and D.B. Webster. 1985. Evaluation of an application of graph theory to the layout problem. *IJPR* **23**, 987-1000.

Harmonosky, C.M., and G.K. Tothero. 1992. A multi-factor plant layout methodology. *IJPR* **30**, 1773-1790.

Hassan, K.M.D., and G.L. Hogg. 1987. A review of graph theory application to the facilities layout problem. *OMEGA* **15**, 291-300.

Heragu, S.S. 1992. Recent models and techniques for solving the layout problem. *EJOR* **57**, 136-144.

Heragu, S.S. and A.S. Alfa. 1992 Experimental analysis of simulated annealing based algorithms for the layout problem. *EJOR* **57**, 190-202.

Kaku, B.K. and R. Rachamadugu. 1992. Layout design for flexible manufacturing systems. *EJOR* **57**, 224-230.

Kouvelis, P., W.C. Chiang and J. Fitzsimmons. 1992. Simulated annealing for machine layout problems in the presence of zoning constraints. *EJOR* **57**, 203-223.

Leung, J. 1992. A new graph theoretic heuristic for facility layout. *Mgmt. Sci.* **38**, 594-605.

Ram R. and N. Viswanandham. 1992. Performance evaluation of cellular flexible manufacturing systems : A decomposition approach. *EJOR* **57**, 287-295.

Rinsma, L., J.W. Giffin, and D.F. Robinson. 1990. Orthogonal floorplans from maximal planar graphs. *Environmental and Planning B: Planning and Design* 17, 57-71.

Rosenblatt, M.J. and B. Golany. 1992. A distance assignment approach to the facility layout problem. *EJOR* **57**, 253-270.

Sarin, S.C., P. Loharjun, C.J. Malmborg and B. Krishnakumar. 1992. A multiattribute decision-theoretic approach for layout design problem. *EJOR* **57**, 231-242.

Sule, D.R. 1988. *Manufacturing Facilities - Location, Planning and Design*. PWS-KENT, Boston.

Tompkins, J.A., and J.A. White. 1984. *Facilities Planning*. John Wiley, New York.

Urban, T.L. 1992. Computational performance and efficiency of lower-bound procedures for the dynamic facility layout problem. *EJOR* **57**, 271-279.

van Camp, D.J., M.W. Carter and A. Vannelli. 1992. A nonlinear optimization approach for solving facility layout problems. *EJOR* **57**, 174-189.

Wascher G., and P. Chamoni. 1987. MICROLAY: An interactive computer program for factory layout planning on microcomputers. *EJOR* **31**, 185-193.

Knowledge Based Hybrid Systems (B-11)
I. Mezgár and P. Bertók (Editors)
Elsevier Science Publishers B.V. (North–Holland)

Management of Flexible Capacity

John E. Lenz
CMS Research, Inc.
Oshkosh, Wisconsin USA

Abstract

A project of developing a knowledge-based model for evaluation of simulation results has evolved into a management system for production cells. The short comings of a rule-based system prompted the development of a mathematical-based model for knowledge representation. This model has proven accurate through several benchmark studies and has evolved into a decision support system for manufacturing cell performance. The decision of capacity allocation which are supported today can be automated through a rule-based system. So the application of a rule-based system for performance measuring has identified the need for a rule-based system in the allocation of capacity. When this is available, we will have production systems that run themselves.

XMAS PROJECT

The purpose of the XMAS project was to develop a knowledge-based model for the evaluation of simulation results. In this project, CMS Research, Inc. provided the expertise in evaluation and a Technical University in the U.S.A provided the knowledge engineer. The results of this effort was a ruled-based system which entitled "XMAS" evaluated simulation results.

The rules in XMAS formed a logical step-by-step procedure for isolating the significant components which contributed to the results. These rules focused at production rates and established whether the actual mix was consistent with what was planned. A second set of rules used station utilization to identify the bottleneck. Rules were developed to isolate whether the transportation system had any influence in the performance. Rules which determined the influence of the inventory level were included.

XMAS operated by reading the results of a simulation and using these rules to determine a "reason" of the production cells simulated performance. XMAS was then benchmarked on 10 different types of production systems. The results of these benchmark tests were disappointing. The conclusion reached in XMAS were too simple.

From the investigation of these benchmarks, XMAS rules were not able to detect integration effects. (Integration Effects are losses within a production system because of material shortage and blocking). Rules were developed which considered various levels of blocking and idle time. These rules took the form of "if bottleneck station," "if blocking greater than 10%'" "if idle time greater than target level," etc.

These rules could identify situations, but they could not identify the cause of the situation. Many situations looked alike from a performance result perspective, but had many different causes. A serious review of this approach to knowledge representation was undertaken.

Representation of knowledge through a set of rules was limited to our current understanding of performance variables and their relationship. Because of this, research changed its direction away from representing knowledge through rules to representation of knowledge through a Mathematical Model.

Manufacturing Integration Model (MIM)

The purpose of the Manufacturing Integration Model is to use mathematical equations to express the relationship between the production variable and performance. These equations could also show the relationship between the variables themselves. Another advantage of representing knowledge through a model over rules was that flexibility could be introduced as a production variable.

MIM has proven to be an accurate prediction of manufacturing cell performance. In this role, MIM provides a capability alternative and identifying which alternative is best. Through the set of equations within MIM, it is possible to interpret the results and explain the "reason" for reaching the results. These reasons include integration effects, and a thorough understanding of MIM can even provide solutions to avoid integration effects.

Knowledge is then transferred by use of a mathematical model and the learning of the model. The decision maker "learns" the relationship between the production variables and how each has impact upon performance. Then, when actual performance is observed, he can apply this understanding.

Decision Support

Today, a decision support system has been developed to assist in the evaluation of manufacturing cell performance. This system has been given the name MASTer and contains an intelligence-design-choice feedback approach.

The decisions have been organized into four specific ways to allocate capacity. This capacity allocation provides a means for the decision maker to identify alternatives which improve performance. The evaluation of these alternatives is performed using MIM. This provides an accurate means to compare performances over all the alternatives.

Selection criteria for identifying the best alternative currently is left to the decision maker. However, recent developments in cost modeling have provided that these will prove an effective means to compare and contrast alternatives.

Monitoring and feedback are provided by the tracking of specific performance variables. These include: production mix, station efficiency, inventory level, and flow time. The actual performance of the production system is then compared to those levels predicted from MIM. This monitoring indicates when/where capacity must be reallocated and the decision cycle repeats.

Future Knowledge-Based Development

The capacity allocation-MIM-performance monitoring decision cycle has proved effective in several benchmark studies. Today, this decision cycle relies on decision support tools and the understanding of the decision maker.

In the future, knowledge-based rules can be applied to the capacity allocation decision. In this application, rules which include the capability of the specific manufacturing cell can be developed. These rules will provide a short-cut to identify the "best" way to allocate capacity for the specific cell. This will reduce the dependency upon the understanding of the production supervisor.

The understanding of the production supervisor is also important in the evaluation of these capacity allocation alternatives. Development in cost modeling will eliminate this need of the supervisor. Today the supervisor considers the benefits of high utilization and meeting production requirements. With a cost-based model, all these issues can be considered at the same time.

Production Systems That Run Themselves

I can see the day that the implementation of production capacity will include the knowledge of its own management. That is, the production system will know how to run itself. This will be accomplished with a combination of rule-based knowledge for capacity allocation, models for evaluation, activity-based costing for selection, and monitoring of specific performance variables for feedback into the capacity allocation decision.

HYBRID SYSTEMS IN SCHEDULING

Knowledge Based Hybrid Systems (B-11)
I. Mezgár and P. Bertók (Editors)
Elsevier Science Publishers B.V. (North–Holland)

ANIMATED AI-BASED SIMULATION IN PRODUCTION SCHEDULING

Oleg Yu. Gusikhin and Anatoly S. Kulinitch

Russian Academy of Sciences,
St.Petersburg Institute for Informatics and Automation,
39, 14th line, St.Petersburg, 199178, Russia,
E-mail: oleg@iias.spb.su & anat@iias.spb.su

Abstract
The paper presents a case study on the integration of traditional scheduling techniques and AI. We propose an animated AI-based simulation as a powerful scheduling tool. An essential feature of animated AI-based simulation is that it uses the power of AI combined with convenient scheduling environment. This approach is illustrated through FMS scheduling. FMS is represented by timed Petri net. The scheduling procedure is based on the interactive heuristic search on Petri net reachability tree. The implementation of the scheduling system in Turbo-Prolog is discussed.

Keywords Codes: I.2.1; I.2.8; I.6.3
Keywords: Artificial Intelligence, Applications and Expert Systems; Problem Solving, Control Methods, and Search; Simulation and Modeling, Applications

1. INTRODUCTION

Production scheduling problems have been intensively studied in the Operations Research. Nevertheless the majority of various approaches developed have a limited applicability in realworld scheduling problems because of their computational complexities. Currently simulation with dispatching rules remains more widely used computer-based scheduling technique.

In simulation approach a scheduler constructs a discrete-event simulator, and then searches appropriate dispatching rules in order to achieve desired changes in operating conditions. This is done by repeating a simulation for several times with variable decision parameters until satisfactory schedule is found and as usual many simulation runs are required to make a decision.

Recent advances in artificial intelligence have created significant new opportunities in production scheduling. From the AI standpoint scheduling could be considered as a search with constraints through a problem space [1] which allows to generate automatically a schedule with required characteristics.

One can recognize a big similarity between simulation control mechanism and data-driven problem space search in AI. This similarity offers possibilities for there useful interrelations, regarding both an improvement of simulation as a convenient scheduling tool and on the other hand - possibilities to exploit the simulation theory results and experience [2] within AI approach of scheduling to improve problem-solving efficiency.

The paper presents a case study on the integration of traditional scheduling techniques and AI. We propose an animated AI-based simulation considered as an improvement of conventional simulation with priority rules [2] on the base of AI and visual interactive modelling techniques [3]. An essential feature of animated AI-based simulation is that it uses the power of AI combined with convenient scheduling environment which enables the user to easily understand and to draw on experience and intuition to guide the schedule generation process toward better schedules.

2. FMS SCHEDULING PROCEDURE

To illustrate the approach being discussed consider the FMS scheduling problem. FMS is represented as a composition of Petri net models of both FMS facilities and technological processes [4].

The manufacturing facilities are represented by state machine Petri net subclass where each transition has exactly one incoming arc and exactly one outgoing arc [5] (fig. 1a):

$R = (T, P; F^-, F^+)$,

where $T = \{t_1, t_2, \ldots, t_n\}$ is a set of transitions corresponding to a set of machine tools;

$P = \{p_1, p_2, \ldots, p_m\}$ is a set of positions corresponding to a set of buffers;

$F^- \subseteq P \times T$ and $F^+ \subseteq T \times P$ are the pre- and post-incidence relations of Petri net R.

The technological processes are represented by acyclic state machine (fig. 1b). Each connectivity component of this Petri net corresponds to the technological process of the particular part type:

$Tp = (C, D; W^-, W^+)$,

where $C = \{c_1, c_2, \ldots, c_k\}$ is a set of transitions corresponding to the set of operation types;

$D = \{d_1, d_2, \ldots, d_l\}$ is a set of positions corresponding to the set of part states;

$W^- \subseteq D \times C$ and $W^+ \subseteq C \times D$ are the pre- and post-incidence relations of Petri net Tp.

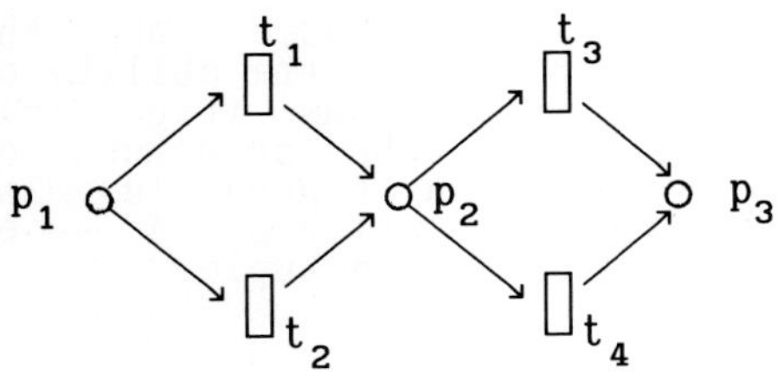

Fig. 1a. FMS facilities model

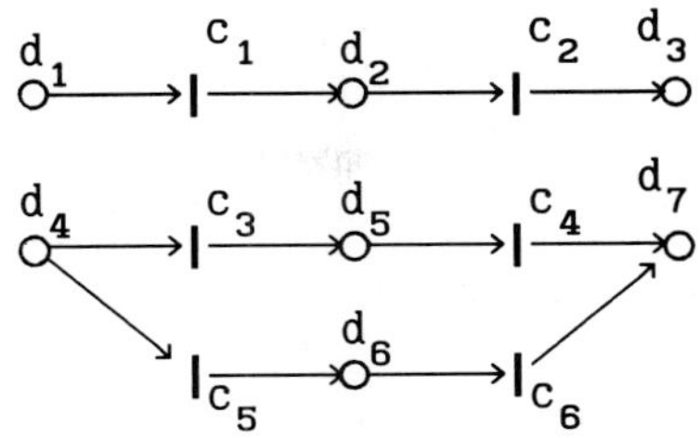

Fig. 1b. Technological processes model

	c_1	c_2	c_3	c_4	c_5	c_6
t_1	+	–	+	–	–	–
t_2	+	–	+	–	+	–
t_3	–	+	–	+	–	–
t_4	–	+	–	+	–	+

Fig. 1c. Compatibility relation

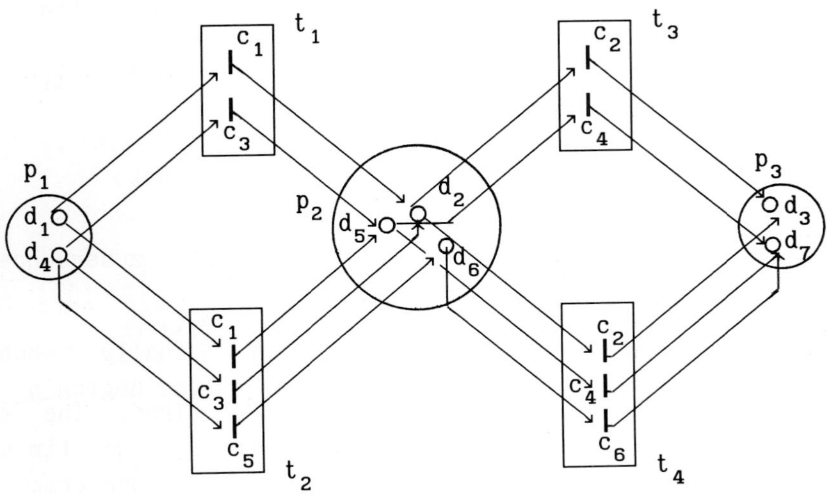

Fig. 1d. FMS production process model

The composition of these Petri nets is based on the compatibility relation: $A \subseteq T \times C$ which defines the ability of machine tools to carry out particular operations (fig. 1c). The resulting Petri net Q which represents the structure of the FMS production process is given in a figure 1d. The durations of operations are defined by a relation $B : A \longrightarrow \mathbb{N}$.

The initial state S_0 of the Q is defined by a tuple:

$S_0 = (M_0, K_0, L_0, H_0)$,

where $M_0: P \longrightarrow \mathbb{N}$ - position marking;

$K_0: T \longrightarrow \{idle, active\}$ - transition states;

$L_0: T \longrightarrow C$ assigns the operations to active transitions;

$H_0: T \longrightarrow \mathbb{N}$ assigns the residual times to active transitions.

The firing rule is defined by the follows:

- Transition $(t_i, c_j) \in A$ is enabled in state S_0, iff $K_0(t_i) = idle$ and $M_0(p_r, d_p) > 0$ for p_r, d_p such that $(p_r, t_i) \in W^+$, $(d_p, c_j) \in F^-$;
- Firing of the transition (t_i, c_j) gives a following state S_i, where

$K_1(t_i) = active$;

$L_1(t_i) = b(t_i, c_j)$;

$H_1(t_i) = c_j$;

$M_1(p_r, d_p) = M_0(p_r, d_p) - 1$.

- If t_i is fired at time τ_0 then during interval $[\tau_0, \tau_0 + B(t_i, c_j)]$ transition t_i will be an active one.
- End of transition t_i firing at time $\tau_1 = \tau_0 + B(t_i, c_j)$ gives a new state S_2, where:

$K_2(t_i) = 0$;

$M_2(p_q, d_s) = M_1(p_q, d_s) + 1$ for (p_q, d_s) such that $(t_i, p_r) \in W^+$, $(c_j, d_1) \in F^+$.

The execution of the model Q is represented as a sequence of steps:

$S_0[G_1, \nu_1 > S_1[G_2, \nu_2 > \dots S_i[G_{i+1}, \nu_{i+1} > S_{i+1}$.

In a given state S the set G_{i+1} of mutually enabled transitions (possibly empty) is selected. At the beginning of the step i+1 the transitions of G_{i+1} are fired. The step duration ν_{i+1} is equal to the minimum of the residual times of the active transitions. A time projection is performed by reducing the residual times of active transitions on value ν_{i+1}. At the end of step i+1 the active transitions whose residual times are equal to zero, end the firing producing the following state S_{i+1}.

The schedule Ps_i corresponding to a given sequence of steps is a set of tuples (τ_p, t_r, c_f) denoting the firing of transition (t_r, c_f) at time τ_p. The schedule Ps is incrementally created by adding at each step i of model execution a set $\{(\tau_i, t_r, c_f)\}$, in one-to-one correspondence to G_i, where τ is a current time equal to $\tau_i = \tau_0 + \sum_{j=1}^{i-1} \nu_i$ and $(t_r, c_f) \in G_i$.

The whole problem state space can be conveniently represented by reachability tree, which defined by the following. The root node X_0 is defined by the initial state S_0, start time τ_0 and empty schedule Ps_0. For a given node X_i corresponding to a model state S_i the set of the immediate descendants is determined by all possible sets G_{i+1}. Each descendant of a given node corresponds to the state produced by applying the step operator determined by the particular G_{i+1} to the current state. Each arc of the tree is labeled by corresponding G and duration of the step.

In terms of reachability tree the scheduling problem could be viewed as a search of the path in the tree from the root node X_0 to the goal node X_g, where every intermediate node X_i satisfies the given constraints on performance parameters (which are functions of corresponding S_i and Ps_i), such as maximal number of parts in the buffers, maximal idle time of machine tool etc.

The large size of the reachability tree even for relatively small test problems and the absence of the efficient algorithm to predict the tree growth in a unsuccessful directions make the AI technique to be an appropriate one. The simplest way to avoid a combinatorial explosion is to use a search with chronological backtracking, which allows to generate a path iteratively , one node at a time, by selecting one operator at each level. If the new state is not satisfactory the search process backtracks and another operator is selected. The efficiency of such approach grows when some heuristic knowledge is used to select the most promising operator at each level.

Suppose that there exists a function $\chi : X_i \longrightarrow \Lambda_{i+1}$, where $\Lambda_{i+1} = \{G^1_{i+1}, G^2_{i+1}, \ldots, G^u_{i+1}\}$ is an ordered set of operators. The choice of direction at each level i made by applying the most left unexplored operator beginning from G^1_i. Thus, in case of backtracking to level i the operator G^{j+1}_i will be applied if the last explored operator was G_i^j. If G_i^j is a last operator in Λ_i then the process backtracks to the previous level of tree.

In order to define a function χ it should be noted that Λ is a subset of the boolean [6] of set of enabled transitions. Therefore we can implicitly define the order on Λ through the order on the set of enabled transitions.

Let Ω be a set of enabled transitions in a given state on which we impose the order $\geq$. It is easy to extend the order $\geq$ on Ω to the boolean $\mathbb{B}(\Omega)$ so that $Y_j \geq Y_i$ for any Y_i, $Y_j \in \mathbb{B}(\Omega)$, if either $Y_i \subseteq Y_j$ or maximal element of $(Y_i \setminus Y_i \cap Y_j)$ not greater than maximal element of $(Y_j \setminus Y_i \cap Y_j)$.

The benefit of such operator selection strategy is that the order $\geq$ on the set of enabled transitions could be defined by the priority functions similar to those used in a conventional simulation [7]. However, in contrast to conventional simulation in this case priority function will define the order of feasible decision trials rather than fixed strategy of conflict resolution. Let us note that the definition of priority functions could be a range from simple static priorities to sophisticated expert systems.

Visualization of the search process is straight-forward through the Gantt chart which is a traditional representation form of the manufacturing schedules (fig. 2). The animated

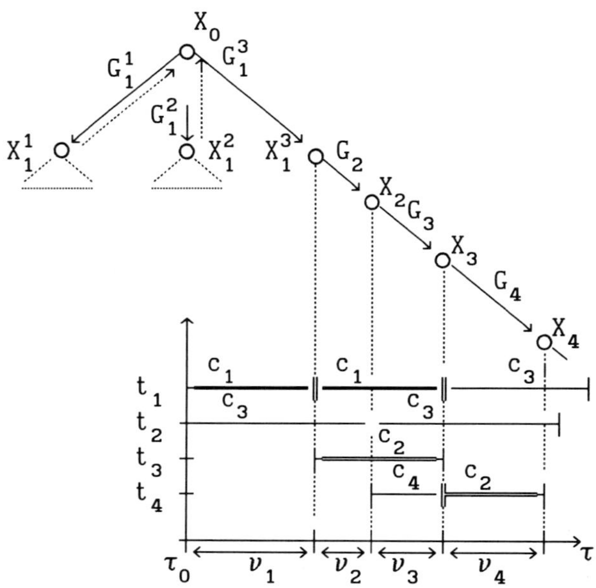

Fig. 2. Reachability tree search

graphical representation of the schedule generation enables the user to assess the current state of the process and if needed directs the search procedure in exploring new alternatives from the desirable level of a tree by indicating a point at a time scale.

3. IMPLEMENTATION

Due to declarative semantics and nondeterminism Prolog seems to be the most appropriate language for the implementation of the scheduling system [8]. The research prototype of the system has been implemented on IBM PC/AT using Turbo-Prolog 2.0. Turbo-Prolog has been chosen because it provides efficient low-cost compiler with useful for our purpose facilities such as powerful graphic capabilities, mouse support and dynamic cutting mechanism.

The Petri net model and schedule are represented by a set of Prolog facts. Corresponding Turbo-Prolog dynamic database declaration is represented in Appendix A.

The schedule generation is realized by predicate 'schedule' and its general format is:

schedule(Time,Top)

in which Time is the current scheduling time and Top is the current pointer to the top of the stack of backtrack points. The 'schedule' clauses are given in Appendix B.

The main process of reachability tree search is performed by nondeterministic recursive procedure in clause (2). At the beginning the list 'L' of enabled transitions ordered in accordance with a given priority function is obtained by calling 'take_list'. The decision procedure is implemented by nondeterministic predicate 'make_step' which takes the list 'L' as an input parameter. Predicate 'check_step' allows to prevent the empty step of the model execution corresponding to a state in which there is no at least one active transition. The following stage is to find a step duration ('find_step_duration') and generate next state ('next_state'). The new state is checked according to the given constraints by calling 'check_constraints'. If the new state satisfies the constrains the tree is further expanded from time Current_Time to time Next_Time at the next search level by calling 'schedule' recursively. When a conflict occurs the procedure explores an alternative decision or backtracks on the previous level of a tree.

Clauses of predicate 'make_step' are presented in Appendix C. This procedure successively checks every transition from a list 'L' by calling 'possible_to_fire' and fires it if it is enabled ('assign'). When 'assign' is called it executes procedure 'assign_made' which performs the operations related to the transition firing, such as changing the transition state and the state of its input position, adding the fact to the schedule database and adding corresponding representation to dynamic display. In the case of backtracking over 'assign' the alternative procedure 'assign_backup' which performs opposite operations will be selected to cancel changes have been made. Thus, if backtracked into the predicate 'make_step' will take a decisions in a sequence according to the order of transitions in a list 'L' (example is given in Appendix D).

The predicate 'next_state' (Appendix E) is realized similarly to predicate 'assign', so that backtracking over 'next_state' results in canceling the changes related to the time projection.

The user control of backtracking is realized on the base of Turbo-Prolog dynamic cutting mechanism, which provides two build-in predicates: getbacktrack and cutbacktrack. The former returns the current pointer to the top of the stack of backtrack points. The later allows to remove all backtrack points above the given pointer.

At any moment the user can interrupt the search process by pressing a key. The predicate 'user_interrupt' (Appendix F) checks the user interruption of the process and recognizes the type of interruption. The user can either stop process to assess the partial schedule represented by Gantt chart and to define a desirable backtrack point or just initialize backtracking on the previously defined point.

In the first case, the procedure 'monitor' is called. The procedure enables the user to observe the static display of the partial schedule. The user can scroll forward and backward through the display using 'mouse' and define the desirable backtrack point. The backtrack point is represented by a fact:

backtrack_point(TIME).

To exit the procedure 'monitor' the user can either initialize the backtracking or give an instruction to continue the search process from the current point.

The user defined backtracking mode is activated by asserting a fact 'backtrack'. The backtracking at a given level of tree is supported by predicate 'user_backtrack' (Appendix G). This predicate performs actions only under the condition of user defined backtrack mode (clauses (19) and (20)). If the 'Time' of user defined backtrack point is less then 'Current_Time' the procedure restores the initial state of the current search level ('backup_state') and jumps to the previous level of a tree preventing the generation of the residual decisions in 'make_step' by calling 'cutbacktrack'. When the backtracking reaches the user defined point the alternative decision of this level of a tree is generated and search renewed.

When the search process reaches the goal state specified by predicate 'goal_state' (clause (1) in Appendix A) procedure 'monitor' is invoked. At this stage the user can assess the suitability of the proposed schedule and direct the computer to explore new alternatives in the search for best schedule.

4. CONCLUSION

The key aspect of animated AI-based simulation as a scheduling tool is that it uses the power of AI combined with convenient scheduling environment which enables the user to easily understand and to draw on experience and intuition to guide the schedule generation process toward better schedules. This approach has been proven to be quite useful in the complicated realworld applications, such as scheduling of the assembly-welding plots in the ship hull production [9]. We argue that the success of this approach is in the rational symbiosis between computer and scheduler that allows to produce quickly the realistic, workable and near optimal schedules.

Additional advantages arising from the use of Petri nets are the follows:
- a graphical and prices nature of Petri net representation scheme provides an efficient and flexible tool for modelling of the concurrent operations in the manufacturing system;
- a hierarchical modelling capability of Petri net models offers the possibility to describe the system at different levels of abstraction and details and hence makes the proposed tool suitable for different hierarchical scheduling approaches;
- an abstract net-theoretic concepts make the proposed technique appropriate for a wide range of applications for scheduling of the discrete concurrent systems different from FMS.

ACKNOWLEDGEMENTS

We would like to thank Irina Podnozova and Tanja Levashova for their help in preparing this paper.

REFERENCES

1 M.S. Fox, Constraint-Guided Scheduling - A Short History of Research at CMU, Computers in Industry, No. 1 (1990) 14.
2 P.A. Newman, Scheduling in CIM System, A. Kusiak (ed.), Artificial intelligence: Implications for CIM, IFS, London, 1988.
3 R.D. Hurrion, Visual Interactive Modelling, R.D. Hurrion, (ed.), Simulation: Applications in Manufacturing, IFS, Berlin, 1986.
4 S. Dwiwedi, A. Leskin, D. Lyons, O. Gusikhin, Flexible Modular Structure for Robot Control System, Proceedings of the IV World Conference on Robotics Research, Pittsburgh, 1991.
5 T. Murata, Petri Net: Properties, Analysis and Applications, Proceeding of the IEEE, No. 4, (1989) 77.
6 P.V. Cohn, Universal algebra, HARPER & ROW Publishers, New York, 1965.

7 K. Wroblewski, R. Krawczynski, Priority Rules in Production Flow Control, Material Flow, No. 2 (1985) 2.
8 I. Futo, AI and Simulation on Prolog basis, C.A.Kulikovski, R,M. Humber, G.A. Ferrate, (eds.), Artificial Intelligence, Expert Systems and Languages in Modelling and Simulation, North-Holland, Amstendarm, 1988.
9 O. Gusikhin, A. Novogilov, A. Spiridonov, O. Fedorov, Prolog Application in Scheduling of Ship Hull Production, Proceedings of International Conference of the Practical Application of Prolog, London, 1992.

APPENDIX A

```
DATABASE - resource
transition(Transition)
position(Position)
transition_position(Transition,Position)
position_transition(Position,Transition)

DATABASE - operations
operation(Operation_Type)
part(Part)
operation_part(Operation_Type,Part)
part_operation(Part,Operation_Type)

DATABASE - compatibility
duration(Transition,Operation_Type,Time)

DATABASE - state
transition_state(Transition,State)
residual_time(Transition,Time)
active_operation(Transition,Operation_Type)
position_state(Position,Part,Number)

DATABASE - schedule
process(Time,Transition,Operation_Type)
```

APPENDIX B

```
schedule(Time,_):-                                        (1)
    goal_state(Time),
    monitor,
    not(backtrack).
schedule(Current_Time,Current_Top):-                      (2)
    take_list(L),
    make_step(Current_Time,L),
    check_step,
    user_backtrack(Current_Time,Current_Top),
    find_step_duration(Duration),
    findall(X,transition_state(X,active),LX),
    next_state(LX,Duration),
    user_interrupt,
    check_constraints,
    Next_Time=Current_Time+Duration,
    getbacktrack(Next_Top),
schedule(Next_Time,Next_Top).
```

APPENDIX C

```
make_step(_,[]).                                                    (3)
make_step(Time,[enabled(Transition,Operation_Type)|L]):-            (4)
    possible_to_fire(Transition,Operation_Type),
    assign(Time,Transition,Operation_Type),
make_step(Time,L).
make_step(Time,[_|L]):-                                             (5)
make_step(Time,L).

assign(Time,T,C):-assign_made(Time,T,C).                            (6)
assign(Time,T,C):-assign_backup(Time,T,C),fail.                     (7)

assign_made(Time,Transition,Operation_Type):-                       (8)
    position_transition(Position,Transition),
    part_operation(Part,Operation_Type),
    duration(Transition,Operation_Type,Duration),
    retract(transition_state(Transition,_)),
    retract(position_state(Position,Part,N)),
    N1=N-1,
    asserta(position_state(Position,Part,N1)),
    asserta(transition_state(Transition,active)),
    asserta(residual_time(Transition,Duration)),
    asserta(active_operation(Transition,Operation_Type)),
    asserta(process(Time,Transition,Operation_Type)),
    add_to_display(Time,Duration,Transition,Operation_Type),!.

assign_backup(Time,Transition,Operation_Type):-                     (9)
    position_transition(Position,Transition),
    part_operation(Part,Operation_Type),
    duration(Transition,Operation_Type,Duration),
    retract(transition_state(Transition,_)),
    retract(residual_time(Transition,_)),
    retract(active_operation(Transition,_)),
    retract(position_state(Position,Part,N)),
    N1=N+1,
    asserta(transition_state(Transition,idle)),
    asserta(position_state(Position,Part,N1)),
    retract(process(Time,Transition,Operation_Type)),
    remove_from_display(Time,Duration,Transition),!.
```

APPENDIX D

EXAMPLE

```
make_step(Time,[enabled(t1,c1),enabled(t1,c2),
enabled(t2,c1),enabled(t2,c2)])
```

RESULT

```
1. {(t1,c1),(t2,c1)}
2. {(t1,c1),(t2,c2)}
3. {(t1,c1)}
4. {(t1,c2),(t2,c1)}
5. {(t1,c2),(t2,c2)}
6. {(t1,c2)}
7. { }
```

APPENDIX E

```
next_state(L,Duration):-                                          (10)
    next_state_made(L,Duration).
next_state(L,Duration):-                                          (11)
    next_state_backup(L,Duration),fail.
```

APPENDIX F

```
user_interrupt:-                                                  (12)
    process_interruption(Mode),
    perform_interruption(Mode),!.
user_interrupt:-!.                                                (13)

perform_interruption(backtrack):-                                 (14)
    assert(backtrack),!,fail.
perforce_interruption(stop):-                                     (15)
    monitor, not(backtrack),!.
```

APPENDIX G

```
backup_state(Time):-                                              (16)
    process(Time,T,C),
    assign_backup(Time,T,C),fail.
backup_state(_):-!.                                               (17)

user_backtrack(_,_).                                              (18)
user_backtrack(Current_Time,Top):-                                (19)
    backtrack,
    backtrack_point(Time),Time<Current_Time,
    backup_state(Current_Time),
    cutbacktrack(Top),!,fail.
user_backtrack(Current_Time,_):-                                  (20)
    backtrack,
    backtrack_point(Time),Time>=Current_Time,
    retract(backtrack),!,fail.
user_backtrack(_,_):-fail.                                        (21)
```

Knowledge Based Hybrid Systems (B-11)
I. Mezgár and P. Bertók (Editors)
Elsevier Science Publishers B.V. (North–Holland)

The hybrid expert system as an approach to the solution of a certain class of scheduling problems.

A. B. Kurzydłowska

Institute for Organization of Production Systems, Warsaw University of Technology, 02-524 Warszawa, ul.Narbutta 85, Poland.

Abstract

The Artificial Intelligence approach is a good solving method for practical applications due to the complexity of scheduling problems. However, the AI methods should be applied when traditional methods do not provide satisfactory answers. On the other hand, both approaches are complementary and to achieve best results both should be applied whenever they may be used efficiently.

The article presents a concept of the hybrid system of solving scheduling problems ORGPLAN3. The system is based on previous version ORGPLAN1 and ORGPLAN2. It employs both numerical and symbolic computation algorithms and as far as knowledge representation of the problem is concerned, both mathematical model and frame-based representation are applied.

Keyword Codes: I.2.1; I.2.4; I.2.8
Keywords: Applications and Expert Systems; Knowledge Representation Formalisms and Methods; Problem Solving, Control Methods, and Search

1. INTRODUCTION

The research work conducted for many years to find efficient methods of scheduling has resulted in defining a number of numerical methods [Graham et al., 1979]. Each of them is efficient (i.e. each provides a good solution in a reasonable period of time) only for problems fulfilling certain specific conditions. To solve virtual problems with these methods an additional assumption that adjusts the real situation to the algorithm assumption should be accepted.

Combinatory complexity[1] and appearing of several, usually impossible to synchronous fulfilling objective functions feature factual scheduling problems [Berry, 1992]. Furthermore,

[1]The computation time for finding an optimal solution to a scheduling problem increases exponentially or worse as a function of a number of subtask and resource under consideration (NP-hard problem).

dynamic changes of parameters and situations are also features of scheduling problems. The dynamic feature of the problem [Graves, 1981] makes optimal solutions pointless due to the continuous changes. Frequently, the stochastic character of input data makes a suitable model impossible to create. The classical methods do not take into consideration the situation at the starting point, either.

Thus, as the practice has indicated many times, a schedule of certain problems is prepared by expert-men with no or relatively small use of computer automatization.

The divergence of theory and practice on one hand and the necessity of automatization of scheduling processes (for instance Computer Integrated Manufacturing) on the other, has resulted in research aimed at substitution of expert-man by scheduling computer system, what is also a part of Artificial Intelligence field of interest.

The AI approach to the solutions of scheduling problems is superior to the traditional one in many respects. Some of them worth mentioning are:

- the possibility of dealing with not only quantity but also quality aspects of the problem;
- the possibility of coping with stochastic and dynamic input data;
- the possibility of an easy modification of a situation description included in the system;
- the possibility of acquiring an acceptable solution in an acceptable period of time;
- the possibility of acquiring the explanations of a solution.

The approach, however, is not void of disadvantages due to the imperfection of an expert's knowledge. An expert system may not find the factual solution, providing only an acceptable solution. In general opinion, however, advantages outweigh disadvantages, as has been confirmed by a large amount of research work in this field. The literature on expert systems in scheduling is very rich [Gilmore et al., 1989, Kanet and Adelsberg, 1987].

In the present paper, it has been assumed that AI methods should be applied when traditional methods cannot be used efficiently i.e. cannot give an acceptable solution in an acceptable period of time. It has also been assumed that the combination of the two approaches into one hybrid system, which could apply either of the methods as necessary, will be most appropriate. The reason for this is the fact that AI methods and methods based on numeric computation are complementary and to reach an optimal solution they should be applied in those aspects and phases of the problem in which they can give positive results. It is an important advantage of the hybrid system that good sides of either method may be used when required. In the same way disadvantages of one method can be counterbalanced by advantages of the other one.

This approach is the subject of the presented paper. The paper describes a concept of the ORGPLAN3 system, which was designed for solving a certain class of scheduling problems. The concept is the result of experience obtained while testing the two previous versions of the system. The first one employed the Operational Research methods, while the other one used heuristic algorithms for constraint-satisfaction problems. The analysis of advantages and disadvantages of both systems resulted in creating the ORGPLAN3 system.

In its first part the article presents the main concept of the ORGPLAN systems and the class of problems for which they were designed. Next, a method for problem description correctness analysis, which is an essential part of the ORGPLAN1 system (realized basing on the mathematical analysis constraint system) has been presented. A subsequent version of the system ORGPLAN2, which is based on a heuristic algorithm with backtracking has been described as well.

The hybrid system ORGPLAN3, presented in the third part of the article combines the advantages of both systems, using numerical algorithms as well as symbolic algorithms. In that system both a mathematical model and frame-based representation are used for the knowledge representation of a problem.

2. THE MAIN IDEA OF THE ORGPLAN SYSTEMS.

The analysis of a number of scheduling problems indicates that four major concepts are sufficient to describe most of the problems [Kurzydłowska and Piasecki, 1988a]:

- concept of task - z_z /What?/;
- concept of element performing the task - e_n /Who?/;
- concept of place where the task is performed - v_s /Where?/;
- concept of time of the task performance - t_k /When?/.

Thus stating a schedule amounts to answering the four questions: Who, What, Where and When. The answer should consider the interdependence of the problem elements defined in the problem description.

It has been assumed that the language whose sentences express the problem constraints will be most natural way of the problem description. The elaborated ORGPLAN language includes:

1. imperative sentences, which define required assignments. For instance, the sentence ZDR[2], $Z=z_z$, $V=v_s$[3] means that the task z_z has to be performed in the place v_s;
2. indicative sentences (ZDO), which describe possible and forbidden assignments;
3. special sentences (ZDS), which define and classify elements due to organizing.

The problem description in the ORGPLAN language can be used as a base for organizing required activity. The ORGPLAN systems achieve this task in two specific phases [Kurzydłowska and Piasecki, 1989]:

I. transformation of a situation description from the ORGPLAN language into mathematical model, which is a constraint-inequality system and checking the correctness of the semantic description;
II. determination of required schedule using an adequate algorithm selected from the method base.

In the first stage, a constraint for binary decision variables is generated for every

[2]The keywords ZDO, ZDR, ZDS are abbreviations of the corresponding polish phrases: ZDO - indicative sentence, ZDR - imperative sentence, ZDS - special sentence.

[3]In the article a simplified form of the ORGPLAN language has been presented. An interactive front-end module with the user is anticipated. It will be designed to translate the more natural form of sentences of the problem description into corresponding sentences in the ORGPLAN language. It was partly realized in the ORGPLAN2 system.

sentence (or a group of sentences).[4] In that way a mathematical model of the problem is obtained on the basis of the linguistic description (see Appendix for an example of generating an mathematical model). The model is a base for correctness analysis of the problem formulation. The assumption that

> Semantic correctness of a text in the ORGPLAN language is equal to the consistency of the constraint system that is generated by the text,

was proven in the work [Kurzydłowska, 1986].

Thus, the method of checking consistency of the constraint system, is also a method of checking the semantic correctness of the problem description in the ORGPLAN language.

In the second phase, the most suitable model and method (algorithm) is selected from the model base and the method base according to the problem description acquired in the first stage.

In the subsequent part the paper presents two versions of the ORGPLAN system, which are concrete implementations of the general philosophy of the ORGPLAN system.

3. THE ORGPLAN1 SYSTEM.

The ORGPLAN1 system was designed so as to allow the user to introduce individually constraints and conditions of a specific organization problem. As a result, the problem description of factual big problems often contain redundant and/or contradictory information. Defined constraints frequently determine some assignments (the decision variables are defined) and in this way a reduction of decision variables or the revealing of contradictions of define conditions are possible.

The concept of ORGPLAN1 system employs *a preanalysis of correctness of a defined problem*. The preanalysis includes those aspects of the problems which can be relatively quickly analyzed and the conclusions result in:

- revealing contradictions of a problem description, as for example imposing constraints, which demand performing two tasks in the same place and at the same time or imposing an impossible to fulfill sequence of a task performance (such contradiction is quite typical for scheduling problems);
- automatic definition of some assignments derived directly from the problem wording (for example, knowing the time of a task performance it is possible to determine other time assignments of the task while defining the exact moment of the beginning of the task performance).

The effectiveness of a preanalysis derives from the fact that it is directed toward a specific characteristics of the scheduling problems, taking into account specific constraints such as continuity of a task performance, one place of performance and frequently occurring in the scheduling problems - sequence constraint.

At the same time, formulating constraints by means of binary variables allows a simple and effective algorithm for checking their correctness. The algorithm is based on a

---

[4]Decision variables express assignments of place for a task (X_{zs}), elements for a task (Y_{zn}), time for a task (T_{zk}).

mathematical analysis of the system of binary constraints derived directly from the problem description. Its effectiveness (see Appendix) is also the result of the method involving structuralization and classification of constraints.

Constraints are grouped in classes with the same consistency principles. Additionally, each class is related to another in the way that the principles of changes of a constraint type within a class are defined. A constraint within a certain class, changes a type as a result of defining qualities of some constraint variables. The change of a type is related to simplification of a constraint form and can result in consequent defining the qualities of some decision variables and, at the same time, in the reduction of problem solving space.

In the approved strategy every assignment of value on a decision variable cause rechecking of fulfillment of the consistency condition by all the constraints containing the variable. If the rechecking shows the lack of consistency, a constraint:

- may change its type (what involves rechecking of the constraint again);
- may appear to be contradictory, which means that a given problem description is contradictory.

The mechanism of checking the problem description correctness provides every constraint with correctness checking within the context of the remaining problem constraints. It means that semantic consistency of each sentence of the problem description within the whole situation description is provided.

A more detailed description of the applied algorithm has been presented in the paper [Kurzydłowska and Piasecki, 1988b]. As a result of the algorithm, the internal consistency of the problem description is examined and in case of the lack of contradiction, the problem description is simplified.[5]

The ORGPLAN1 system was implemented by means of IBM PC computers. An important conclusion derived from testing the ORGPLAN1 system was the ascertainment that the analysis of the problem description correctness before running the solution algorithm is an essential element of the system effectiveness. The analysis helps discover some of the contradictions (and in this way to avoid the situation in which the algorithm works very long due to the lack of an acceptable solution). It also allows a considerable decrease increase in the solution space (by decreasing the number of decision variables and the constraints related to them). In some cases the achievement of a reduction of the problem quantity makes effective application of the optimalization methods possible, whilst the revelation of contradictory constraints has frequently allowed the avoidance of time-consuming work of an algorithm.

The experience of the ORGPLAN1 system work has confirmed the usefulness of a preanalysis of the wording correctness. At the same time, it has been concluded that application of mathematical programming methods to problem solving (even in its reduced form) is in most cases inefficient due to NP-hard complexity of the problem.

---

[5]The presented algorithm does not find out all the contradictions. The whole consistency check-up is equal, as far as computing complexity is concerned, to finding the solution of a problem.

4. THE ORGPLAN2 SYSTEM

The ORGPLAN2 system was created on the basis of a critical analysis of the ORGPLAN1 system. The transformation of problem description into a binary constraint system and at the same time the preanalysis in the shape it was employed in the ORGPLAN1 system were relinquished in the ORGPLAN2 system. The solution algorithm was replaced by the *heuristic algorithm.*

The main concept of this algorithm is assigning the following parameters to the tasks $z_z \in Z$: time of the beginning $t_p \in T$ and ending $t_k \in T$, place $v_s \in V$ and elements $e_n \in E$, in such a way as to minimize the number of algorithm returns [Partyka and Peszyńska, 1987]. It is basically a branch and bound algorithm with backtracking, in which the heuristic method of defining sequence of tasks (to be assigned) is essential to minimize the number of failures while defining the parameters.

Priority, which is dependent on the degree to which a task's parameters are defined is computed for every task. The lower the lack of definition, the higher the priority. If a task has got assigned place, time and performers, its parameters' indefiniteness reaches the zero mark and the task will be introduced into schedule first of all.

The priority of assigning organization elements to tasks is defined in the same way. The priority of considering tasks is characterized by the following elements:

- demanded time of performance (the beginning and the end);
- a demand for sequence (performing before or after another task);
- a constraint on the beginning or the end;
- a demand for the time of performance;
- prohibition of performing other tasks at the same time;
- demanded place of performance;
- a demand for performance in certain periods of time;
- prohibition of performance by certain performers;
- prohibition of performance in certain places.

5. THE HYBRID ORGPLAN3 SYSTEM

Summing up both of the presented systems, it should be pointed out that the analysis of problem description correctness is a great advantage of the ORGPLAN1 system. On the other hand, the heuristic algorithm of the ORGPLAN2 system seems to be most suitable for scheduling this type of problem. The concept of the hybrid ORGPLAN3 system was created while drawing conclusions from both of the systems. The new system combines advantages of the previous ones and employs frame representation in the created knowledge base.

The problem analysis in the ORGPLAN3 system is a two-phase analysis. Each phase can employ different kinds of algorithms.

In the first phase an analysis of problem wording correctness is carried out, next one of the problem solving algorithms is selected and run based on specific characteristic of the problem description acquired in a simplified form. The selection is made according to the rules forming a knowledge base of the method base system. If it is impossible to apply one of the optimization algorithms due to the complexity of the problem and a small degree of the mathematical model reduction in the first phase, one of the heuristic algorithms (for example, one of the ORGPLAN2 system) will be selected. It is worth mentioning that information about the problem conditioning is included into the frame knowledge base in the ORGPLAN3 system. More detailed description of both phases is shown in the following.

Phase I. An analysis of problem wording.

As in the ORGPLAN1 system, a mathematical model of the problem is generated in this phase based on the description in ORGPLAN language. The algorithm checks the semantic correctness of the problem wording by means of a mathematical analysis of acquired constraint system.

The preanalysis algorithms are based on numerical computing. In the ORGPLAN3 system an additional analysis related to the frame knowledge base created in the system is carried out. The frame representation possesses built-in mechanisms to maintain the consistency and integrity of the knowledge base (compare the procedures of *if_added* type described later on).

Phase II. Problem solving.

In this phase, a solution in the form of a schedule is assigned. Needless to say, the second phase is performed when no contradiction of description has been found in the first phase.

A simplified description of the problem excluding most of the redundant information but including assignments directly derived from the problem description becomes input data. The model of the problem situation obtained in this way may differ from the output one. In particular cases, the model can be as simple as to allow application of one of the optimalization algorithms. However, if it is impossible to apply a numerical algorithm due to the size of solution space (the number of constraints and variables) the system will select from corresponding algorithms of symbolic computation. The algorithm employed in ORGPLAN2 system may be an example of that.

Generally speaking, the modulus has the form of the rule knowledge base, which contains rules reflecting the knowledge of applications of different optimalization methods to the problem with specific features (size and conditioning of the problem, form of constraints and decision variables) [Graham et al., 1979], also the knowledge of application of priority rules [Panwalker and Islander, 1977; Johnson et al., 1990] or solving methods of constraint-satisfaction problems [Kumar, 1992].

5.1. Frame-based representation.

The situations, which may be described by means of the ORGPLAN language are typical problems characterized by the importance of the structure of input data and their interrelations. Therefore, frame-based representation is most suitable for that type of problem [Fikes and Kehler, 1985].

Three basic types of frame correspond to the main concepts of the ORGPLAN language. They are: frames representing tasks, frames representing elements and frames representing places of a task's performance. The organization theory axioms, which define the main concepts, are connected with frames in the form of attached procedures. The procedures are automatically started during the adequate operations on the frame.

For instance, the axiom ensuring that each task is performed only in one place, is expressed by the procedure (demon *if_added*) launched at the moment when the value is given to a place slot (slot representing a place) of the task frame (frame representing a task). The procedure's work is basically checking whether a place slot of the frame has already been given value. If so, the system asks the user whether he wants to change the place of the task's performance. This compares to the place frame which has got an attached procedure checking whether only one task is being performed in a certain place at a certain moment. The *if_added* type procedure is also set when an attempt of assigning place to a task with a given time of performance or an attempt of assigning

The constraints of the specific problem caused assigning the value to certain slots of frames. For example, an imperative sentence defining that the task should be performed in the certain moment of time determine a value of a slot requested time of the task frame representing this task. But, before the value is assigned, a procedure (demon) is activated checking the consistency of requested time with acceptable time of task's performance.

The frame-based representation allows the knowledge base to keep its integrity and consistency due to the mechanism of the attached procedures. Thus, it is complementary to the first phase of analysis in this aspect. However, as it was emphasized, the application of several complementary methods of correctness checking is in accordance with the hybrid nature of the ORGPLAN3 system.

At the same time, the frame-based representation is suitable for a number of heuristic algorithms, including the algorithm presented in the paper. The attached *if_checked* type procedures make automatic the checking the admissibility of successive steps of the heuristic algorithm.

5.2. Interrelations among the ORGPLAN3 system models.

In the ORGPLAN3 system, in accordance with its hybrid character, there are three models of problem situation under consideration:

1. linguistic model;
2. mathematical model;
3. AI model.

The linguistic model is a text in the ORGPLAN language, which describes a certain problem. The model exists only exactly at the moment of creating the two other models.

The mathematical model is a constraint system of decision variables with related constraint types and the rules of constraint correctness and constraint type changes. The model is generated automatically by the system and is used in the first phase of analysis. The model is also a base for a number of numerical algorithms.

The third model is a frame-based model describing the problem knowledge base, which is used by the algorithms based on the Artificial Intelligence methods.

The interdependence of the models results directly from the sequence of phases of problem analysis. During the creation of the problem description in the ORGPLAN language by the user, a constraint system together with a frame representation of knowledge is generated automatically. The attached *if_added* procedures check simultaneously the consistency of newly introduced sentences with the existing description of a problem situation. If all the sentences of the text are accepted (as consistent ones) by the system, at that exact moment there will be two semantically equivalent models of the problem - mathematical and a frame model.

As a result of the analysis of the mathematical model, the reduction of decision variables and problem constraints may occur, providing there is no contradiction. It causes a reduction and/or change of slots representing constraints within the frame-based representation and giving the value to corresponding slots representing tasks, place and elements. Automation of the process is possible due to the mutually explicit conformity of the problem constraints and mathematical constraints-inequalities.

The required schedule will be generated after using the selected algorithm of solution (if there is a solution). The determined place, elements and time assignments to perform tasks are defined by the values of corresponding slots of the task frames.

6. CONCLUSIONS

The application of various methods to a problem solution should be done in such a way as to allow the use of the advantages of each of them. Thus the methods should be applied to those aspects of the problem in which they can give positive results.

Due to the wealth of classical methods of scheduling and a growing number of systems involving methods based on symbolic computing [Fox, 1990; Burke and Posser, 1990] a wide-spread discussion on the best approach to the problem has been taking place [Grant, 1986; Kusiak, 1990; O'Keefe et al., 1986]. It seems that no extreme attitude in the OR versus AI question is justified. As it has been pointed out before, both of the approaches have their advantages and disadvantages. The only reasonable way of selecting an approach should be a pragmatic criterion of a method's usefulness.

The concept of the presented scheduling system conveys the principle that AI methods should be applied when the application of numerical methods is either ineffective or impossible. The ORGPLAN3 system is a hybrid system, which employs both numeric and symbolic computation methods in order to solve a problem, based on various models (representation) of the problem situation.

It should be emphasized that there are two ways of applying different methods to find a problem solution:

1. application of some *complementary methods* of checking the correctness of problem wording (which takes place in the first phase of analysis);
2. *selection of a solution method* from the method base; a method adequate to the problem specificity and the character of a model acquired after the first phase of analysis (which take place in the second phase of analysis).

After translating a problem description into mathematical expressions and applying the rules of checking correctness of the expressions, their (part) correctness can quickly be checked automatically by the application of mathematical methods for finding redundant and contradictory information. The effectiveness of the algorithm has been confirmed in practice.

At the same time, it should be emphasized that the second phase leading to assigning a schedule is, in most cases, solved by means of different methods using symbolic algorithms. The main reason for that is the size of a given problem. Nonetheless, the acquired reduced model can be the basis of an optimalization algorithm in those problems, in which most of the assignments are determined in advance (for example the moment of the beginning of the task performance or settled assignments of equipment to the task) after preanalysis, which should take out settled assignments and some of the constraints.

The created expert system possesses some more advantages resulting from philosophy of the AI systems:

- a possibility of acquiring the explanations of decisions taken, description of contradictions etc.;
- a possibility of an easy adaptation of the system to the change of problem conditioning.

In the presented system the AI methods are used in three ways:

1. as an aid in numerical algorithm applications;
2. as an aid in checking integrity, redundancy and consistency of the problem description;
3. as a method of scheduling (replacing the expert-man).

It is expected that in a developed version of the ORGPLAN3 system, AI approach will also be used for:

4. relaxation of constraints in case of their contradiction;
5. altering and updating of an already existing schedule for the sake of applying it to changing conditions.

7. REFERENCES

1. Graham, R.L.; Lawler, E.L.; Lenstra, J.K. and Rinnooy Kan A.H.G. 1979. Optimization and Approximation in Deterministic Sequencing and Scheduling: A Survey. *Annals of Discrete Mathematics 5: 287-326.*

2. Berry, P.M. 1992. A Predictive Model for Satisfying Conflicting Objectives in Scheduling Problems. *AI Magazine, vol. 13, No.1.*

3. Graves, S.C. 1981. A Review of Production Scheduling. *Operations Research 29(4): 646-675.*

4. Rodammer, F.A. and White, K.P. 1988. A Recent Survey of Production Scheduling. *IEEE Transaction on Systems, Man and Cybernetics 18(6): 841-851.*

5. Gilmore, J.F.; Williams D.L. and Thornton S. 1989. A Knowledge-Based Approach To Planning and Scheduling. *In Proceedings of the SPIE Applications of Artificial Intelligence VII, Orlando, FL, March 1989.*

6. Kanet, J.J. and Adelsberg, H.H. 1987. Expert systems in production scheduling. *European J.Operational Research 29, No.1: 51-59.*

7. Kurzydłowska, A. and Piasecki, S. 1988a. Formalization and Computer representation of Organizing Problems for Purposes of Computer-Aided Problem Resolution. *Computers in Industry 10: 21-33.*

8. Kurzydłowska, A. and Piasecki, S. 1989. ORGPLAN - An Information-Decisive Aid System to Resolving Organizing Problems. *Computers in Industry 11: 275-285.*

9. Kurzydłowska, A. 1986. The Language Problem-Oriented to Organizing Problems and its Employment in Computer System for Automatic Determination of Schedule.. Ph.D. Thesis, *Systems research Institute, Polish Academy of Sciences (in Polish).*

10. Kurzydłowska, A. and Piasecki, S. 1988b. A Semantic Analysis of ORGPLAN. *Computers in Industry 10: 247-260.*

11. Partyka, A. and Peszyńska, M. 1987. The concept and development of organization decision support system. *Proceedings of Conference "Computer system and decision support methods", Warszawa (in Polish).*
12. Panwalkar, S.S. and Iskander, W. 1977. A Survey of Scheduling Rules. Operations Research (25)1: 45-61.
13. Johnson, L.M.; Dileepan, P. and Sen, T. 1990. Knowledge based scheduling systems: a framework. Journal of Intelligent Manufacturing 1: 117-123.
14. Kumar V. 1992. Algorithms for Constraint-Satisfaction Problems: A Survey. *AI Magazine (13)1: 32-44.*
15. Fikes, R. and Kehler, T. 1985. The Role of Frame-Based *Representation in Reasoning. Communications of the ACM 28(9).*
16. Fox, M.S. 1990. Constraint Guided Scheduling A Short History of Research at CMU. *Computers in Industry 14(1-3): 78-89.*
17. Burke, P. and Prosser P. 1990. Distributed Asynchronous Scheduling. *Technical Report, University of Strathclyde.*
18. Grant, T.J. 1986. Lessons for O.R. from A.I.: A Scheduling Case Study. *Journal of Operational Research Society (37)1: 41-57.*
19. Kusiak, A. 1990. Manufacturing Systems: A Knowledge- and Optimization-Based Approach. *Journal of Intelligent and Robotic Systems 3: 27-50.*
20. O'Keefe, M.R.; Belton, V. and Ball, T. 1986. Experience with using expert systems in O.R. *Journal of Operational Research Society, (37)7: 657-668.*

APPENDIX: AN EXAMPLE OF GENERATING A MATHEMATICAL MODEL.

Fixing a schedule of lessons with respect to class rooms.

The problem concerns fixing a time table for school work. It has been presented in a simplified exemplary version which concerns only one group of students. Three kinds of occupations are considered: lectures, classes and laboratories. The planning concerns only two of the lectures with the corresponding classes and one laboratory. The four class rooms have been taken into consideration and incidentally not each room is suitable for each kind of classes. The planning period considers only two days: Monday and Tuesday. Each day the occupations can be started at 8 a.m. and end at 2 p.m. Of course, a continuation of the occupations which were started a day before is inadmissible. The individual classes have the defined duration. The detailed conditions of the problem will be specified by means of ORGPLAN sentences.

Using ORGPLAN to describe the problem, one should identify the concepts of the language. The occupations (lectures, classes, labs) will be identified with the *tasks* (z_z); the class room in which the occupation will be performed will be treated as the *places* of the realization of the tasks (v_s).

The description of the problem should be started from the definition of the data structure of the problems, i.e. from the definition of vocabularies:

1. VOCABULARY Z: MATH, PHYS, CLASSES_MATH, CLASSES_PHYS, LAB.
2. VOCABULARY R: LECTURE, CLASSES, LAB.

3. ZDS, R=LECTURE, Z=MATH,PHYS.
4. ZDS, R=CLASSES, Z=CLASSES_MATH, CLASSES_PHYS.
5. ZDS, R=LAB, Z=LAB.
6. VOCABULARY V: 314, 321, 204, 325.
7. VOCABULARY M: LECTURE_ROOM, CLASSES_ROOM, LAB_ROOM.
8. ZDS, M=LECTURE_ROOM, V=314, 321.
9. ZDS, M=CLASSES_ROOM, V=204.
10. ZDS, M=LAB_ROOM, V=325, 314.
11. VOCABULARY T: MON_8, MON_9, MON_10, MON_11, MON_12, MON_13.
12. VOCABULARY T: TUE_7, TUE_8, TUE_9, TUE_10, TUE_11, TUE_12.

The sentences with the identifier VOCABULARY define the vocabularies, the sentences with identifier ZDS specify the relations between the vocabularies. The third sentence, for instance, defines that the occupations MATH and PHYS are the lectures.

The vocabularies defined in this way specify the following binary decision variables:

$$z=1,2,3,4,5; \qquad s=1,2,3,4; \qquad X_{z,s}=(0 \text{ or } 1); \tag{1}$$

$$z=1,2,3,4,5; \qquad k=1,2,\ldots,12; \qquad T_{z,k}=(0 \text{ or } 1). \tag{2}$$

The essential constraints of the decision variables have the following form:

A. Each occupation has the defined class room for realization

$$z=1,2,3,4,5; \qquad \sum_{s=1}^{4} X_{z,s} = 1. \tag{3}$$

B. Each occupation is realized without breaks

$$z=1,2,3,4,5; \qquad \sum_{k=1}^{11} (T_{z,k} - T_{z,k+1})^2 + T_{z,12} + T_{z,1} = 2. \tag{4}$$

C. At each time interval at most one occupation is realized

$$k=1,2,\ldots,12; \qquad \sum_{z=1}^{5} T_{z,k} \leq 1. \tag{5}$$

The constraints of the problem will be specified by means of the following sentences:

1. ZDO, R=LECTURE, M=LECTURE_ROOM, O=2.

This sentence specifies that the lectures are to be in the lecture room and their duration is 2 hours. This sentence generates the constraints in the form:

$$\sum_{k=1}^{12} T_{1,k} = 2 * X_{1,1} + 2 * X_{1,2} \tag{6}$$

$$\sum_{k=1}^{12} T_{2,k} = 2 * X_{2,1} + 2 * X_{2,2} \tag{7}$$

$$X_{1,3} = 0, \qquad X_{1,4} = 0, \qquad X_{2,3} = 0, \qquad X_{2,4} = 0. \tag{8}$$

2. ZDO, N, R=CLASSES, M=LECTURE_ROOM.
R=*, M=LAB_ROOM.

These sentences specify the places in which the task cannot be performed. In this case, the classes cannot be performed in the lecture rooms and not in the lab room. These sentences generate the constraints in the form:

$$X_{3,1} = 0, \qquad X_{3,2} = 0, \qquad X_{3,4}=0, \qquad X_{4,1} = 0, \qquad X_{4,2} = 0, \qquad X_{4,4} = 0. \tag{9}$$

3. ZDO, R=CLASSES, O=2.

This sentences specifies that the duration of classes is 2 hours. The following constraints are generated:

$$\sum_{k=1}^{12} T_{3,k} = 2, \qquad \sum_{k=1}^{12} T_{4,k} = 2. \tag{10}$$

4. ZDO, R=LAB, M=LAB_ROOM, O=3.

This sentence specifies that each laboratory should be performed in the labs room and its duration is 3 hours. This sentence generates the constraints in the form:

$$\sum_{k=1}^{12} T_{5,k} = 3 * X_{5,1} + 3 * X_{5,4} \tag{11}$$

$$X_{5,2} = 0, \qquad X_{5,3} = 0. \tag{12}$$

5. ZDR, Z=MATH, V=314.

This sentence commands that the occupation MATH is to be performed in the room 314. The following constraints are generated by this sentence:

$$X_{1,1} = 1, \qquad X_{1,2} = 0, \qquad X_{1,3} = 0, \qquad X_{1,4} = 0. \tag{13}$$

6. ZDR, Z=LAB, V=314, T=TUE_8:TUE_11.

This imperative sentence specifies that the occupation LAB is to be performed in the room 314, Tuesday, 8-11 hours. This sentence generates the following constraints:

$$X_{5,1} = 1, \qquad X_{5,2} = 0, \qquad X_{5,3} = 0, \qquad X_{5,4} = 0, \tag{14}$$

$$T_{5,8} = 1, \qquad T_{5,9} = 1, \qquad T_{5,10} = 1. \tag{15}$$

7. ZDR, Z=CLASSES_PHYS, TMIN = MON_12.

This sentence specifies that occupation CLASSES_PHYS will be started at Monday 12.00. The following constraints are generated:

$$T_{4,1} = 0, \qquad T_{4,2} =0, \qquad T_{4,3} =0, \qquad T_{4,4} =0, \qquad T_{4,5} =1. \tag{16}$$

8. ZDO, N, R=LECTURE, T=TUE_7.
ZDO, N, R=CLASSES, T=*.
ZDO, N, R=LAB, T=*.

This sentences specify that the lectures, the classes and the labs (i.e. all the occupations planned) cannot be performed Tuesday 7 a.m. In this way the condition is realized that the occupation which is started one day cannot be continued to the next day. These sentences generate the constraints in the form:

$$T_{1,7} = 0, \qquad T_{2,7} = 0, \qquad T_{3,7} = 0, \qquad T_{4,7} = 0, \qquad T_{5,7} = 0. \tag{17}$$

The constraints (1)-(17) form the mathematical model of the problem. This constraints system contains 67 constraints and 80 decision variables.

As a result of *the semantic analysis* (which is based on a mathematical analysis of this system of binary constraints) this model has been reduced to 17 constraints and 30 decision variables.

KNOWLEDGE BASED SIMULATION

Knowledge Based Hybrid Systems (B-11)
I. Mezgár and P. Bertók (Editors)
Elsevier Science Publishers B.V. (North-Holland)

Control Automation of a Multioperational Section for the Flexible Manufacturing of Highly-Diversified Products

G.M. Acaccia, M. Callegari, R.C. Michelini, R.M. Molfino, P. Peri, A. Ricci

Industrial Robot Design Research Group, University of Genova, Istituto di Meccanica Applicata alle Macchine, via all'Opera Pia 15A, 16145 Genova, Italy

Abstract

An object-oriented expert-simulator is used to develop the control logic for the automation of the central multioperational section of a plant, manufacturing highly diversified thermoplastic components. Continuous and discrete time processes run simultaneously, for material blending and moulding, for components and tools procurement, for part machining and assembling, for production storing and shipping, etc. For the plant's automation, the control setting of the sorting and storing system is a critical requirement, and the expert simulation is a valuable aid at both the design and the management stage.

Keyword Codes: I.2.1; I.6.3
Keywords: Artificial Intelligence, Applications and Expert Systems;
Simulation and Modelling, Applications

1. INTRODUCTION

Automation in the manufacturing of low cost thermoplastic components presents pacewise trends, with sophisticated isles (e.g. moulding machinery, assembly stations, etc.) and extended manual operations inbetween, in order to preserve flexibility. The paper deals with the assessment of the control logic supervising the multioperational section that needs be purposely developed and included for the overall factory automation. This section is characterized by an automatic buffer, for parts sorting and storing operations, between the (continuous) moulding stations and the machining, assembling and packing stations; it superintends to the shop logistics in order to provide, both, the part-and-products and tools-and-fixtures balanced flows.

The modelling uses object-programming [1] to exploit hierarchic inheritance between resources' properties. The control logic exploits distributed knowledge, with hybrid (algorithmic and heuristic) relational structures, and concurrent processing capabilities. The resulting simulator embeds granularity and expandability [2] and can operate in real time, according to options provided by the AI shell G2 [3] that has been used.

The approach was developed in the past years [4,5] and is hereafter discussed considering an existing manufacturing environment; the comments provide details on the physical facilities and on the modelling and programming developments performed for simulational purposes.

2. REFERENCE FACILITIES

The customer-oriented manufacturing moves the products shipping and packing requirements back to the parts flow (and, then, to the material procurement) and to the fixtures scheduling. Parts are suitably handled into containers (according to the prescribed assortments). The section to be investigated is composed by the following facilities:

- an interoperational buffer, for storing and sorting the parts outcoming from the moulding shop and forwarded, into containers, to subsequent workstations;
- a travelling lift robot, for moving the containers within and out of the buffer;
- two machining centers, for the technological operations required by a given subset of components;
- six assembly stations, for the requiring subset of final products (valves, etc.);
- a bag filling machine, for packaging products (valves, fittings, etc.) with diameters not exceeding 32 mm;
- two thermoforming machines, for packaging products with diameters exceeding 32 mm.

The buffer is organized in two opposite stands with superimposed shelves and has pre-established I/O gates, according to the interfacing requirements, namely:

- the plant-storage gate, with superimposed conveyors that accommodate five containers each;
- the machining-centers gate, with in and out roll-ways;
- the assembly stations gate, with multiple locations in order to present the correct parts assortment for the requested final products;
- the bag-filling machine gate, providing 18 locations (two arrays of 9 places) directly accessible from outside;
- the thermoforming machines gate, with 4 locations for each machine, presenting an external port, with elastic external protection.

The robot moves horizontally along the stands and vertically up to the addressed shelf. At the interfacing gates, special purpose manual stations are left for the presetting of the pallets to be fed to the machining centers and for the preparation of the part assortments to be fed to the assembly stations. Then the machining centers perform the milling, boring, facing, etc. operations, as scheduled.

The bag-filling machine automatically executes the products packing, as soon as the 18 locations at the assigned gate are full; when the production changes, the resetting of the feeding service is assumed to be performed manually. The thermoforming machines, as well, operate automatically provided that an initial parts positioning is assured; these machines require a different presetting, following to the changes on the scheduled final products.

3. PROGRAMMING DEVELOPMENTS

The modelling is implemented with generality, based on object-classes and rules; the specialization follows through instanctiation. The programming distinguishes the physical and the logical object-classes.

The previously described facilities are standing physical resources not-modified through manufacturing, and are associated with correspondingly stable objects. The containers, the (element) parts and the (final) products are transient physical objects, that are originated, modified and destroyed at given locations of the manufacturing section. The consideration of actual transient objects faces important programming drawbacks, and a different modellization has, therefore, to be used [6]. The details on the running production can be kept at the level of containers (since the assortment requirements are fixed by the scheduling) and the related attributes (parts or products types and amounts) can accordingly be transferred to the manufacturing facilities that temporarily have in charge the individual containers.

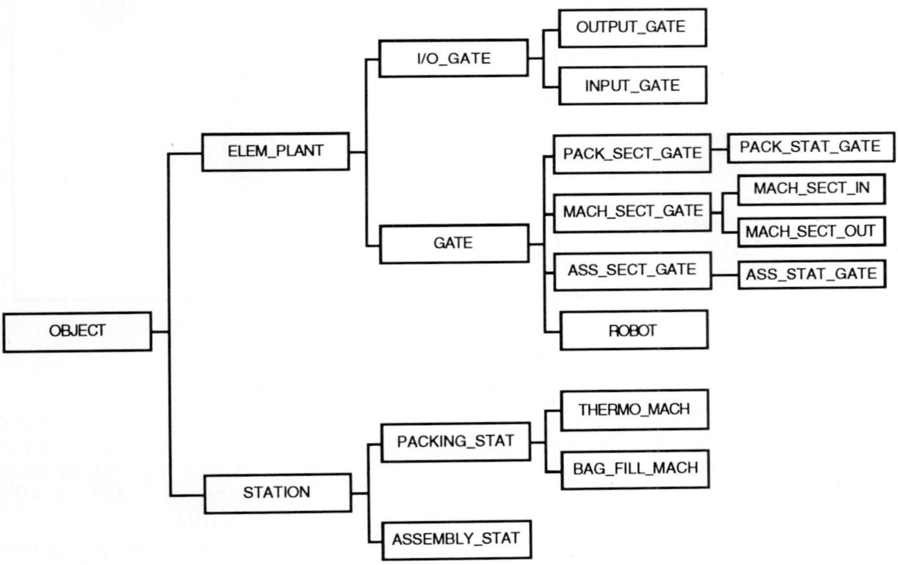

Figure 1. Classes hierarchy

Each object belongs to a class with the hierarchical organization [7] shown by Fig. 1. The object definition specifies an icon and an attribute list. The basic level distinguishes between buffer shelves and workstations; then location and stations are further detailed following to the related operational requisites. The typical icons of the buffer locations are given by Fig. 2; each position can be void, can have an empty container, or a full container.

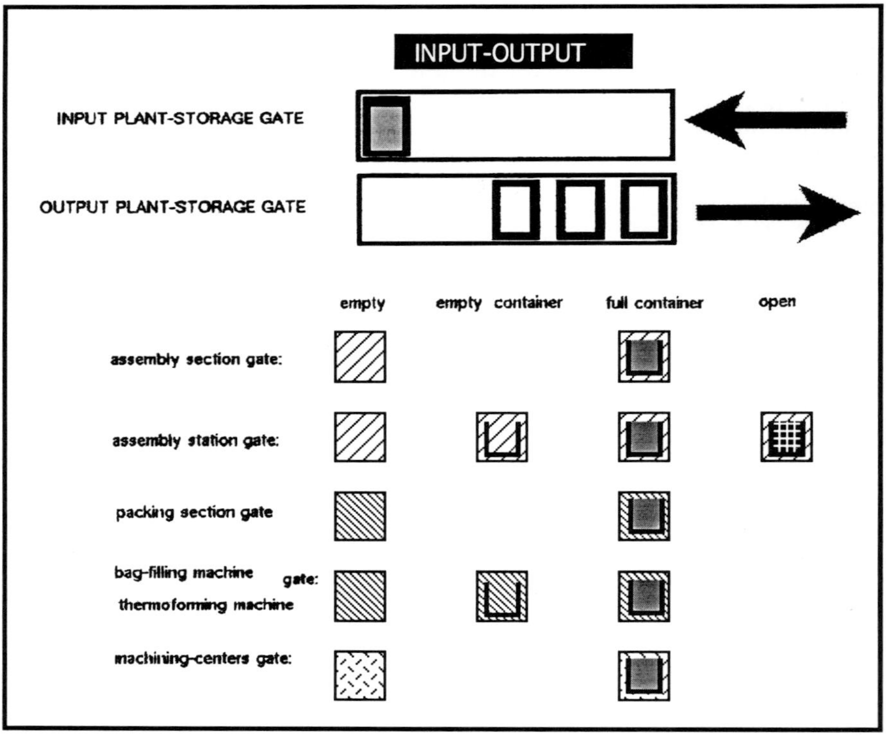

Figure 2. Buffers locations' icons

The icons of Fig. 3 describe the workstations: the considered assembly stations, for example, require the input of nine parts that must be combined for delivering the requested products; the central portion of the individual icon has, furthermore, a colored flag: *green*, if the station is operating; *white*, if off-line; *transparent* (not distinguishable from the icon body), if idle.

A second set of objects defines the logical resources. The correspondence bounds among the physical resources are also defined, through connections between objects that may be switched on or off by an action defined by a rule or by a procedure [7]. The option is useful, for example, for associating the product type-code (and relating the manufacturing operations) timely transferred to a given workstation, and, in general, for expanding the programming flexibility, in order to assure the functional specialization of the individual rule and procedure.

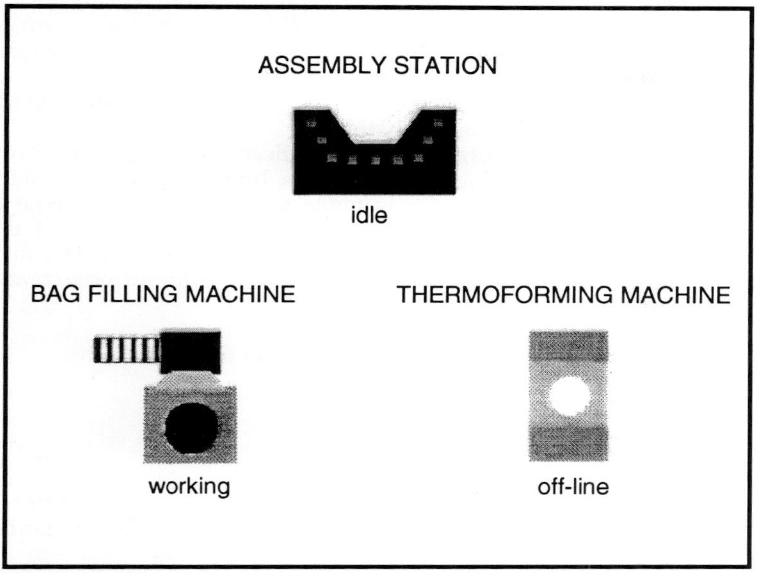

Figure 3. Workstations' icons

4. CONTROL REQUIREMENTS

The control of the logistic section is governed by transient objects, corresponding to the transportation requests generated each time that a container should be moved between two locations of the buffer, with focus on the I/O gates. The generated requests are conveniently grouped into specialized lists (with the priority scale corresponding to the order listing), such as:

- entry list, for processing the input service of the containers in the buffer;
- remove list, for managing the withdrawal of empty containers, whichever location is actually occupied;
- reset list, for moving back to the buffer entry the half empty containers that come from the assembling stations (typically at products changes);
- forward list, for moving the containers from the machining centers or from the dedicated buffer gates, to the assembling stations;
- delivery list, for planning the transport of the containers from the buffer to the machining centers and to the packaging stations.

The control logic aims at establishing just-in-time policies, therefore conditioning, through the assembly lots, the moulding processes, according to the packing batches, that properly match with the customers' requests.

The moulding batches define the agenda of the new arrivals to the buffer, and allow for the correct feeding, through the (two) machining centers, of the (six) assembly stations and of the final packaging stations. The backward actions of the shipping requests slightly affect the packaging segment; it, rather, requires a careful analysis of the machining centers behavior. These are, as a matter of fact, flexible cells, but appropriate refixturings are needed according to the product type to be processed; the refixturing operations are complex and time consuming, and the corresponding specialized manufacturing schedules require a purposely related module, that is enabled to run in parallel with the main simulation program, with due conditioning effects on the assembling stations schedules.

The control strategy is, hereafter, shortly described referring only to the main program, and considering a worksession limited from an initial and a final simulation time [8]. Reference is made to an occurrence-driven agenda [9]; the situational updating is connected to a list of conditional occurrences that are dealt with sequentially. Each occurrence appears in the form of a structured step within the (C-coded) process. Seven different occurrence types are defined, namely:

- entry of a new container (from the moulding section);
- draw of a container (from a buffer shelf or from a workstation);
- unloading of a container;
- end of a workshift;
- end of an assembly batch;
- activation of a transport request;
- end of a simulation run.

The driving module is a set of C-functions, assembled in a library, that perform the management of the lists, by means of ordered commands (item insert, extract, or cancel) according to the agenda requests.

5. SIMULATIONAL ACHIEVEMENTS

The simulational package is built exploiting the concurrency of a set of mutually cooperating parallel processes: one set employs a standard programming language (actually the C language) and provides an efficient modelling of the dynamic behavior of the plant physical resources; the second uses the expert shell G2 and assures the implementation of the control logic.

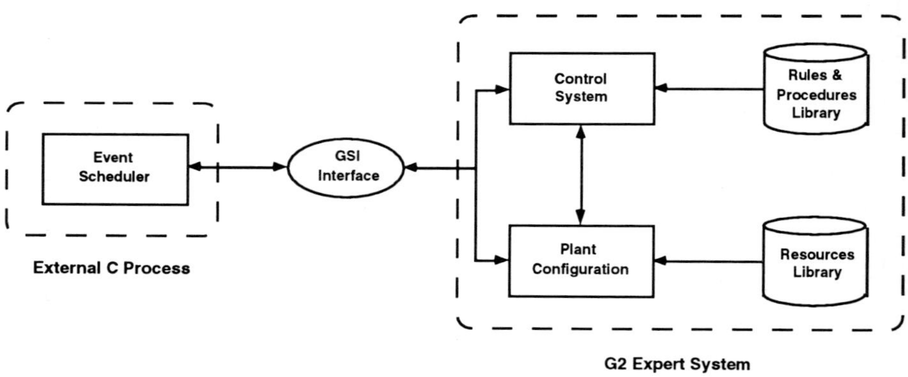

Figure 4. Simulator overall architecture

In order to enable the recalled software architecture, the communication between the two sets of processes has been developed through specially tailored modules and appropriate interfaces [10]. Fig. 4 schematically shows the principal features of the overall architecture, distinguishing:
- the G2-block, performing the expert-control,
- the interfacing protocol, assuring the data communication,
- the C-block, duplicating the plant behavior.

In view of the control integration for the factory automation, the third block could be directly replaced by the actual plant, and the expert-controller would directly operate in real time on the manufacturing processes. The first and second modules are already written for on-field operations, and the simulational results have shown the feasibility of the proposed control.

It should be put stress on the fact that the knowledge structure of the expert-controller is based on two different data-bases: a stationary data-base and a dynamic one.

The stationary data-base is fixed for every situational updating (specifying the system actual behavior, according to the driving agenda). The overall knowledge architecture is presented in the specialised *workspace*, corresponding to physical frames of the display, into which the objects, connections, procedures, display-elements, etc. can be created. A knowledge architecture can contain any number of *workspaces*, and the fact is properly exploited for orderly grouping the elements into specialized layers (namely following to the related class, relation, rule, procedure, etc. definition).

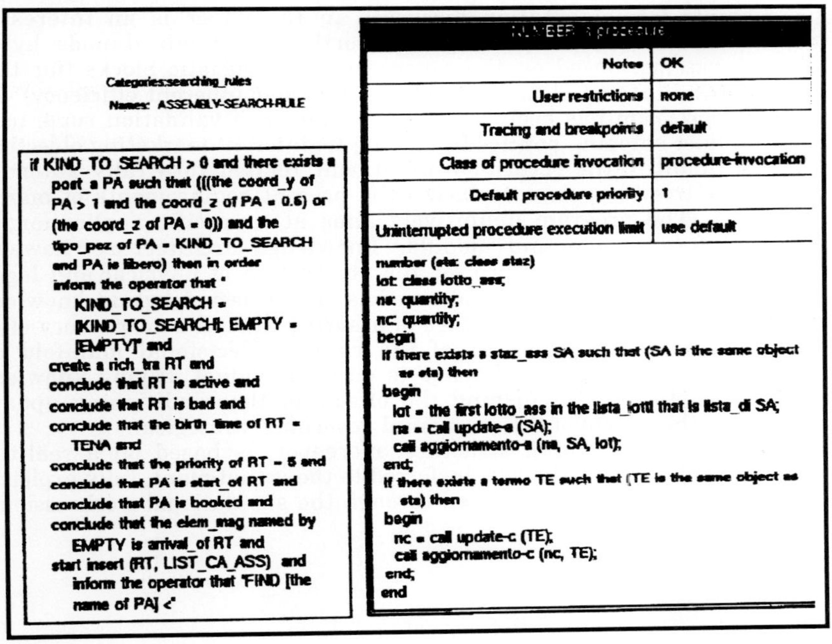

Figure 5. Example of a rule and a procedure

The dynamic data-base is interfaced to the controller, which appears to be a decision-keeper that runs the plant, using the established production-planning strategies, conditioned by the situational knowledge. These strategies are coded: with rules, for mainly heuristic frameworks (e.g. for the quickest selection of the appropriate container, from the sorting-and-storing system); and with procedures, for properly established algorithmic contexts (e.g. for the sequencing of operation cycles); examples are provided by Fig. 5.

The used interfacing protocol imposes, by now, synchronization restrictions [11]; if the remote G2 procedures, for example, have to give back updated values to the algorithmic block, these need be transferred when the calling C-process is still operating, since a single initialized block may only allow for legal communications.

6. CONCLUDING COMMENTS

The full automation of the traditional manufacturing plants, engaged in the production of highly diversified (due to the extended customisation) low-price components, faces important technical and economical hindrances, since the return on investments is a rather hypothetical figure. The market conditioning factors can very quickly switch a solid enterprise into a marginal one, and good plant efficiency markers should be used to forecast in due time the innovation pace. Expert-simulation is being more and more considered by production engineers as a powerful evaluation instrument.

The example implementation discussed in the paper is an interesting prototypal development, exploiting the worthiness of mixed-mode hybrid knowledge systems, based on the concurrency of heuristic blocks (for their expanded flexibility) and algorithmic blocks (for their inherent efficiency). The simulational program has been tested by a number of validation runs, using different product shipping policies (according to different marketing objectives) in order to quantify the return-on-investment related to a fully automatic storage and retrieval system replacing the present requirements of manual interventions. The program definitively aims at a specific application; its object-based formulation and modular knowledge architecture, however, represent a noteworthy reference for the abstraction of physical and logical facilities into consistent mixed-mode models. The selected hybrid knowledge architecture has the further advantage of assuring the full transparency of the control logic, so that the effects of the running decisional manifold can modularly be assessed with reference to each individual driving knowledge element (separately distinguishing the effects of the conditioning updated situations and the effects of the procedural governing strategies).

The distributed governing structure, moreover, is based on a real-time expert shell, and can directly be transferred to the actual plant, fully exploiting the experiences previously completed through the simulational worksessions.

7. ACKNOWLEDGEMENTS

This work has been partially supported by the Italian National Council of Researches (CNR) through the contract number 92.01083.PF67 with the *Progetto Finalizzato Robotica* (PFR).

8. REFERENCES

[1] Eldredge D. L., MacGregor J. D., Summers M. K. "Applying the object-oriented paradigm to discrete event simulations using the C++ language", Simulation, Vol. 4, N. 2, February 1990, pp.83-91

[2] Kiran A.S., Schloffer A., Hawkins D. "An integrated simulation approach to design of flexible manufacturing systems", SCS Simulation, vol.52, n.2, San Diego, CA, USA, February 1989, pp.47-52

[3] Gensym Corporation: "G2 Reference Manual: ver 3.0", Cambridge, MA, July 1992

[4] G.M.Acaccia, R.C.Michelini, R.M.Molfino: "Design of intelligent governors for the material handling equipment of automated factories" IFIP Workshop 'Software for Factory Automation', Tokyo, oct. 19-21, 1987, pp. 184-208.

[5] G.M.Acaccia, M.Callegari, R.C.Michelini, R.M.Molfino: "Simulation facilities for the development of computer-integrated manufacturing" Int. J. Advanced Manufacturing Technology, vol. 7 n° 4, 1992, pp.238-250

[6] Townsend M.A., Lamb T.W. "Detailed simulation of a real world job shop with subassembly requirements", Simulation, Vol.57, N.2, August 1991, pp.114-128

[7] Alasuvanto J., Eloranta E., Fuyuki M., Kida T., Inoue I.: "Object oriented programming in production management - two pilot systems", Int. J. Production Research, Vol.26, N.5, May 1988,pp.765-776

[8] Fox M., Sathi N., Baskaran V., Bouer J. "SIMULATION CRAFT: an expert system for discrete event simulation", Eastern Simulation Conferences, Norfolk, VA, March 10-12, 1986, pp.186-191

[9] Banks J., Carson J. S., "Discrete-event system simulation", Prentice-Hall nc., Englewood Cliffs, New Jersey 07632, 1984

[10] Foret J.M. "Connectivity of G2", Gensym User Society Europeen Meeting, Paris, France, September 25-27, 1991

[11] Sugars M. "Interfacing to G2", Gensym User Society Annual Meeting, Houston, Texas, USA, March 6-8, 1991

Knowledge Based Hybrid Systems (B-11)
I. Mezgár and P. Bertók (Editors)
Elsevier Science Publishers B.V. (North–Holland)

Knowledge Based Modelling, Simulation and Evaluation of Flexible Manufacturing Systems

S. Kopácsi[a], N.T. Koussoulas[b] and D. Gavalcová[a]

[a]Computer and Automation Research Institute, Hungarian Academy of Sciences, CIM Research Laboratory, 1111 Budapest Kende u. 13-17., Hungary

[b]University of Patras, Electrical Engineering Department, Laboratory of Automation and Robotics, Rio Patras 26500, Greece

Abstract:

In this paper a brief description of Flexible Manufacturing Systems (FMS) will be given, then the knowledge based simulation of FMS will be discussed. We will emphasise the methods of automatic model generation, decision support during simulation and result evaluation. Proving our concepts a rapid prototype system will be introduced.

Keyword Codes: J.6; I.2.1; I.6
Keywords: Computer-Aided Engineering; Applications and Expert Systems; Simulation and Modelling

1. INTRODUCTION

Flexible Manufacturing Systems (FMS) have continuously increasing importance not only in big factories but in the small and medium serial production, as well. The reason is that in a well-organised FMS the production costs and delivery times are relatively low.

Planning and operating of FMS need careful work, because of the high value of the system and in consequence of the precious production on it. Computer aided *simulation* is very helpful both in planning and operating of FMS. Simulation can be used in many parts of the life cycle of a manufacturing system, like capacity planning, bottle-necks detection, creating and testing manufacturing schedules, etc.

Although several different simulation languages, that can be suitable for FMS simulation, are available on the software market, the knowledge representation and management power of these languages are usually not satisfactory. The knowledge that is necessary for planning and controlling of an FMS as well as for preparation of the simulation model of the FMS to be examined and evaluating the results of the simulation can be more easily built into *knowledge based* or *expert systems*.

2. FLEXIBLE MANUFACTURING SYSTEMS

Flexible Manufacturing Systems (FMS) are computer controlled complexes of NC machines, automated material and tool handling units, transportation and storing devices, measurement and inspection appliances, etc. FMSs are able to operate with minimal human control fulfilling the lowest times of delivery.

FMS - depending on its size and structure - can be divided into different categories. *Flexible Manufacturing Units* (FMU) usually consist of one CNC machine or a manufacturing centre. *Flexible Manufacturing Cells* (FMC) have usually two or three CNC machines or manufacturing centres, automated workpiece handling and buffering system (e.g. AS/RS = Automatic Storage and Retrieval System) and transportation facility (e.g. AGV = Automatic Guided Vehicle). *Flexible Manufacturing Systems* (FMS) are built up from FMUs and FMCs, as well as, other cells (assembly, measurement, storage cell, etc.)

Several criteria of *flexibility* have been defined in the literature (e.g. [Browne 1984]) some of them are as follows: modular extendibility of the system, possibility of manufacturing different workpiece types, potentiality of handling breakdown situations, manufacturing in different operation orders.

The manufacturing process in a manufacturing system is based on the *manufacturing schedule* that is created from a *process plan*. For every workpiece type the process plan consists of the operation order, the identifiers of the machines to the operations, tools, NC program identifiers, etc. and the time period of each operation. The manufacturing schedule contains the previous information moreover the start and finish times of the operations, because a scheduler, that generates the schedule, can take into account the finish time of the preceding operations, waiting time and the duration of transportation. Other important input data for the scheduler are the due dates or due times which always should be complied with.

3. SIMULATION OF FLEXIBLE MANUFACTURING SYSTEMS

Computer based simulation can be used in many parts of the life cycle of an FMS. One goal of FMS simulation is the *initial configuration* and *capacity planning*. The simulation can check whether the machines, distances, transporters, etc. were suitably chosen. The bottlenecks of the system can be detected before the expensive real operation of the FMS starts. After evaluating the simulation results the number of machines, the transport facilities, the sizes of the buffers, etc. can be modified toward a better overall system.

Another domain of FMS simulation is when simulation is used for *creating manufacturing schedule*. There are different methods (mathematical, heuristic, etc. approaches) for scheduling, each of them has advantages and disadvantages. Scheduling with simulation is usually fast and reliable. During the simulated manufacturing process, that is based on the process plan, the start and finish time of each operation can be recorded, which can later serve as a schedule. A simulation based scheduler can include *quality assurance* conditions as well, like pre-determined in-process and off-process measuring. This results in a manufacturing schedule ensuring better quality of the product.

Simulation can be used for *testing schedules* created by other schedulers or human experts. A complex simulation-evaluation-scheduler system can detect errors or drawbacks of the given schedule, and can suggest improved ones. This mode of the simulation can be combined with *simulation of breakdown situations* to check the effects of the errors in the given system and/or schedule.

A simulation software of a system can be written in several languages like traditional, object oriented or special simulation languages. It is usually difficult to choose the most suitable approach. The special simulation languages offer a controlled environment but little flexibility, while the programming languages can satisfy every need of the user, present and future, but they entail the often unwelcome cost of software development. If an available simulation language is acceptable for our purposes, then it is the most advisable to choose. For example SIMAN [Pegden 1990] is advised to be used for simulating manufacturing systems.

4. KNOWLEDGE BASED SIMULATION OF FMS

In traditional languages, and in the most of the simulation languages, decisions can be made by using the well-known *if-then-else* structure only. If we have a large number of conditions to satisfy at the same time it is very difficult to write decision rules this way.

In a complex simulation system where not only the simulation itself, but the preparation of the simulation model and the evaluation of the results of the simulation are also automatic, big amount of knowledge is to be handled and many decisions are to be made. By means of model preparation the most appropriate simulation model should be generated. In the act of simulation, especially if we use simulation for scheduling, several decision points occur. Let us consider as an example the routing problem, where the scheduler has to choose the next resource. Automatic evaluation of the simulation results is based on using evaluation knowledge. The latter knowledge base contains the evaluation rules of statistics and manufacturing schedule in order to state whether the schedule and utilisation is feasible or not.

In the past few years the so-called *knowledge based systems* became widespread. The common properties of these systems are, that they contain facts that reflect general and domain specific knowledge, contain general and domain specific rules describing the relationships between the facts, and contain a so-called inference engine which applies the rules for the facts. The result of this knowledge processing is new facts or rules and certain side effects. Expert Systems (ES), Production Systems (PS), Inference Systems (IS), Rule Based Systems (RBS), etc. belong to this category.

Combining these two kinds of different technologies (i.e. traditional simulation systems and knowledge processing methods) seems to be a good idea to solve the problem of decision making in a simulation environment. This combination will result in the so-called *knowledge based simulation*. Consequently knowledge based systems can be helpful prior to, during, and after the simulation.

In the following sections we will first investigate the theoretical backgrounds of the application of knowledge based systems for model preparation, decision support and result evaluation, then a prototype of Simulation Based Evaluator and Scheduler will be introduced.

4.1. Model Management

To illustrate the importance of the concept of the model management, let us examine the following issue. The theoretical backbone of the simulation system will consist of a rich collection of models, which will have to be easily retrieved and manipulated. In addition, the system must make the correct choice of a model for the task at hand. Thus, on one hand, the efficient representation and management of this information (perhaps an ES) become a point of major importance for the success of the project. On the other hand, equally crucial is the establishment of rules and procedures for identifying the most suitable model.

Selecting a model for simulating, or simply studying an FMS, depends largely on the questions asked. Investigating the logical performance of an FMS, e.g. absence of deadlocks, requires a totally different model than the analysis of the FMS throughput under varying forms of loads and task complexities. Thus, for example a Petri net model is capable of analysing the correctness of the process plan for the various tasks, while a queuing network model may be necessary for determining the behaviour of the system when there are uncertainties regarding delays, complexity of the process plans, machine failures, etc. Therefore, users must be suitably guided to express their exact needs for either qualitative-microscopic and quantitative-macroscopic analysis (or both), and appropriate corresponding models must be provided.

The set of different models, along with their properties, assumptions, and requirements constitute an appreciable body of knowledge that has to be efficiently handled. Similar issues have been the subject of research in the past within the context of Decision Support Systems and come under the general term *Model Management Systems* (MMS). Knowledge representation for such systems is treated in [Dolk 1984].

The requirements for the MMS can be summarised as follows:

- The MMS must be capable of containing a large number of different classes of models, with all necessary information for applying them, including the set of the appropriate objective functions for the ensuing optimisation.
- The MMS must be able to offer alternative ways for modelling the same real system, even at the same level of analysis. For example, generalised semi-Markov processes and perturbation analysis are two approaches to performance analysis through simulation that are both valid at the macroscopic level.
- The MMS must have a suitable structure and operation that will facilitate and support the operation of the inference engine.
- The MMS should support aggregation within the same model or through combinations of models.
- The MMS should contain enough knowledge to be able of assisting in the interpretation of the simulation results in a consistent way.
- The MMS should have facilities for adding, removing, or modifying a model in an easy and transparent manner.

A partial list of the models that the MMS must know about includes Petri nets, timed Petri nets, dioids, finite state machines (automata), finite recursive processes, communicating sequential processes; mathematical programming methods; Markov chains, open and closed queuing networks with all kinds of distributions and disciplines at the nodes, infinitesimal and large perturbation analysis, generalised semi-Markov processes, stochastic Petri nets; and heuristic methods that time has proven successful.

An inference engine, which can be called a *Model Selection System* (MSS), is necessary for selecting the appropriate model or models after receiving input or exchanging information with other parts of the simulation system.

The requirements for the MSS can include the following:

- Given the difficulty of the task the MSS must be frame-based. However, some fundamental guidelines at the very top of the decision tree, such as the selection of appropriate classes of models, may be more efficiently encoded in a rule-based way.
- The MSS must be easily extensible, flexible, and re programmable.

- The MSS must be able of detecting and appropriately reacting to conflicting requirements.
- The MSS should be able to sort stated user preferences in order of their overall bearing to the simulation.

It is easily realised that implementing immediately all the features and the full scale of the MMS and MSS is impossible and may prove unnecessary, at least regarding some of the model categories for which theoretical and/or practical unpopularity seems to gain ground. However, the implementation of a small number of well-regarded methods are possible now, while a knowledge based environment is likely to bring out all their power.

4.2. Decision Support During Simulation

Basically two different approaches can be imagined to help the decision making in simulation with knowledge based system:

- creating a special knowledge based simulation language,
- connecting traditional simulation system to a knowledge based system

The first approach is the less widespread, while usually it results in a less powerful system. It will have many limits both in simulation properties and in knowledge processing facilities. The advantages of these systems that they are compact. One example for this is the *CS-PROLOG* (Communication Sequential Prolog) language [Futó 1987]. In this language we can define processes with Prolog clauses. These processes can communicate with each other through messages. We can also define decision rules in Prolog clauses which will make the necessary decisions.

Connecting a traditional simulation system to knowledge based systems is more promising, because we can connect the most suitable simulation system to the most appropriate knowledge based system for our purposes. The connection usually happens with so-called user written routines (usually C) which write and read files or by means of local network. Therefore it is very important that both systems should have interface for user written routines. When simulation reaches a decision point, that decision is to be made in the connected knowledge based system, the simulation stops, sends its message to the knowledge based system and waits for the answer.

4.3. Evaluation of Simulation Results

First we have to distinguish the evaluation of the simulation itself, which can be called model verification, from the evaluation of the results of the simulation. The latter one in our case is the evaluation of the utilisation statistics and the manufacturing schedule. We can see the difference clearly if we keep in mind that the simulation is a tool for the investigation of a system. From this aspect the evaluation of the simulation means determining the measure of how successfully the tool we have built approaches the real system, its behaviour and properties. Therefore the evaluation of the simulation means much more evaluation of our attempt to build a tool simulating a system than evaluation of the amount of results that we receive after the simulation has proceeded.

In this way, when we want to manage the results of the simulated system, we must consider the proper domain and according to its criteria the results can be evaluated.

Our interest is focused on the FMS simulation and the evaluation of the results gained from it. Since we can evaluate something only if we know what it is supposed to be and to do, evaluation depends on our demands on the FMS simulation.

Researchers give the greatest emphasis on criteria such as production rate, machine and system utilisation and flow time, while for most companies among the most important criteria one can find meeting due data, maximisation of machine and system utilisation and minimisation of in-process inventory.

The results of the simulated system gain their importance when one tries to improve the performance of the system. Since there does not exist a universal and satisfactory method for determining the connection between the criteria used in scheduling rules directing the performance of the system and the results of the system, it is advisable to test various scheduling rules against the performance measures, and thus find the most appropriate set of scheduling rules, providing such a performance of the system, which is the most acceptable one for us.

The Evaluation System can be connected to (or can be the same as) the Preparation system (Model Manager). If corrections are suggested by the Evaluation System then the Preparator can modify the condition system of the simulation. For example the workpiece priorities, the number of pre-defined events or the manufacturing route can be modified if the manufacturing schedule is not acceptable according to the Evaluator. The Evaluator can suggest that the Advisor should give other advises in scheduling decisions in a new simulation run, as well.

5. PROTOTYPE OF THE SYSTEM

In order to prove our concepts we have implemented some prototype versions of the Simulation Scheduler System. The latest prototype system investigates the experimental Flexible Manufacturing System located at the Technical University of Budapest. The structure of this pilot plant can be seen on the animation layout (Fig.1.)

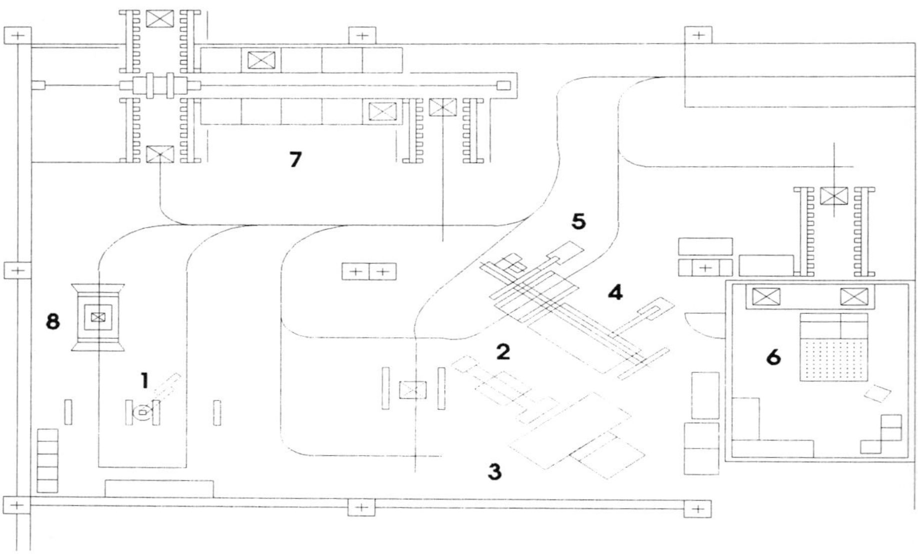

Fig.1. CIM Pilot Plant at the Technical University of Budapest

This FMS consists of four main cells (numbers in bracelets identify the elements in Fig.1.):

- storage cell: Roboplan storage (7)
- manufacturing cell: - EEN 400 CNC Lathe (4),
  - Polyax TC3 Manufacturing Centre (3),
  - FANUC M3 Robot (2)
  - IGM Rekard MR08 Portalrobot (5)
- assembly cell: Unimate Puma 760 assembly robot (1)
- measuring cell: Opton UMC850 measuring machine (6)
- Automatic Guided Vehicle (8)

The prototype configuration of the Simulation Scheduler System can be seen in Fig.2. Five main modules build up the system.

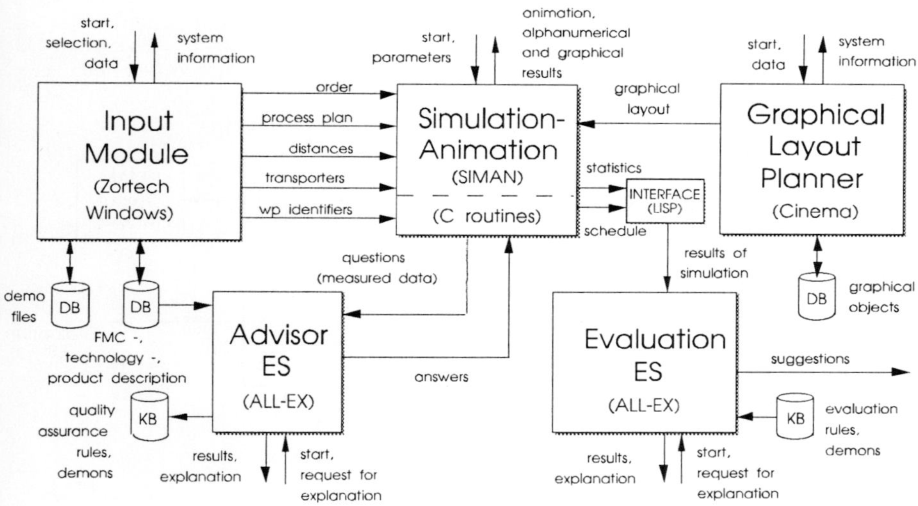

Fig.2. Prototype of Simulation Scheduler

The user can give the parameters of the cell (e.g. velocity of the AGV), manufacturing order (number and type of workpieces to be produced, due dates, etc.) and process plan (operation order with time period) to the *Input Module*. This menu driven program has been written in Turbo C using the Zortech C++ Windows library. It runs on IBM PC AT under DOS operating system.

The *Graphical Layout Planner* is the part of the *SIMAN/Cinema* simulation software package running on IBM PC/AT (VGA), under DOS operating system. The user can edit the graphical layout of the manufacturing cell, and can define connections between the graphical symbols of the resources and the resource identifiers in the SIMAN simulation program. The output of the Graphical Layout Planner can bee seen in Fig.1.

The *Simulation-Animation* subsystem receives the prepared information from the Input Module and from the Graphical Layout Planner. This simulation-animation module, which is also based on the *SIMAN/Cinema* simulation package, has five main parts (see Fig.3.). The *Model Frame* describes the logical structure of the model of the simulated system, while the *Experimental Frame* contains the parameters of the simulation model. According to our future plans these frames will be generated by the *Model Manager*, but in the present version it is done manually yet. The manufacturing order, the process plan and some information of the cell configuration (distances between the resources, properties of the transporters) belong to the parameters of the simulation model. When the Model and Experimental Frame are prepared, they are compiled and linked by the *SIMAN compilers* and *linker* before being executed by the SIMAN *Run Processor*. The SIMAN Run Processor has been extended with C routines. These routines keep the contact with the *Advisor ES*. If the simulation process reaches a decision point (what should happen to the workpiece next), then the simulation stops, sends a message to the Advisor and waits for the reply. Using the SIMAN *Output Processor* (part of the SIMAN/Cinema system) the user can get the graphical representation of the statistics of the simulated manufacturing process.

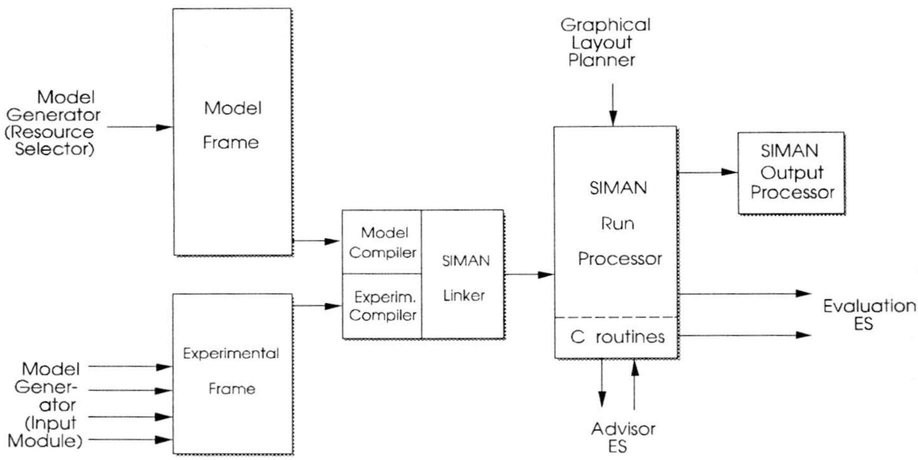

Fig.3. Simulation-Animation Module in SIMAN

In this prototype version the knowledge base of the *Advisor ES* (written in ALL-EX expert shell [Multilogic 1991], under OS/2 operating system) is filled with quality control knowledge only. Using this knowledge the Advisor can decide what should happen after a simulated measuring. The Advisor waits for the questions of the Simulation system, which are the simulated measured value of certain workpieces, and decides whether the workpiece is good, faulty but repairable or waste. Scheduling rules are not implemented in the knowledge base yet. They should help in the decisions like which workpiece can leave the storage first, or which resource is more advisable for a certain operation, etc. In the prototype version these decisions have been built into the simulation program (for example the storage leaving control is the well-known Round Robin strategy).

The Evaluation Expert System is related with the program which simulates the working of the manufacturing cell. Its task is the evaluation of the simulation s results. As input data it processes the requirements of the manufacturing cell, data and coming from the simulation respectively their preliminary elaboration (statistics). The evaluation procedures of the EES evaluate the input data set and put the qualification at the disposal of the preparation module (in the lack of the preparation module to the programmer setting up the simulation model) that make change on the initial conditions of the simulation model conrrespondently to the results of the evaluation. The Evaluation Expert System consists of the knowledge base and the inference machine. The knowledge base contains first all necessary data concerning the cell, second the relationships existing between the data, third the solution strategies leading to the goal. The inference machine is such a program which knows effectively use the knowledge base and gives an answer to the questions raised. We implemented our Evaluation Expert System in ALL-EX expert system frame [Multilogic 1991]. The objects of the system are represented by *frames*. There is a frame which contains the identifier of a workpiece, among its attributes there are average flow time, due date, another frame contains the identifier of a machine and its attribute is utilisation. The knowledge is represented by *rules* and *demons*. There is a rule which compares the due date with completition date for every workpiece and then concludes whether the due date was met or not. There is a demon which fires whenever the utilisation of any machine receives a value and determines its relative level. The inference engine has *goals* which are reached from rules by backward chaining making use of demons, too. There is a goal to determine if the due date of any workpiece was met or not. This goal can be reached from the rule mentioned above.

An interface program (written in LISP, under DOS) supplies the connection between the Simulation module and the Evaluation ES.

The data transfer in this version has been realised by using data files which is sometimes a bit uncomfortable, but it was the quickest way in the developing phase.

6. CONCLUSIONS

With the working prototype version of the Simulation Scheduler System we have proved that knowledge based systems can efficiently help the work of traditional simulation systems. It is important to complete the Advisor's knowledge base with scheduling knowledge, because now the scheduling knowledge has been built into the simulation model.

Although the SIMAN/Cinema is a simulation system of general purpose, with some special features (e.g. special modelling possibilities of transportation systems) it is very efficient in simulating manufacturing systems. There are many SIMAN applications and therefore it is worth dealing with developing standard interfaces between SIMAN and knowledge based systems.

We are going to change our computer platform from IBM PC we used for the prototype to *SUN Sparc Station*. The SIMAN/Cinema can run on SUN machine under UNIX operating system, and a real-time expert shell called *G2* [Gensym 1991] will be used for knowledge processing. The development of this new prototype has been started.

This work has been partially supported by the Hungarian State Committee for Technological Development (OMFB), partly by OTKA fund (No.T4122) and by the Greek Ministry of Industry and Technology under the Terms of a Greek-Hungarian Scientific Co-operation

7. REFERENCES

[Dolk 1984] D. R. Dolk and B. R. Konsynski: *Knowledge Representation for Model Management Systems*, IEEE Trans. Software Engineering, Vol. SE-10, No. 6, pp. 619-628., Nov. 1984.

[Browne 1984] J. Browne, D. Dubdis, K. Rathmill, S. P. Sethi, K. E. Stecka:Classification of Flexible Manufacturing Systems, The FMS Magazine, pp. 114-117., April 1984.

[Futó 1987] Futó I.: *AI and Simulation on PROLOG basis*, Int. Symp. on AI, Expert Systems and Languages in Modeling and Simulation, Barcelona, Spain, June 2-4., 1987.

[Gensym 1991] *Proceedings of Gensym Users Society European Meeting*, Paris France, September 25-27.,1991.

[Multilogic 1991] *ALL-EX Reference Guide*, Multilogic Computing Ltd, 1991.

[Pegden 1990] C.D. Pegden, R.E. Shannon, R.P. Sadowski: *Introduction to Simulation Using SIMAN*, McGraw-Hill,Inc. New York, 1990.

Knowledge Based Hybrid Systems (B-11)
I. Mezgár and P. Bertók (Editors)
Elsevier Science Publishers B.V. (North–Holland)

Composing Simulations by Expert Rules: Modeling Plasma Sprayed Films

G.Jacucci, M. Marchese and C. Uhrik

Laboratorio di Ingegneria Informatica, Dip. di Informatica e Studi Aziendali,
Università degli Studi di Trento, 38050 Mesiano-TN, Italy
email: gianni@itnsun4.cineca.it, marchese@itnsun4.cineca.it, carl@itnsun4.cineca.it

Abstract

A hybrid knowledge-based simulation model of a plasma sprayed film is presented. It represents an example of a multiple part simulation model coordinated by an expert system.

Films are formed by spraying plasma drops. With the simulated arrival of each plasma drop at its target area on the film, a knowledge-based program tests a set of conditions which take into account various characteristics of the drop and of the target impact area on the film deposited so far. Based on the conditions, one of multiple simulation scenarios corresponding to different physical situations is activated. Thus, one simulation generates the spray and travel through space to the target area; an "expert system" chooses which of many simulation scenarios to run as each drop makes contact with the film; and the process continues until the completed film is built up. Statistics are run over the whole film volume to determine quality features which are of interest to engineers such as the percentage of crack-like pores in the various layers of the film. Subsequently the simulation model result is used in a finite element residual stress analysis of the film, which serves as a validation of the technique since useful design analysis has come out of the study.

Keyword Codes: I.6.3; I.6.5; I.6.8
Keywords: Simulation and Modeling, Applications; Model Development; Types of Simulation

1. INTRODUCTION

The past 10 years have seen an explosion in developments in the field of knowledge-based simulation. This field deals with the broad topic of "making simulations more intelligent" by linking simulations with expert knowledge. Round [1989] suggests a natural motive for the linkage in that both knowledge-based programming and simulation predict complex behaviors by constructing computer models of systems. However the two methods differ in the way the system is modelled and used to make predictions:

1. simulation employs mathematical equations and probability distributions to predict future states by propagating values of system variables through time,

2. knowledge-based programming uses symbolic representations of states to infer facts about a current state or show how to achieve a future state without explicit reference to time.

Whereas numerically driven simulations can give more precise results, the knowledge-based method can be more explanatory of the reasoning that produced a result. Thus, linking the two methods overcomes difficulties inherent in either method alone. O'Keefe [86] and Mezgar [90] give interesting taxonomies and examples of many such systems.

Engineering has long recognized simulation as an alternative to purely analytical models. In especially complex systems, the analytical solution in closed form may be prohibitively complicated, such that simulation offers the only feasible solution technique. A primary question then would be "whether the integration of numerical and rule-based components greatly extends the range of complex systems that can be simulated." Round [1989] concludes that until present, lack of validation has made the answer to this question unclear.

Herein, we present a particular style of combining simulation and rule-based approach shown in Figure 1. The problem studied is the modeling of plasma sprayed films by breaking the simulation into 2 components: 1. generation of the plasma and tracking a stream of droplets through space until they arrive at the target spray area, 2. treatment of the plasma drops according to one of several possible simulation scenarios according to the precise condition of the droplet and the impact point where it will be deposited. In effect, a knowledge based system which takes into account the physics of the plasma deposition is used to pick which of the scenarios in 2. is to be invoked.

In section 2, the numerical simulation to track the thermal and mechanical history during the deposition of a protective plasma-sprayed coating is presented. The simulation is presented in 2 parts: splat formation and combining splatting events into the film; then the statistical analysis of simulation output is considered. Subsequently, section 3 describes the results of the simulation and how it proves useful for a stress analysis of the film. Finally, section 4 gives some perspectives and advantages of using such an approach.

2. COMPUTER MODELING OF PLASMA SPRAYED COATINGS

Plasma spraying has been used for more than 3 decades to make coatings of one material on another [Zaat 1983]. The quality of the coating obtained is dependent on a large number of parameters: the power and design of the torch, its position relative to the substrate, the type of powder used and how it is introduced into the plasma, the nature and preparation of the substrate. Traditionally, these parameters have been optimized empirically. In recent years, much progress has been made in understanding the physics of plasma

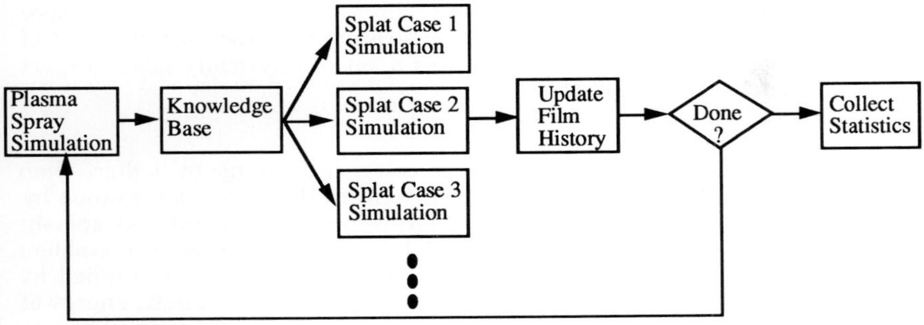

Figure 1 - Knowledge Based Approach to Plasma Sprayed Film Simulation

spraying, particularly the physics of the plasma and the plasma-particle interactions (see, for example [Vardelle 86, Fauchais 89]). The microstructures of the resulting coatings have been studied by a number of authors. This subject has recently been reviewed by McPherson [89]. The objective of such studies is to understand how the properties of the coatings correlate with the process parameters used to obtain them. Here, we wish to calculate the effective properties of the coating and the temperature distribution produced by the deposition process.

Information about the detailed microstructure can be obtained by analysis of micrographs. However, such information is tedious to obtain and only available for one or two cases. On the other hand, simulation models can also give this information [Cirolini, et. al 91, Ferrari, et. al. 91]. Such a model should not be considered as a replacement to experiments since, as we shall see, a number of simplifying assumptions are necessary. However, the results of such modelling can be used to suggest where particular experiments are likely to be useful.

The microstructure of the coating results from the details of the spraying technique and in turn determines the properties of the coating. The production of the microstructure is not a thermodynamic process at (or anywhere near) equilibrium and thus any model of the coating must emerge from a model of the deposition process. The coating is built up from individual particles hitting the substrate, splashing and forming thin laminae. The freezing time of the particles (10^{-6} - 10^{-8} seconds) is sufficiently fast that each particle impact may be considered to be independent of all the others. We can therefore split the task of modelling the deposition process of a thermal barrier plasma-sprayed coating film into two parts:

1. spraying the plasma and description of the behaviour of the particle on arrival at the film,

2. combining of plasma droplets to obtain the film.

The first part takes account of experimental data for the plasma temperature and enthalpy content and the velocity, temperature and size distributions of the sprayed particles. The second part uses a set of physically based expert rules to construct the coating.

2.1 Splat Formation

When a droplet hits a surface at low velocity, it flattens to a disc, then rebounds to form a cone and finally spreads to form the shape determined by surface tension as shown in Figure 2. At high velocity, the liquid may splash; the surface tension forces are not great enough to hold it together. The problem of the impact of liquid droplets on a cold, flat surface has been studied by Madejski [76]. He made four main assumptions: first, that the kinetic energy of the particle goes into viscous dissipation and changes in surface tension due to the change of shape; second, that the particle forms a thin spherical disc; third, that heat flow is everywhere normal to the surface; fourth, the simplest velocity distributions that satisfy the continuity equation. If we further assume that solidification can be ignored, an equation for the ratio of the radius of the thin disc to that of the original particle ($\xi = D/d$) may then be obtained as

$$\xi = D/d = 1.2941\,\Re^{\,0.2} = 1.2941\,(\,\rho V d/\eta\,)^{0.2} \tag{1}$$

where ρ, V and η are the density, velocity and viscosity of the particle, and $\Re$ is the Reynolds number. This formula has been investigated experimentally and is reasonably successful [Madejski 76, McPherson 80] and has been used in our model. Once the particle has spread into a thin sheet, it will rapidly solidify. We have approximate the solidification of the splat as a Stefan problem and solved it by standard numerical methods [Cirolini, et. al 91].

If we assume a splat thickness of about 3μm we get a temperature drop across the splat as the solidification front moves to the top of about 50-100°K. The calculation also suggests that the solidification time is a few microseconds, at least an order of magnitude less than the mean time between the arrival of particles at a particular place. This justifies the assumption of the independence of each splat formation. When the splat arrives, it first spreads across the surface. However, as it cools it will tend to shrink and curl up to relieve the stress. Once it is within the film, it clearly cannot do this and the stress is locked into the coating. To model this effect we make two assumptions. First, we assume that the curling is caused by the temperature drop across the splat when the solidification front just reaches the top. That is, we assume that until the solidification front reaches the top, the splat is free from stress and only when all the splat is solid does the differential thermal contraction begin. Second, we assume that when the next particle arrives, it pushes the curled-up splat onto the substrate provided that it has not been covered over by material arriving from elsewhere. Thus pores and interlaminar cracks arise because the splat curls up, and the porosity so formed is buried before it can be firmly attached to the substrate by the next arriving particle.

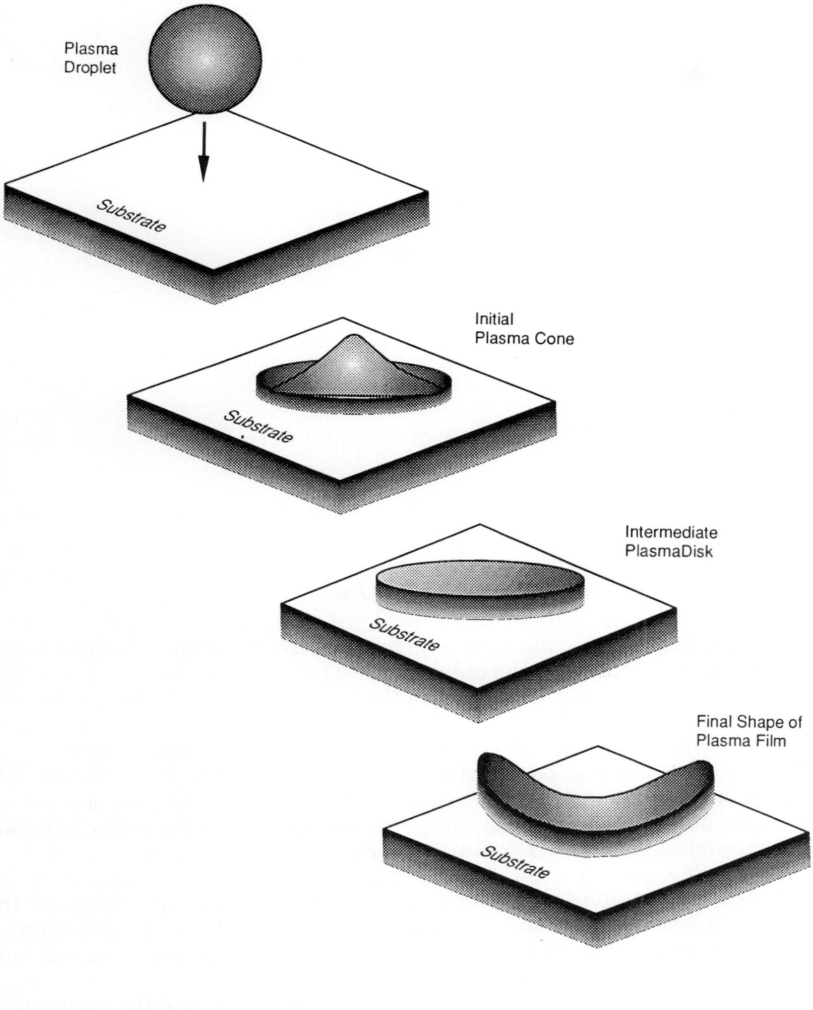

Figure 2. Splat Formation from a Plasma Droplet

We can estimate the amount of curling as follows. If the thermal expansion coefficient is γ and ΔT is the temperature drop across the splat, the top will contract $\gamma\Delta T$ more than the bottom. For a splat of radius *d* and thickness *a*, the

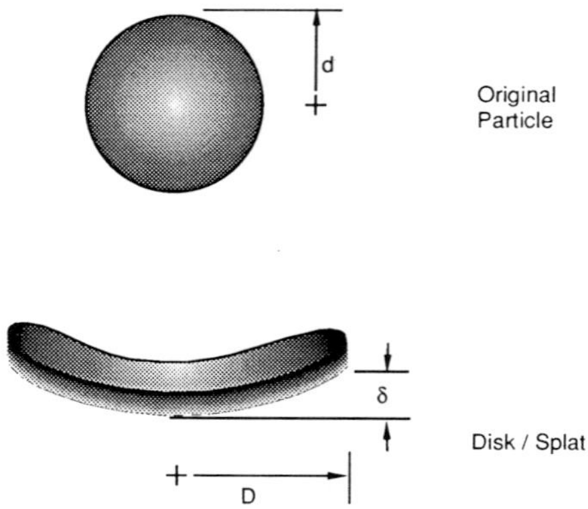

Figure 3 - Geometry of the Splat Formation

radius of curvature is given by $d\,/\gamma\Delta T$, about 5mm. Thus, the amount of lift at the end (see Figure 3) is given by

$$\delta = d^2\,\gamma\,\Delta T\,/\,4a \qquad (2)$$

which is the order of a few microns. The formula assumes the splat retains its cylindrical shape throughout; the only response to cooling being a curling effect. Thereby, it relates the splat behaviour to the temperature of the substrate.

2. 2 Combining Splatting Events according to a Knowledge Base

Splats form the coating by linking together in a complex fashion. The knowledge based component of the program attempts to model this using a few simple rules as illustrated in Figure 4. We shall here list the rules but refer their discussion to reference [Cirolini, et. al 91]:

A. The program assumes that the splat follows exactly the shape of the underlying layer where this is under the impact region of the original particle (i.e. out to the radius of the particle from the impact point in both directions)

B. If the program detects porosity one splat thickness below the uppermost layer under the impact region, it hammers the uppermost layer to the next layer below it destroying the pore.

C. The splat beyond the impact region initially follows the surface of the layer below and is constructed from the impact point outwards in both directions. It can therefore grow under existing splats if there is a continuous open space permitting this (interlocking).

D. If the splat encounters a dead end, it fills the space available and then flows over the outer surface above.

E. If the splat comes to a vertical drop, it falls straight down until it finds the surface again.

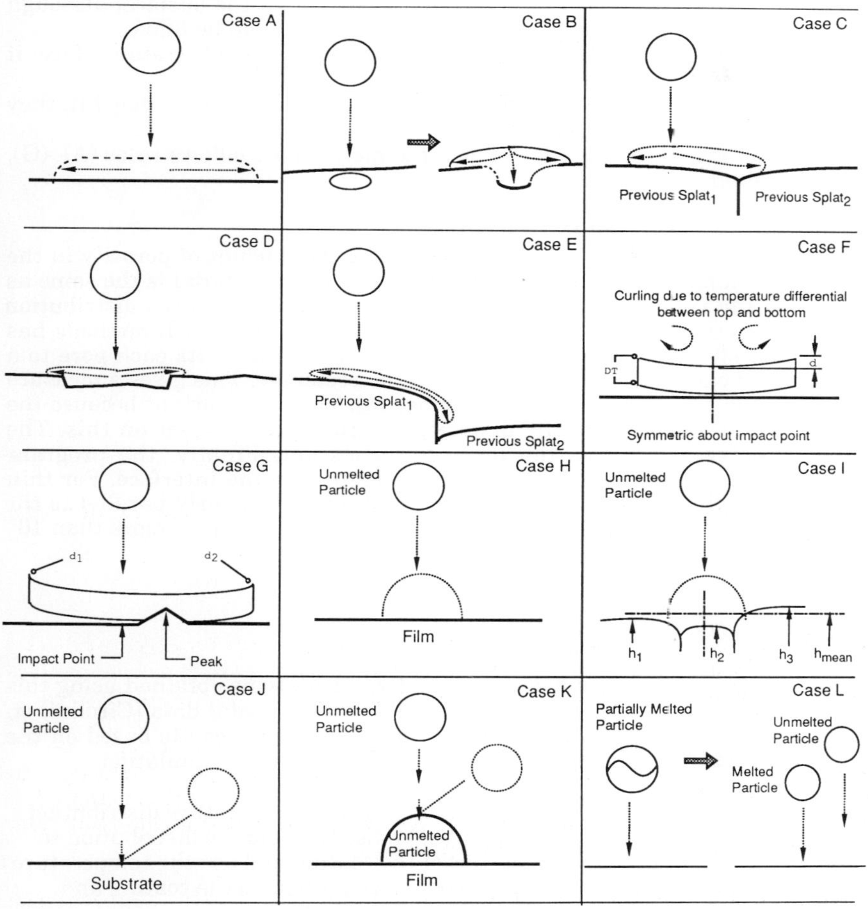

Figure 4 - Rules for Simulation Scenarios of Splatting on Film Surface

F. The splat is then permitted to curl up provided that it is the top most layer. The total amount of curling is estimated as a function of distance from the impact region using equation (2). above
G. If the underlying region contains a large peak, this can pin the splat and then the curling is calculated with respect to the peak.
H. Unmelted particles form hemispheres on the coating.

I. Hemispheres are assumed to lie on the mean line of the surface, although very wide (wider than the particle diameter) cavities will be filled.
J. Unmelted particles are assumed not to adhere to the substrate surface; if they hit, they bounce off.
K. Unmelted particles are assumed not to adhere to each other; if they hit, they bounce off.
L. A partially melted particle is divided; the melted part follows rules (A)--(G), the unmelted part follows rule (H)-(K).

2.3 Derivation of Summary Statistics
The statistical analysis program first calculates the amount of porosity in the coating. It assumes that the ratio of the area of void to material is the same as the ratio of the volumes. Methods exist for transforming the area distribution of voids to a volume distribution [Wichsell 25]. The use of such methods has only a small effect on the total porosity. The program also fits each pore to a rectangle and calculates the aspect ratio of the rectangle. This gives a measure of whether the pores are crack-like or spherical. This is important because the effective elastic constant of the material is strongly dependent on this. The more crack-like the pores, the lower the constant. Finally, the program calculates the orientation of the pores with respect to the interface. For thin coatings, it is obvious that pores of any length must be nearly parallel to the interface. The calculation suggests that the angle is never much more than 10° except for pores that are nearly spherical.

3. RESULTS OF THE MODEL

Two levels of validation are available. Firstly, simulations obtained using this model can be compared directly with available experimental data [Cirolini, et. al 91]. In addition, we can derive further computational results based on the mechanical and thermal history of the film indicated by the simulation.

First we consider the calculation of the porosity and temperature distribution within the growing coating. Subsequently we use the porosity distribution to calculate the effective elastic properties of the coating and use the temperature distribution to calculate the residual stresses generated as the coating and substrate cool to room temperature.

3.1 Results of the Deposition Model
The rules for constructing the deposited coating are implemented by growing the coating on a mesh. This is necessary if the evolution of the splats is to be followed. Figure 5 shows a typical coating produced by the model. This figure is produced using a 500x500 mesh, which is large enough to show the features of the model, but small enough to be printed out. It is important to ensure that the simulated structure of the coating is independent of the details of the mesh on which it is simulated. Meshes of up to 2000x2000 are required to check this. Most of the results presented here were obtained using such large meshes. Although absolute values are dependent on mesh size, the trends are not.

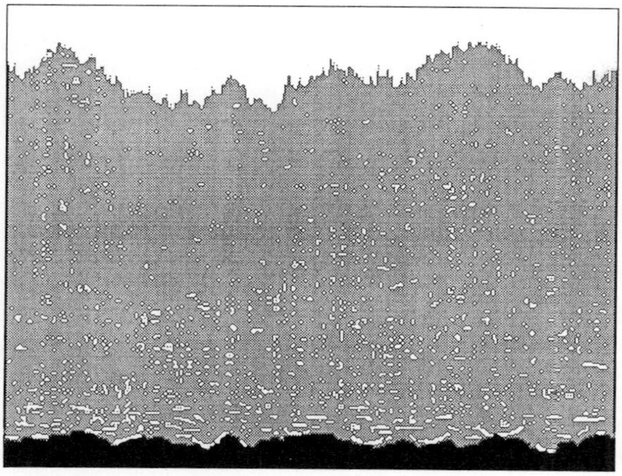

Figure 5 - Typical Output of the Deposition Model

The porosity levels calculated are typical of those observed in coatings (of the order of 10 %). For our later work, it is not enough to evaluate the bulk porosity, we require also an estimate of what kind of porosity it is; whether in small spherical inclusions or long cracks. The latter kind of porosity has a far larger effect on the elastic properties of the coating. Table 1 shows the variation of the pore aspect ratio within the coating.

The Table suggests that a large proportion of the porosity (greater than 50%) is in the form of thin crack-like objects. A further point of interest here is the suggestion that the interface region is significantly different from the bulk of the coating. The porosity due to the crack-like objects falls away from the interface. The porosity due to spherical objects remains approximately constant. This correlates with the observed fact that the coatings usually break just above the interface rather than along the interface itself.

Table 1
Variation of pore aspect ratio

	Aspect Ratio				
	0-25%	5-50%	50-75%	75-100%	
Thickness (μm)	Porosity (%)				Tot. porosity
0-10	10.6	4.0	1.4	3.5	19.5
10-85	5.4	2.9	1.3	2.3	11.9
85-160	3.2	2.7	1.0	3.1	10.0
160-210	1.8	2.3	0.9	2.6	7.6

3.2 Application to Residual Stress Computation

The results from the model can be used to calculate the effective elastic and thermal properties of the coating and hence to calculate the stresses within the coating for some simple configurations [Ferrari et.al. 91]. The computed thermo-mechanical properties can then be used in a thorough investigations of the behaviour of the protective coating under service conditions.

The finite element method was selected to determine detailed stress states in a cylindrical thermal barrier coating specimen, for the linear thermo-elastic case.

The geometric data of the problem are the following: cylinder's height and diameter are 50 mm, and the coating's thickness, assumed uniform, is 0.2 mm. The finite element mesh consists of 1661 nodal points and 1500 four-nodes axi-symmetrical elements. A very fine mesh was introduced to model both the ceramic-metal interface and the ceramic coating. In these regions the elements sizes are in the 1-2 μm range (additional refinements were found not to influence the results).

For all of the numerical studies, the initial reference time was taken to be the instant at which the spraying process ends. The corresponding ceramic state is assumed to be stress-free. Two kind of information are required to perform the thermo-elastic computation: (1) the temperature distribution across the specimen, (2) the thermo-mechanical properties.

The temperature field was calculated from the deposition model described above and therefore it is assumed to be varying only in the axial direction. The effective thermo-mechanical properties needed for the thermo-elastic problem were calculated starting from the simulated micromorphological portrait, and using the Mori-Tanaka approach as described in [Mori 73, Ferrari 89].

The microstructural gradient is accounted for, upon by dividing the coating into four layers, the thermo-mechanical properties of which are calculated with the above mentioned homogenization techniques. A summary of the porosity volume fraction portrait used for the calculation of the effective elastic properties has been given in Table 1. The computed elastic properties are given in Table 2, for the specific case of a macroscopically isotropic 8% Yttria-stabilized Zirconia coating sprayed on a Nickel superalloy. The computed thermal expansion coefficient is uneffected by the porosity.

Table 2
Effective elastic constants for the isotropic case

	Young's mod.	Poisson's ratio	Therm. exp. coef.
Metal	137 GPa	0.250	15.0 e-6
Layer 1	131 GPa	0.239	7.0 e-6
Layer 2	164 GPa	0.234	7.0 e-6
Layer 3	177 GPa	0.224	7.0 e-6
Layer 4	190 GPa	0.214	7.0 e-6

The basic finite element program was used to study the response of the coating to different initial temperature profiles [Ferrari 91]. One weak point of the total composite coating system seems to lie at the metal-ceramic interface. Even if this layer exhibits neither the highest stresses nor the greatest temperature gradient, most alternating compressive and tensile loads are found there. The effect of the substrate temperature on the interface stresses is studied in Figure 5. It is noted that, in order to minimize the corresponding transverse stresses, the substrate temperature should be kept in the 300-400°K range. However this induces tensile transverse stresses at the free surface, and may thus generate surface cracking.

4. CONCLUSIONS

The knowledge based simulation is a hybridization between a "parallel-integrated system" and "front-end system" [O'Keefe 86]. In parallel-integerated systems, information continuously passes back and forth between the knowledge based and simulation components whereas, in a front-end system, the knowledge-based component acts as a "front-end" for defining a numerical simulation which is subsequently used on its own. According to Mezgar's classification [Mezgar 90], the system may be seen as a shallow coupled composite intelligent hybrid model.

The approach is particularly conducive to validation since it provides a plausible model from which complex analyses useful to production engineering are preformed. The same simulation implemented without the knowledge-based method would involve an order of magnitude more work for development and maintenance of the computer code written or possibly be infeasible from the number of special cases handled.

The model captures the main qualities of trends to be found in the microstructure observed in the manufacture of a plasma sprayed coating. The simulation shows how the microstructure depends on details of the deposition process and provides a deeper understanding of the interrelations between material properties and coating behavior. This serves the ultimate goal of predicting the performance of the coating under various service conditions, enabling an inexpensive "empirical route" to optimization.

The developed model provides an innovative and much needed design tool that overcomes the limits of traditional empirical optimization procedures, such as high costs and difficulty in investigating every combination of production technique and service condition.

5. REFERENCES

Cirolini , S., Harding, J. H. and Jacucci, G., Surf. Coat. Tech.,48 (1991), p137;

Fauchais, P. , Grimaud, A. , Vardelle, A., and Vardelle, M. , Ann. Phys.Fr.,14, (1989) p261;

Ferrari , M., and Johnson, G.C., Mechanics of Materials, 8 (1989), p67;

Ferrari, M., Harding, J.H. and Marchese, M. , Surf. Coat.Tech.,48(1991), p147;
Madejski, J. , J. Heat Mass Transfer,19,(1976) p1009;
McPherson, R. , J. Mater. Sc .,15,(1980) p3141;
McPherson,R. , Surf. Coat. Tech., 39--40, (1989) p173;
Mezgar,I., The Evolution of Models for Flexible Manufacturing, Modelling the Innovation: Communication, Automation and Information Systems, IFIP Intl. Conf., Rome, 1990.
Mori, T. , and Tanaka, K., Acta Metallurgica, 21 (1973), p571;
O'Keefe, R., Simulation and Expert Systems - A Taxonomy and Some Examples, Simulation , 46(1):10-16, 1986.
Round, N., AI Handbook, Barr, et. al. eds., Volume 5, 1989.
Vardelle, A. , Vardelle, M. and Fauchais, P., Rev. Int. Hautes Temp. Refract, 23 (1986) p69;
Wicksell, S. D. , Biometrika, 17,(1925) 85;
Zaat, J. H. , Ann. Rev. Mater. Sci., 13 (1983), p9.

HYBRID TECHNIQUES IN MONITORING AND MAINTENANCE

Knowledge Based Hybrid Systems (B-11)
I. Mezgár and P. Bertók (Editors)
Elsevier Science Publishers B.V. (North–Holland)

Failure recovery of flexible production systems through cooperation of distributed agents

E.D. Adamides and D. Bonvin

Institut d'Automatique, Ecole Polytechnique Fédérale de Lausanne,
CH-1015 Lausanne, Switzerland

Abstract

This paper reports on the design and development of a novel distributed AI-based framework for the automatic failure recovery of Flexible Production Systems (FPS). A distributed frame-based knowledge representation scheme of the physical production system, the production process and the products being processed is used by a set of intelligent control agents with local perception and action capabilities to recover a system through cooperation. The recovery process is being carried out in an optimal way by optimizing numerically, in a decentralised fashion, the trajectory from the error state to a required safe state.

Keyword Codes: I.2.4, I.2.11, J.2
Keywords: Knowledge Representation Formalisms and Methods, Distributed Artificial Intelligence, Computer Applications in Physical Sciences and Engineering.

1. INTRODUCTION

Although modern large-scale sophisticated production processes increase production rates simultaneously over a wide range of products, using "flexible" machinery and advanced computational techniques, failures do happen and cost much. Effective management of these failures requires complex organisational and operational changes in the production line. Fortunately, in today's high-tech factory, it is possible to implement these with the required speed by using the information available in order to choose the best alternatives in terms of product range, product quality and line productivity. Since modern production technologies are not deterministic [1], once a failure has been encountered in the shop-floor, a large number of decisions have to be made by a number of hierarchically structured and interdependent decision. To make these decisions, to communicate them, and to synchronise the actions to be taken, all in real-time, without the use of some sort of automation is an impossible scenario.

From the operational point of view, what is required is to construct a *recovery plan*, i.e. to obtain a sequence of actions which have to be taken from the moment a failure alarm is signaled until the moment the system is able to resume normal operation. Since during this transient period, the safety of operators, as well as expensive machinery and unfinished products may be at stake, this transition has to be made with the greatest possible care. At the highest decision levels, decisions are concerned with whether complete recovery of the system is possible, and if not, with how to keep productivity and product quality measures as high as possible. Due dates are also of concern. At the lower levels, decisions have a more reactive and opportunistic nature.

Technically, automatic failure recovery of large flexible production systems is a difficult task to accomplish for two main reasons: first, it is very difficult to assess the state of the entire system in very short time intervals and to provide correctly alternative solutions and set points for the recovery plan (e.g. recovery states), and secondly, the task to find a sequence of actions which will recover the system in a cost-effective way (meeting the production constraints) is of combinatorial complexity. From the computation point of view the problem of failure recovery is hybrid, i.e. of both symbolic and numeric nature. Symbolic because the level of abstraction of the dynamics of the production process is of qualitative (logic) nature requiring a dynamic model of the system to function at this level; numeric because recovery plans can be optimised by quantifying all the decisions made with respect to equipment usage, safety, product quality, line productivity, etc.

In this paper, by taking a holistic view of the problem, we present the architecture, the elements and the operation of a failure-recovery system for flexible production processes based on a hybrid distributed artificial intelligence and distributed optimal control approach. Frames are used to encapsulate all the knowledge about the processing, storage and transportation equipment, the production processes and the products on the production line, and they are distributed among the recovery-planning agents. Relations among the planing agents themselves are also included. Reactive and reparative actions are triggered by the presence of specific events. *Matching* and *inheritance* are the basic inference mechanisms. After two stages of bidding and negotiation, a recovery plan at-large is constructed and distributed among the agents. Then, the plan is refined and optimised by cooperative effort. In order to quantify their decisions for the construction of the final cost-effective detailed recovery plan, the agents use a distributed high-level Petri net model in connection with a dynamic programming algorithm.

2. A HOLISTIC APPROACH TO FAILURE RECOVERY IN FPS

Due to their increasing importance, production failure localisation, identification and processing have attracted the interest of many researchers from various disciplines and with different perspectives. Depending on the flexibility of the application domain, various approaches to recovery have been proposed. However, most of the work published so far, as well as most of the prototypes and industrial products currently available, address the problems of fault diagnosis and recovery through localised repairs and pre-defined reactions limiting the application domain to metal processing industries [2-5]. For the case of production processes with low flexibility and minimal availability of on-line information to the monitoring system, simple optimisations of the recovery process seem to be sufficient [6]. However, for processes where batches of different products are fabricated using the same raw materials on a limited number of equipment, a holistic approach to the recovery process is required. By using the information offered on-line, large and expensive production processes do not have to be halted or diverted to reduced-rate processes only because of a small equipment failure. The availability of information imposes a more demanding view of the issue. The view of a reactive opportunistic procedure is not enough. Recovery has to be considered as a minimisation of the transient period from a failure state to a recovery goal state that guarantees the less disturbed process at the factory level. This means that the production system must move in the fastest possible way at minimum cost towards a state which guarantees that the production schedule is followed in both quantitative and qualitative terms. During this period, some part(s) of the system may be responsible for maintaining the original production schedule (or a reduced-rate schedule) through diverted operations, whilst other part(s) of the system are trying to recover. In a more analytical view, the whole operation involves:

- identification and localisation of the disturbance
- execution of high-priority emergency actions
- identification and (re)organisation of the resources of the reduced potential system
- identification of safe recovery states and rough estimation of their reachability potential and costs
- determination of the maximum production potential of the reduced-configuration system (probably using redundant resources)
- determination of the effects of the reduced-potential production system on the quality of the final product
- determination and execution of a detailed plan for the complete recovery of the system.

The overall problem can be decomposed into three consecutive stages/sub-problems, namely:
i) find the best state that the system has to be brought to by its controls,
ii) decide which parts of the system will do what during the recovery process,
iii) find the best decision sequence that will bring the system to the required state.

The first two sub-problems are principally of symbolic (knowledge-based) nature whereas the last involves numerical optimisation using a discrete-event system model. In addition, the size and the distributed and modularised nature of the production system itself, as well as, the popularity of distributed control systems put forward the requirement for a distributed approach to recovery.

To meet the above objectives, the shop-floor control system should operate on a very rich and complex structural, functional and behavioural model of the production system which encapsulates all the knowledge needed to meet the individual objectives. The necessary heterogeneous information cannot be represented solely by a mathematical formalism such as production flow equations or Petri nets. Neither can a knowledge-based representation scheme per se include the dynamic attributes needed for the evaluation of decisions.

In the system presented here, *frames* are used to encapsulate all the (structured and unstructured) knowledge about the production schedules and the physical production equipment. Frames act as data structures or schemata in a simple meta-model which encapsulates both unstructured domain knowledge and discrete-event dynamic models such as Petri nets. The modified (after-failure) high-level Petri net representing the production flow model (schedule) is generated by the agents using the pre-defined master schedule, the frame-based model of the production system and information on the products that can be processed on the line (production recipes). The structure of the frames used in our system is described, in more detail, in the following section.

The data and procedures that form the recovery agents are assumed to be distributed on the nodes of the FPS's monitoring system. All interaction strategies among agents, as well as among agents and the production and monitoring systems are incorporated as an integral part in the design of the recovery system.

3. FRAME-BASED MODELLING OF FPS

The concept of *frames* can be used as the *at-large* formalism for structural and (partly) behavioural knowledge representation of a production system, with the dynamic components being triggered by the presence or absence of specific values in specific slots. Frames seem to be a suitable means of knowledge representation in manufacturing systems, but they have been used mostly in the context of scheduling and simulation of FMS (e.g. [7-8]). Moreover, frames have already been used in manufacturing fault management, but most of the effort has been concentrated on fault detection and localisation issues with statically predefined maintenance and system re-configuration actions [2-3].

In the system presented here, frames are used as a means of organising the knowledgebases of the general monitoring agents. Inference and coordination of the agents takes place by consulting and updating frame structures. There are two major frame superclasses: EQUIPMENT (physical system equipment/devices and their attributes, including failure information) and PRODUCTS (all products that can be produced by the system, their characteristics (including quality information) and required production steps.

The EQUIPMENT frames are distributed on the control agents according to their physical and operational proximity in relation to system devices or domains. Their attributes include device dynamics and related physical constraints as well as the effect of various processing alternatives on different products. More specifically, a frame structure of a system equipment includes the following:

- general characteristics and relations for the frame hierarchy (to inherit properties)
- coloured-Petri net model of the equipment dynamics (physical constraints)
- physical properties of device
- logical properties of the device in terms of products on the line and degree of importance of device for various production processes
- connectivity with other devices for various products

- sets of immediate actions for recovery after specific faults have been identified
- weights and costs associated with every action or change of state of the device and/or its actuator
- knowledge about the state of the production equipment with corresponding knowledge about the state of the products being processed or being carried out by the equipment
- a machine-product state and product quality relation, i.e. contribution of a specific processing stage to the total product quality.

Frames of the PRODUCTS class hierarchy exist in the knowledgebases of all agents with the information relevant to the specific agent. A frame structure for a product includes:

- general characteristics and relations for the frame hierarchy (to inherit properties),
- physical characteristics of the product and its parts,
- production recipe(s),
- quality characteristics in relation to various production processes,
- pointers to other products with quantitative relations (e.g. reduced quality by-product, etc.)

4. ARCHITECTURE AND OPERATION OF THE RECOVERY SYSTEM

4.1 Agent distribution and internal architecture

Various domain-specific approaches, ranging from locality-dependent to object-oriented [9], have been proposed in the literature for problem decomposition and task assignment to agents. Since efficient operation of a production system means efficient management of its resources, and since the interactions between the various resources are known before the problem solving activity starts, the planning/recovery agents of the system proposed are directly associated with specific limited domains (resources) of the system. The agent closest to the failure takes, by default, the role of the coordinator (SUPERVISOR mode) of the recovery. The agents are assumed to be connected directly to sensors and actuators to determine the state of the system and, after planning, to implement their control actions. They are all connected to each other through point-to-point connections, but their logical connectivity depends on the particular situation. Focused addressing disconnects some of the agents from the network. Point-to-point addressing, although cumbersome and more expensive, implements more easily the semantics of synchronised communications.

Figure 1 shows the overall structure of a simple reference production system consisting of three processing stations and two intermediate in-process limited size buffers. The areas of responsibility as well as the connectivity of the recovery planning agents is also shown. Products are assumed to be processed and be circulated within the system in batches. Routing of batches is by means of programming the transportation medium (conveyor belts).

Since failure recovery management is only a part of the production system's supervisor, and since the recovery process must be as fast and flexible as possible, agents have to be simple and fine-grain incorporating only the essential reasoning, optimisation and communication mechanisms. In line with this, our system consists of fine-grain agents with application specific characteristics of the general "standard" autonomous-agent-definition characteristics (privileges, intentions, limited rationality and acting capabilities, etc.). [9]. All agents are assumed to be able to access with the same privileges the enterprise's global information resources (databases, knowledgebases). No single agent has sufficient expertise, resources and information to solve the overall recovery problem.

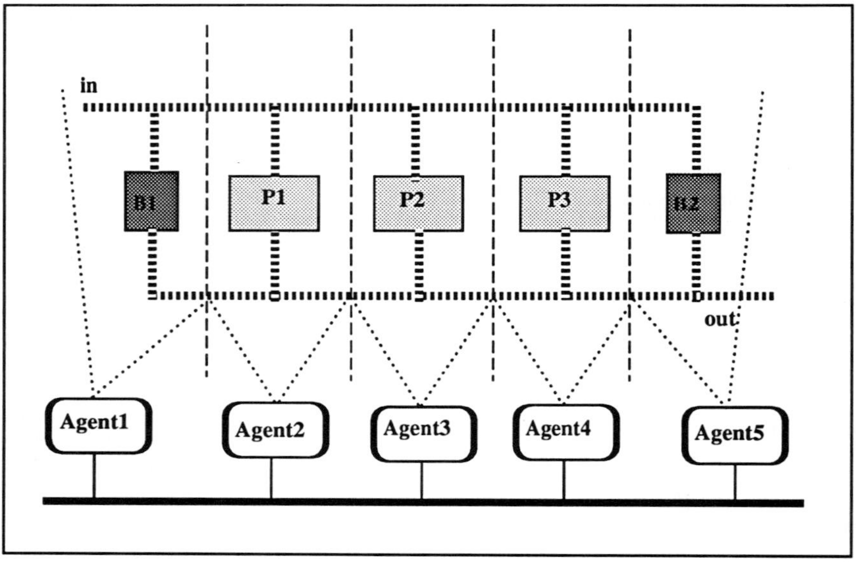

Figure 1. A Flexible Production System with its controlling agents

The internal architecture of the agents is shown in figure 2. The *knowledgebase* stores the frame structures relevant to the specific agent, as well as predefined heuristic rules and relations between the agents which are used in the recovery. The *numerical routines* module implements numerical functions such as a dynamic programming algorithm. In addition, it stores representation transformation procedures, e.g. algorithms for the descolorisation of the coloured high-level Petri net (hl-PN), that is, the transformation of the hl-PN into algebraic representation of a simple Place/Transition net [11]. Finally, this module includes some other ad hoc procedures, e.g. routines to quantify and evaluate bids offered by other agents. The *inference engine* module implements a simple matcher for the frame slots and uses the inheritance properties of slots to infer about the target safe state, the suitability of the particular agent to the recovery process, etc. The *mode controller* oversees the operation of the agent and it is in synchrony with the mode controllers of the other agents. The *communications* module handles all the interactions with the physical system and the other agents. Finally, the *working model* is a dynamic list-like data structure implementing a hl-PN which describes the system dynamics at the discrete-event level of abstraction.

The total recovery processes, as it is seen from the point of view of a single agent, can be split into four consecutive stages which correspond to four agent modes. More specifically:

1. location and identification of the nature of the fault and determination of the target state and operational constraints (PROBLEM_SETTING),
2. estimation of the agents position and role in the total recovery process (SELF_CHECKING),
3. bidding and negotiation for task assignment (BIDDING_NEGOTIATION),
4. construction of detailed optimal plan (OPTIMAL_PLANNING).

In addition, there exists a SUPERVISOR mode that the coordinator of the recovery process enters from the moment of failure identification until the recovery of the system has been completed, for example to initiate, coordinate and terminate the operations and actions of the other agents. The operation of the SUPERVISOR mode is explained in more detail in the following section.

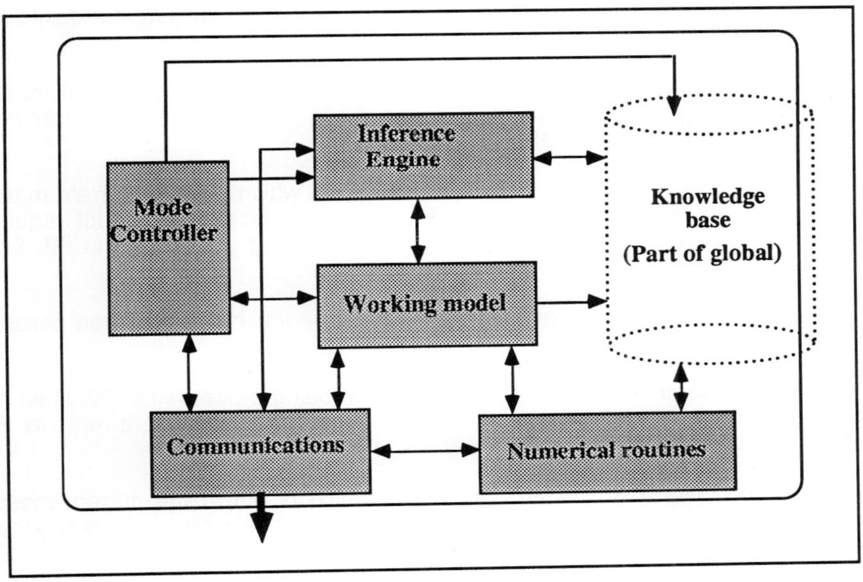

Figure 2: Internal architecture of a control agent

4.2 Operation of the recovery system

After failure location and identification by the shop-floor monitoring system, the global error state signal triggers specific slots of the distributed knowledgebase which result in immediate recovery actions. Consequently, the production system is brought into a new global (error) state. Then, the multi-agent system reasons about the goal state for which it has to plan. The nature of alternatives that have to be considered and the rationale behind the reasoning and selection mechanism is illustrated by means of the following simple scenario.

Assume that a failure has been encountered in the processing station P2 whilst a batch of 400 parts of product A were being processed. As a result of the failure, the batch has been processed for 4 min. instead of the required 10 min. This resulted in thinner coating of the products. At first, since the processing station has a fault, the control system has to empty P2; this constitutes one element of the final goal. Then, it has to decide what to do with the faulty products. By searching through their EQUIPMENT frame structures, the agents have to decide whether this batch can be reprocessed in any form or has to be thrown away.

After examination of its PRODUCTS frame hierarchy, the agent responsible for P1 (*Agent2*) evaluates three possible alternatives:

i) to processes the parts of Product A which being processed in P2 when the failure happened. Processsing in P1 will take 20 min, and after that, it requires P3. End-products will be of 2nd level quality.

ii) To process them for end-product B which is not in the current master schedule). Processing in P1 will take 10 min. Then processing by P3 is required. End-products will be of 1st quality.

iii) To process them for end-product Product C which is in the current master schedule. Processing in P1 will take 5 min. To produce this product requires removal of previous applied coating, i.e., requires fixing of new tool in P3. End-products will be of 2nd level quality.

Accordingly, the agent responsible for P3 (*Agent4*) evaluates three possible alternatives:

i) to processes the parts of Product A which being processed in P2 when the failure happened. Processsing in P3 will take 2 min, and after that, it requires P1. End-products will be of 2nd level quality.

ii) To process them for end-product B which is not in the current master schedule). Processing in P3 will take 10 min. Then processing by P1 is required. End-products will be of 1st quality.

iii) To process them for end-product Product C which is in the current master schedule. Processing in P3 will take 15 min. To produce this product requires removal of previous applied coating, i.e., requires fixing of new tool in P3. End-products will be of 2nd level quality.

In order to decide for which alternative they have to plan, agents *Agent2* and *Agent4* initially calculate the costs of ready-product inventories, processing, buffering, delaying first quality products, etc. for every alternative, and then negotiate their best choice with the rest of the agents. A product is finally chosen by forming a consensus of opinions based on a market-oriented logic, i.e. the solution chosen has to be the least compromising for all agents. Further, the solution has to satisfy the time constraints imposed by the coordinating agent (*Agent3* for the above example). Given these parameters, and in relation to the reference example, the choice of Product B may be a wise decision, if finished-product inventories are cheap. By the end of this stage each agent knows the goal state for which he has to plan the best reachability trajectory. In addition, at this stage, the agents can reason about the possibility of some deadlock situations that are known a priori, but mostly deadlocks are treated the moment they happen by considering them as failures and by restarting the recovery process with different parameters. These initial actions constitute the PROBLEM_SETTING mode of the agents. By the end of this stage, all the agents enter the SELF_CHECKING mode in order to estimate their position and role in recovering the system from the encountered fault. The outcome of this stage is the filling of a "bidding form" that includes the following information:

AGENTS_IDENTITY
PROXIMITY_TO_CAUSE_OF_FAILURE
APPROPRIATENESS_FOR_RECOVERING_SYSTEM
through availability of ACTIONX, ACTIONY, ... ACTIONZ
PROXIMITY_TO_AGENTS (names of agents)
ACTIONS_RELATED_TO_FAULT
PROCESSING_OF_PRODUCTS (names and processing times and due-dates
PHYSICAL_SELF_CHECK_STATUS
TOTAL_ACTIONS_SPEED_INDEX
ACTIONS_REQUIRED
AGENTS_REQUIRED

The second bidding and negotiation process (BIDDING_NEGOTIATIONS mode) is to resolve some conflicts and redundancies that the coordinator might have observed after the target selection. This stage follows the architecture of the CONTRACT NET protocol [12], however, with all the agents acting both as *bidders* (offering a bid) and *managers*. The agent close to the failure, constantly in SUPERVISOR mode, and having very limited action capabilities by itself (due to the failure), acts as a *meta-manager,* i.e. coordinator of the bidding process. During this time, the meta-manager calculates the *appropriateness indices* for the agents, which are weighted functions of the values of the items presented in its bidding form with the requests filled by the responding agents. In addition, it registers the *relations* among the agents which are to be used in the next phase of the recovery. The forms presented to the coordinator are filled in an optimal way by the bidding and negotiation process among the agents which eliminate redundancies and fuzzy answers. The structure of the problem-solving team which is the outcome of this phase, defines the final (distributed) working PN model by modifying some of its transitions The agents chosen to lead the system to the desired state enter their OPTIMAL_PLANNING mode, which consists of implementing an open-loop distributed optimisation algorithm operating using the algebraic (state-space) representation of the PN. The supervisor agent being in the SUPERVISOR mode coordinates their actions. The firing of all transitions in the distributed Petri net is synchronised externally from the model by a system clock with a period large enough to allow for communication delays between the various agents. In this way, temporal conflicts can be resolved by predefined operators.

To find the optimal trajectory from the failure state to the desired target state, a recursive multi-stage cost function of the form

$$J_p(\mu_p, k) = \min_{q_p(k-1)} \Big\{ C_{1\,p}^T(k-1)\big[\mu_p(k-1) + R_p g_p(k-1)\big] + \big[C_{1\,p}^T(k-1)D_p + C_{2\,p}^T(k-1)\big] q_p(k-1) + J_p(\mu, k-1) \Big\} \quad (1)$$

is used by the agents (p is the agent index, k is the stage index, q is the control vector, R is the structural coupling between the constituents of the distributed model, g is the control influence, and C_1 and C_2 are the cost vectors formed from the EQUIPMENT and PRODUCT frame classes). For each agent/component of the system, the distributed model of the production process, based on the matrix representation of the decolorised Petri net, has the form:

$$\mu_p(k) = \mu_p(k-1) + D_p \cdot q_p(k-1) + R_p \cdot g_p(k-1) \quad (2)$$

In order to perform the global optimisation in this large-scale system formulation, the principle of *interaction prediction* [13] is used. According to this, the global system optimum is achieved when the predicted interactions $\tilde{g}_p(k)$ are correct, i.e., equal to the real interactions $g_p(k)$ taken from

$$\hat{q}(k) = \left[\hat{q}_p^T(k) \; \vdots \; g_p^T(k)\right]^T \quad (3)$$

where $\hat{q}_p(k)$ is the decision vector that minimizes eq. (1) with the predicted interactions $\tilde{g}_p(k)$. This comparison is done by the supervising agent which, in addition, provides a new prediction (if the interactions have not predicted correctly with the required accuracy). Since the structural couplings between the various parts of the distributed hl-PN model are known, the coordinator has to supply, to every agent, predicted optimal decisions (*q*'s) which are used for the solutions of the local optimisation problems. The predictions are made based on the system state, the previous predictions, predefined strategy heuristic rules, as well as on relations between agents.

Strategy heuristic rules, indirectly, define operators on the plans of the agents in order to resolve conflicts, and thus, to limit the space of possible combined decision alternatives, e.g. "empty buffer to the processing unit with the highest speed of processing for this specific batch size". Constraints are applied directly on the transitions of the models of the corrected agents. For the prediction of interactions, relations are used in two ways: first, they provide operators that update the space of possible predictions based on old values and, secondly, they rate the possible values so that the most suitable are selected first. Both positive and negative relations [10] exist among the plans of the agents. For example, if we assume that there exists a *favor* relation between two transitions of agents P2 and P4, for some specific colours, then the coordinating agent does not have to consider, for both agents, the common alternatives, since one of the two can do the work for the other. Relations are also used when defining the cost vectors of the agents.

This leads to the following two-level regulator-coordinator algorithm which is executed by the control and the supervising agents at the final stage of the recovery planing process. The algorithm terminates when a predefined minimum level of "optimility" has been achieved as demonstrated by an appropriate discrepancy between the values of the predicted the real interactions.

```
SUPERVISOR:
for every stage k until the required marking is reached
  begin
    for each legal state μ(k-1)
      begin
        Obtain enabled transitions
        by interpreting the state of the hl-PN;
        Supply sets of enabled hl-PN transitions to
        subsystem controllers;
        Check strategy heuristics and inter-agent relations;
        Check interaction vector;
        Predict interaction vector from current state
        AND heuristic rules defined by strategy;
  CONTROL AGENT(i):
          parbegin
            Put predictions in the objective function;
            Find firing vector that minimizes it;
            Send optimal value to coordinator;
          parend;
         Find real interaction;
         Compare real and predicted interactions;
         if satisfy required precision exit
         else
         endfor;
endfor;
```

5. CURRENT STATE OF THE PROJECT AND FUTURE PERSPECTIVES

The framework presented in this paper attempts, for the first time, a consideration of the error recovery issue in FPS from a holistic, non-reparative point of view. This is achieved by appropriately structuring and distributing the information available in the shop-floor, as well as, by distributing the additional intelligence required for the recovery task among the monitoring devices of the production system. Although most of the features of the multi-agent system have been developed, the system is still at the implementation and incremental testing phase. The frame-based modelling of a medium size FMS has been implemented using Borland C++ over a network of PCs [14]. Inference mechanisms over frame structures have also been implemented. C++ provides a more flexible platform for both the implementation of the agents communication functions and the numerical optimisation routines embedded into the agents. Furthermore, it is a necessity for real-time speeds of operation. Simple agents with static, cooperation protocols have been implemented and tested. The bidding and negotiation functions, together with the numerical optimisation routines, have also been coded. Development of Petri-net-related functions (dynamic definition, descolorisation, etc.), and integration of the various components represent the final stages towards complete system implementation.

REFERENCES

1 Derwin, D. and Kolondy, H., "Management of Advanced Manufacturing Technology. Strategy, organisation and Innovation", Wiley, 1992.

2 Arnoux, M., Becker, G. and Thomas, M.C. "Un système expert de gestion des aléas en production", CIIAM'86, Proceedings of the 2nd International Conference on Artificial Intelligence, 1986, pp. 15-30.

3 Pau, L.F. "Survey of expert systems for fault detection, test generation and maintenance", Expert Systems, Vol. 3, no. 2, 1986, pp 100-111.

4 Zhou, M.C. and DiCesare, F. "Adaptive Design of Petri Net Controllers for Error Recovery in Automated manufacturing Systems", IEEE Trans. on Systems, Man and Cybernetics, Vol. 19. No.5, 1989, pp.963-973.

5 de Bonneval, A., Courvoisier, M. and Combacau M., "F.M.S. real-time monitoring: Decision making aspects in automatic recovery" in: Doumeingts G., Browne J. and Tomljanovich M. (editors), Computer Applications in Production and Engineering (CAPE '91), Proc. of the Fourth International IFIP TC5 Conference, 1991, pp. 409-416.

6 Paul, C.J., Holloway, L.E., Yan, D., Strosnider, J.K. and Krogh, B.H., "An Intelligent Reactive Monitoring and Scheduling System", IEEE Control Systems Magazine, June 1992, pp. 78-85.

7 De, S. "Representing FMS production planning knowledge in a knowledge-based system", Managerial Decision Support Systems, M.G. Singh, K.S. Hindi, D. Salassa (Editors), Elsevier Science Publishers B.V. (North-Holland), 1988.

8 Fox, M.S. and Smith, S.F. "ISIS - A Knowledge-based System for Factory Scheduling" in: J. Allen, J. Hendler and A. Tate (editors),*Readings in Planning*,, Morgan Kaufmann Publishers, 1990, pp. 336-360.

9 Kamel, A. and Syed, A., "An Object-oriented Multiple Agent Planning System", in: L. Gasser and M. N. Huhns (editors), *Distributed Artificial Intelligence*, Volume II, Pitman Publishers, 1989, pp. 259-290.

10 von Martial, F., "Coordinating Plans of Autonomous Agents", Lecture Notes in Artificial Intelligence, Vol. 610, Springer-Verlag, 1992.

11 Murata, T., "Petri Nets: Properties, Analysis and Applications," Proc. of IEEE, Vol.77, No.4, 1989, pp. 541-580.

12 Smith, R.G., "The Contract Net Protocol: High-Level Communication and Control in a Distributed Problem Solver", IEEE Trans. on Computers, C-29(12), 1980, pp. 1104-1113.

13 Mesarovic, M.D., Macko, D. and Takahara, Y., "Theory of Hierarchical Multilevel Systems", Academic Press, 1970.

14 Oeqvist, P., "Frame-based Modelling and Control of Discrete-Event Dynamic Systems", Diploma Dissertation, Institut d'Automatique, EPFL, 1991.

Knowledge Based Hybrid Systems (B-11)
I. Mezgár and P. Bertók (Editors)
Elsevier Science Publishers B.V. (North–Holland)

Monitoring of the milling process through neural network and fuzzy techniques

S. Markos[a], L. Monostori[b], J. Nacsa[b], G. Szöllősi[a]

[a]Manufacturing Technology, Technical University Budapest
Egry J. u. 1, Budapest, H-1111 Hungary

[b]Computer and Automation Institute, Hungarian Academy of Sciences
Kende u. 13-17, Budapest, H-1502 Hungary,

Abstract

Real-time nature, uncertainty handling and learning ability are essential features of knowledge representation and processing techniques to be applied at lower levels of intelligent manufacturing systems. The paper demonstrates, that artificial neural networks and fuzzy systems are possibly approaches to comply with these requirements. Their applicability for modelling and monitoring of manufacturing processes is illustrated and compared. At the end of the paper the combined use of the neural and fuzzy techniques is suggested to produce adaptive fuzzy systems, and the integration of such hybrid system in an intelligent manufacturing environment is investigated.

Keyword Codes: I.2.1; I.2.6; I.5.4
Keywords: Artificial Intelligence, Applications and Expert Systems, Learning; Pattern Recognition, Applications

1. INTRODUCTION

Four main evolution phases of manufacturing systems can be distinguished [12]:

- *direct numerical control (DNC)* of groups of machine tools (in the late sixties),
- *flexible manufacturing systems (FMS)*, where machines are equipped with automatic workpiece and tool exchangers, and some kind of on-line computer based scheduling is incorporated (starting in the seventies, and still spreading),
- *computer integrated manufacturing (CIM)*, which can be characterized by the synthesis of CAD, CAPP and CAM techniques (in breakthrough from experiment into industrial practice),
- *Intelligent manufacturing systems (IMS)*, capable of solving, within certain limits, unprecedented, unforeseen situations and problems on the basis even of incomplete and imprecise information.

The complexity of highly integrated manufacturing systems is beyond a certain threshold limit, and the enhanced requirements can hardly be fulfilled with conventional methods, new - mostly artificial intelligence (AI) - techniques are needed [20].

In the paper three important requirements for knowledge representation and processing techniques to be applied at lower levels of intelligent manufacturing systems are outlined, i.e. *real-time nature, uncertainty handling* and *learning ability*.

Artificial neural networks (ANNs) and *fuzzy systems* are two approaches of artificial intelligence which seems to comply with these requirements.

The fundamental aim of the paper is to demonstrate and compare the applicability of neural network and fuzzy techniques for modelling and monitoring of manufacturing processes, especially of milling.

At the end of the paper the combined use of the neural and fuzzy techniques is suggested to produce *adaptive fuzzy systems*, and the integration of such hybrid system in an intelligent manufacturing environment is investigated.

2. CUTTING EXPERIMENTS, PREPROCESSING OF SIGNALS

Many efforts have been made to develop reliable and effective sensing technologies, which guarantee the reliability of manufacturing and product quality, as it was surveyed e.g. in [21]. In *direct tools wear sensing* the actual progress of the physical wear while cutting is being sensed on-line by radioactive, pneumatic, electrical resistance, optical and other sensing techniques. None of them proved to be reliable and rugged enough.

Consequently, *indirect methods* are usually used, which rely on monitoring other signals (tool temperature, cutting force, spindle torque, spindle power, rotational speed of the spindle, mechanical vibration, sound, acoustic emission, etc.), which are expected to be fairly correlated to the tool condition.

However, all the efforts to find a unique indirect signal, which is applicable from different (technical, economic) points of view and reliable over a range of operating conditions, failed. Therefore *sensor fusion or integration* is a feasible approach for modelling and monitoring of manufacturing processes [9].

Milling is considered as one of the most complicated intermittent cutting operation. The difficulties arise from the fundamental nature of milling, i.e. cutting edges of the cutter move in and out of cut while tool is rotating and chip thickness varies through the immersion angle for each cutting edge. Moreover, in multi-edge cutting more then one cutting edge may be active at the same time. Process dynamics is affected by the intermittent interaction of the multi-tooth tool and workpiece. During modelling and monitoring of the process, varying cutting parameters and workpiece geometry must be also considered.

Milling experiments were carried out on a VF 22 vertical knee type milling machine at the Technical University of Budapest. Six-tooth 125 mm diameter WIDAX M40 cutter with ISG Böhlerit TPAN 1603 PPN R121 inserts (without coating) was used.

In the *face milling investigations* structural steel (C45 - AISI 1045- Rm=770 N/mm^2) was machined under varying operating conditions /cutting speed (v = 96 - 239 m/min), tooth feed (s_i= 0.018 - 0.22 mm), axial depth of cut (a = 0.9 - 4 mm)/ and different stages of tool wear. Average wear of teeth was determined by measuring the flank wear on each of the cutting edges by tool room microscope.

Prismatic (width 92mm, length 500mm) test workpieces were mounted on a KISTLER 9281B board (supplement table), which generated electrical signals proportional to cutting force components (F_x, F_y, F_z). Vertical component of the vibration signal (R_z) was also measured using accelerometer fixed on the KISTLER board (Fig. 1).

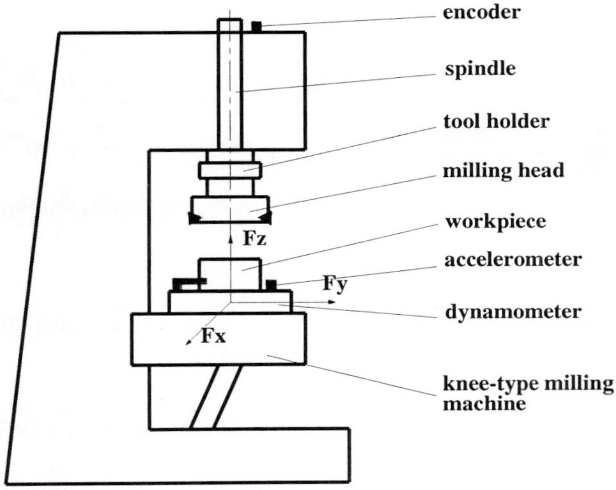

Figure 1. Experimental setup

The signals were first passed through a KISTLER 9863A charge amplifier, and then fed into a GOULD 1604 digital oscilloscope for digitalization. They were sampled with 2 kHz (Fig. 2). Start of sampling was synchronized to the rotational position of the spindle by an encoder mounted on the machine's spindle. Through an IEEE 488 GPIB interface system, the digitized data were conveyed to a 386 PC/AT with mathematical coprocessor for acquisition and further processing.

Numerous (100 - 150) statistical (e.g. mean value, quadratic mean value, variance, standard deviation, effective value, third and forth central moments, skewness, and excess, etc.) and spectral features (e.g. absolute and relative power in frequency bands, frequency and power of the largest peak, normalized moment of the spectrum, and normalized central moment of the spectrum, etc.) were computed from the measured signals.

Because of the great number of considered features and in order to enhance the learning, estimation and classification performances of the monitoring process, SFS (sequential forward search) *feature selection* was accomplished in all phases of investigations, using the criterion

$$J = \frac{trace(S_b)}{trace(S_w)} \tag{1}$$

where S_b is the between class and S_w is the within class scatter matrix respectively [8].

Through this feature selection the relative importance of measured signals in various modelling and classification assignments could be also evaluated.

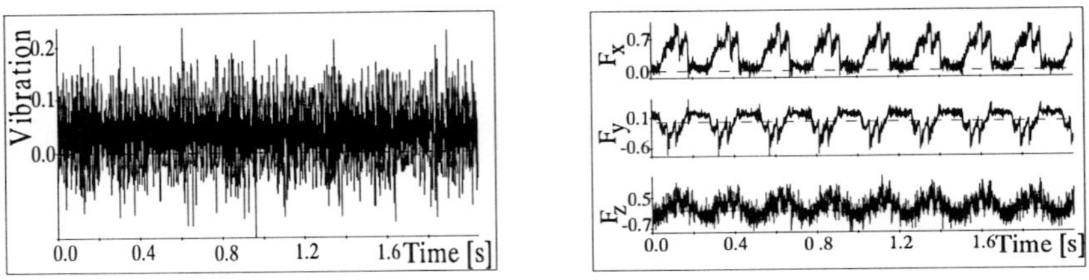

Figure 2. Mechanical vibration (left) and force components (right) during milling (normalized values)

3. APPLICATION OF THE ANN TECHNIQUE FOR MODELLING AND MONITORING OF THE MILLING PROCESS

In the last few years *artificial neural networks (ANNs)* or *connectionist systems* came into the limelight and divided to some extent the AI society [7,11,13,18,29]. As a subsymbolic approach, connectionism has some true benefits over symbolicism exactly in those fields (real-time feature, quantitiveness, uncertainty handling, effective learning ability), which were considered as fundamental requirements for intelligent manufacturing.

For multisensor integration, process modelling and state classification the *back propagation (BP)* technique [29] was applied, which being the most frequently used ANN model usually serves as basis of comparisons with other approaches. Some accelerations of the BP procedure were used, in order to get reasonable training times [23]. Three layer networks were investigated with varying number of input (pattern features) and hidden elements.

3.1 Inverse modelling of the milling process

The usual way to apply pattern recognition or neural network technique in monitoring of manufacturing processes is to use *signal features as inputs and wear parameters of the tool as targets*. Mostly the state of the wear is ranged in some, usually two (sharp, worn) intervals. The published results refer generally to one combination of process parameters, and the investigations show, that these models have limited validity over a broader range of process parameters.

The *incorporation of process parameters in the model inputs* can theoretically overcome the above difficulties [26,4]. However, a reliable model requires a great number of experiments with different combinations of process parameters, and what the main problem constitutes, with a great number of wear states.

The approach to be demonstrated here for milling relies on the *modelling of the normal operation*, i.e. cutting with sharp cutting tools [25,26]. In contrast to the more common solutions, where process parameters act as inputs and features of measured signals as outputs, *inverse models* are used, that is, signal features serve as input and the process parameter to be modelled acts as output. In order to assure applicability over a broader range of process parameters, model inputs are supplemented with the other process parameters not used as target elements. As an example, the left side of Fig. 3 illustrates this approach for the case of modelling of the axial depth of cut (a). Similar structures can be used for the modelling of other process parameters, e.g. cutting speed (v) or tooth feed (s_i).

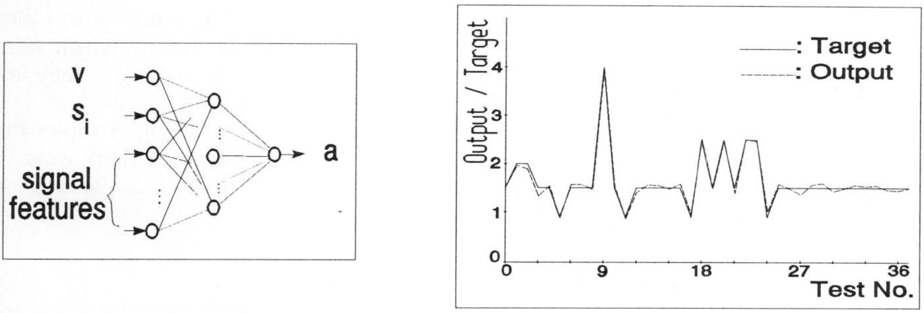

Figure 3. Structure of the networks for axial depth of cut modelling and performance of such a BP network with 14*4*1 structure, sharp milling tools, varying cutting conditions (v, s_i, a), target values and network outputs for training patterns [mm]

During the investigations, inverse models of the milling process, i.e. separate models for three process parameters (axial depth of cut, cutting speed, and tooth feed) were generated. Having computed 112 signal features, they were supplemented by the values of the two remaining process parameters, i.e. those which were not modelled (Fig. 3). In each variation of process parameters, feature selection was accomplished using the SFS procedure with criterion (1), with the restriction that the best signal features were searched for in addition to the two process parameters serving as inputs in the actual modelling assignment.

For the actual two process parameters serving as inputs, the best 6, 12, 20 signal features were selected. Consequently, BP networks with 8*x*1, 14*x*1 and 22*x*1 structures were applied. The number of hidden elements (x) was varied between 2 and 16 (2, 4, 6, 10, 16). The patterns composing the training sets were selected in such a way as to cover a relatively large range of process parameter combinations. The networks using force and vibration features as well gave very good estimation of process parameters in investigations where all the three incorporated process parameters were changed.

In the following, some of the results of axial depth of cut modelling will be described. The other modelling assignments resulted in similar performance. It must be emphasized that these investigations were accomplished with tools, which can be regarded as sharp, with average flank wear equal to or less than 0.1 mm.

The right side of Fig. 3 shows the estimation results of this network for training patterns. The estimation with test patterns is certainly worse, but at the same time surprisingly good, with a maximum relative deviation of 38.8% (left side of Fig. 4) and with an average relative error of 8.2%. It must be mentioned that the maximum relative error appears at such combinations of process parameters, which were hardly represented in the training set.

3.2 Application of inverse models for tool wear monitoring

The applicability of process parameter models for tool wear monitoring was also investigated in the experiments. The main question was: how do models, which were trained with patterns generated during machining with sharp tools, behave if they are applied for patterns gained from machining with different *worn tools*.

Remaining at the axial depth of cut modelling, the right side of Fig. 4 shows the actual and estimated axial depth of cut values for tools with an average flank wear of about 0.2, 0.3 and 0.4 mm respectively containing the values of 8 tools, 3, 2, 3 tools from the above wear classes in sequence.

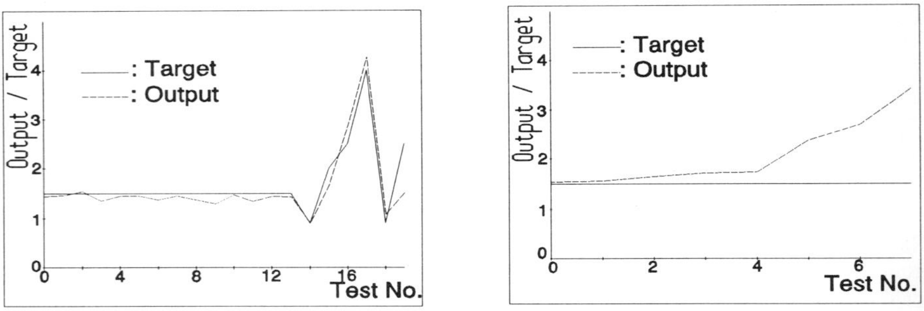

Figure 4. Axial depth of cut estimation with a 14*4*1 BP network, varying cutting conditions (v, s_i, a), target values and network outputs for test patterns [mm], sharp (left) and worn (right) milling tools

Estimation results of the network of Fig. 3 for these patterns is considerably worse, than for test patterns computed from signals generated by cutting with sharp tools (left side of Fig. 4). The deviation between the actual and estimated axial depth of cut values grows with increasing wear and reaches a maximum relative deviation of 127.8% with an average relative error of 38.7% (right side of Fig. 4). These deviations (average flank wear of 0.3, 0.4 mm) significantly surpass the deviations of test patterns characterizing sharp tools.

Modelling other process parameters (cutting speed, tooth feed), similar phenomena were experienced. The results indicate that there is a definite hope of success in applying inverse models of the cutting process for tool wear detection.

4. APPLICATION OF FUZZY TECHNIQUE

Fuzzy systems [30,27,31] lie in some sense between traditional AI systems using structured knowledge representation in symbolic manner and neural networks which cannot directly encode structured knowledge. Fuzzy systems also estimate sampled functions from input to output. They may use linguistic (symbolic) or numeric samples. The fuzzy approach combines the purely numerical approaches of neural networks and mathematical modelling with the symbolic, structure-rich approaches of AI.

Fuzzy and AI systems are often compared: the latter deals with the deep knowledge of a narrow field which is exact, quantitative, precise, no contradicting, while the former represents the wide knowledge in the real world which is empirical, qualitative, general, "fuzzy", etc.

Fuzzy systems directly encode structured knowledge but in a numerical framework. Fuzzy associations between fuzzy properties (e.g. HIGH, SEVERE) can be entered in a *FAM-rule matrix* (FAM: fuzzy associative memory), Fig. 5. The system maps input fuzzy sets A to output fuzzy sets B. Separate FAM rules are stored separately. For each input these FAM rules are parallel fired. In the fundamental case the w weights are set to 1 or 0. Output fuzzy sets are usually defuzzified (are converted into single scalar outputs) as illustrated on Fig. 5.

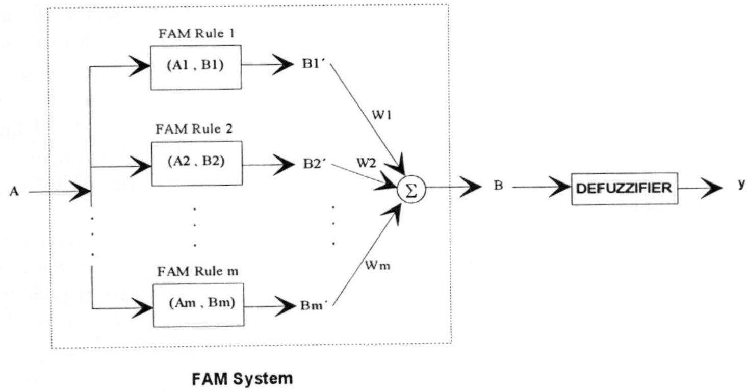

Figure 5. Fuzzy system architecture [15]

4.1 Fuzzy recognition of tool states

Fuzzy technique was used for wear state classification and for detection of serious changes in cutting depth. The developed fuzzy logic software acts as an inference engine, the input parameters are some calculated features of the force and vibration signals while the output parameters describe possible diagnoses. The FAM-rule matrix is made up by rules of the following type:

IF FEATURE1 IS fuzzy_property
AND FEATURE2 IS fuzzy_property
...
THEN
THE FIRST DIAGNOSIS IS DIAGNOSIS1
(WITH fuzzy_property)
AND THE SECOND DIAGNOSIS IS DIAGNOSIS2
(WITH fuzzy_property)
...

Here FEATURE1,... and DIAGNOSIS1,... are given features and events/pieces of information accordingly. The possible "fuzzy properties" together with the membership functions are also given in advance.

The inference process works with a few min-max operations instead of searching through the rules. The efficiency does not depends on the number of rules this way.

The number of rules is about 30, they describe sufficiently the cutting circumstances and the possible events. The developed system showed good classification performance.

5. HYBRID USE OF NEURAL AND FUZZY APPROACHES; NEURO-FUZZY SYSTEMS

Beside learning ability, model-free estimation is, after all, the central computational advantage of neural networks. The cost is system inscrutability. ANN is limited to unstructured parameter and state estimation, pattern recognition, and cluster formation.

Neural and fuzzy systems encode sampled information in a parallel-distributed framework. Both frameworks are numerical. In contrast to the fuzzy approach, knowledge representation in neural networks is incomprehensive for users. Using fuzzy systems, the user can determine which FAM rules contributed how much membership activation to a concluded output. FAM-structured knowledge can be added or deleted without disturbing stored knowledge.

Fuzzy systems store and process FAM rules in parallel. Mathematically they associate output fuzzy sets with input fuzzy sets, and so behave as associative memories.

Earlier investigations confirmed [2,24], that - similarly to our present conception on biological systems - ANN techniques seem to be the most viable solution for the lower level of intelligent, hierarchical control and monitoring systems. Since the higher levels of the control/monitoring hierarchy require mostly symbolic knowledge representation and processing, the integration of symbolic and subsymbolic methods can be predicted.

Fuzzy systems are considered to be natural link between symbolic and subsymbolic approaches. On the one hand they can work in real-time circumstances and handle uncertainties as artificial neural networks, on the other hand they can manage both symbolic and numerical information. However, fuzzy systems usually do not incorporate automatic learning ability.

It seems, best performance can prospectively be reached by combining neural and fuzzy approaches, integrating their benefits. The resulted *neuro-fuzzy system*, is a *hybrid system*, where the system architecture remains fuzzy, but using neural training techniques, it can be trained automatically.

Training procedures originating from the ANN world, can be used in two levels. First, unsupervised, competitive technique [29] can generate FAM rules from training data. At this stage, the input-output space is clustered with neural network techniques, and each cluster estimates a FAM rule. The experts need not articulate these rules, only generate input-output data for learning. Secondly, further training procedure (e.g. back propagation) modifies the w weights in the system (Fig. 5).

Therefore, neuro-fuzzy systems comply with the requirements enumerated in the introduction, i.e. *real-time nature, uncertainty handling and learning ability*, combining the purely numerical approaches of neural networks and mathematical modelling with the symbolic, structure-rich approaches of AI. Consequently, they can provide the sufficient framework for the solution of numerous problems in intelligent manufacturing, and can contribute the future realization of intelligent manufacturing systems.

6. CONCLUSIONS

The fundamental aim of the paper was to focus some requirements of knowledge representation and processing techniques to be used in intelligent manufacturing. Artificial neural networks and fuzzy systems seem to be able to comply with these requirements. Their applicability was demonstrated and compared on modelling and monitoring of the milling process which belongs through its intermittent nature to the most complicated cutting operation. It can be established that

- ANN models can be advantageously used for developing various monitoring strategies,
- applying process parameters as inputs in addition to signal features, models can be produced which work over a great variety of operating conditions,
- with inverse modelling techniques, parameters can be individually modelled with a high degree of reliability,
  - applying other process parameters as inputs, models can be produced which work over a great variety of cutting conditions,
  - these models show increasing deviations between actual and estimated process parameters with growing tool wear, indicating their applicability for tool wear detection,

The outlined approach of inverse modelling requires the modelling of the normal operation, i.e. cutting with sharp tools. The models work over a broad range of process parameters. These features enable their on-line use during machining and let the networks adapt themselves to actual machining conditions (machine tool, material and cutting tool).

An experimental fuzzy system was developed and applied for wear state classification and for detection of serious changes in cutting depth.

At the end of the paper the combined use of the neural and fuzzy techniques was suggested to produce *adaptive fuzzy systems*, and the integration of such hybrid system in an intelligent manufacturing environment was investigated.

Further research was initiated to develop of such hybrid systems and to circumscribe more precisely their application fields within intelligent manufacturing.

7. ACKNOWLEDGEMENTS

The work was partially supported by the National Research Foundation, Hungary, Grant No.: 2626. The cutting experiments were accomplished at the Technical University of Budapest. The analysis of the measured signals was partly performed in the Institute of Electrical Measurement, University of Paderborn, FRG. Special thank to the *Humboldt- Foundation* and *Prof. D. Barschdorff* head of the Institute of Electrical Measurement, Paderborn, allowing L. Monostori to carry out scientific research work in Germany.

8. REFERENCES

1. Barschdorff, D.; Monostori, L.: Neural networks, their applications and perspectives in the intelligent machining, Computers in Industry, Elsevier, Spec. Issue on Learning in Intelligent Manufacturing Systems, 1991, pp. 101-119.
2. Barschdorff, D. and Monostori, L.: Symbolicism and connectionism: rivals or allies in intelligent manufacturing?, Proc. of The 23rd CIRP International Seminar on Manufacturing Systems, 6-7 June, 1991, Nancy, France, Section 4, pp. 14-28.
3. Barschdorff, D.; Monostori, L.; Ndenge, A.F.; Wöstenkühler, G.W.: Multiprocessor systems for connectionist diagnosis of technical processes, Computers in Industry, Elsevier, Spec. Issue on Learning in Intelligent Manufacturing Systems, 1991, pp. 131-145.
4. Barschdorff, D.; Monostori, L.; Kottenstede, T.; Warnecke, G.; Müller, M.: Cutting tool monitoring in turning under varying cutting conditions via artificial neural networks, The sixth International Conference on Industrial & Engineering Applications of Artificial Intelligence & Expert Systems, June 1-4, 1993, Edinburgh, Scotland (in print)
5. Chryssolouris, G and Domroese, M.: Sensor integration for tool wear estimation in machining, Proc. of the The Winter Annual Meeting of The ASME on Sensors and Controls for Manufacturing, Chicago, Illinois, Nov. 27 - Dec. 2, 1988, pp.115-123.
6. Chryssolouris, G. and Wright, K.: Knowledge-based systems in manufacturing, Annals of the CIRP, Vol. 35/2/1986, pp.437-440.
7. DARPA neural network study, AFCEA International Press, 1988

8. Devijver, P.A. and Kittler, J.: Pattern recognition - a statistical approach, Prentice Hall, New Jersey, 1982
9. Dornfeld, D.A.: Neural network sensor fusion for tool condition monitoring, Annals of the CIRP, Vol. 39/1/1990, pp. 101-105.
10. Dornfeld, D.A.: Unconventional sensors and signal conditioning for automatic supervision, AC'90, III. CIRP International Conference on Automatic Supervision, Monitoring and Adaptive Control in Manufacturing, 3-5. Sept, 1990, Rydzyna, Poland, pp. 197-233.
11. Fahlman, S.E. and Hinton, G.E.: Connectionist architectures for artificial intelligence, IEEE Computer, Jan. 1987, pp. 100-108.
12. Hatvany, J.: The efficient use of deficient information, Annals of the CIRP, Vol. 32/1/1983, pp. 423-425.
13. Huang, W. Y., Lippmann, R. P.: Neural net and traditional classifiers, in: Conference on Neural Information Processing Systems - Natural and Synthetic, IEEE, Nov. 1987.
14. Isermann, R.: Process fault detection based on modeling and estimation methods, Automatica, Vol. 20, 1984, pp. 387-404.
15. Kosko, B.: Neural networks and fuzzy systems, A dynamical systems approach to machine intelligence, Prentice Hall, 1992.
16. Markos, S., Raifu A. W., Szalay, T. and Szöllősi, G.: Monitoring face milling using the fuzzy logic technique, Joint Hungarian-Japanese Symp. on Fuzzy Systems and Applications, 1991, Budapest, Hungary, pp. 114-117.
17. Markos, S., Szalay, T., Szöllősi, G., Raifu, A. W.: Tool and process monitoring of milling operation, 2nd DAAAM Symposium, 1991, pp. 119-221.
18. Minsky, M. and Papert, S.: Percetrons: an introduction computational geometry, MIT Press, Cambridge, 1969.
19. Monostori, L.: Learning procedures in machine tool monitoring, Computers in Industry, Vol. 7., 1986., pp. 53-64.
20. Monostori, L.: Intelligence in machine tool monitoring and diagnostics, Proc. of the 11th Triennial World Congress of the International Measurement Confederation, Plenaries, Invited papers, Computers and Intelligent Systems, Houston, Texas, 16-21 October 1988, pp. 267-273.
21. Monostori, L.: Signal processing and decision making in machine tool monitoring systems, Proc. of the Manufacturing International'88 Symposium, Vol. I., Product and Process Design, Atlanta, Georgia, April 17-20, 1988, pp. 277-284., and: The Journal of Condition Monitoring, Vol. 3, No. 1, 1989, pp. 1-20.
22. Monostori, L. and Nacsa, J.: On the application of neural nets in real-time monitoring of machining processes, Preprints of the 22nd CIRP International Seminar on Manufacturing Systems, 11-12, June, 1990, Enschede, the Netherlands, pp. 6A/15-27.
23. Monostori, L., Bothe, A.: Convergence behaviour of connectionist models in large scale diagnostic problems, in Belli, F. and Rademacher, F.J.: Industrial & Engineering Applications of Artificial Intelligence and Expert Systems, Proc. of the 5th International Conference, IEA/AIE-92, Paderborn, Germany, Lecture Notes in Artificial Intelligence, 604, Springer, 1992, pp. 113-122.

24. Monostori, L. and Barschdorff, D.: Artificial neural networks in intelligent manufacturing, Robotics and Computer-Integrated Manufacturing, Pergamon Press, Vol. 9, No. 6, 1992, pp. 421-437.
25. Monostori, L.; Markos, S.; Krämer, S.; Nacsa, J.; Szöllősi, G.: Neural networks for modelling and diagnosis of manufacturing processes; Milling, Proc. of the 8th International IMEKO Symposium on Technical Diagnostics, Sept. 23-25, 1992, Dresden, Germany, pp. 502-512.
26. Monostori, L.: Intelligent manufacturing through neural networks; Monitoring of machining processes, Proc. of the IAS-3, International Conference on Intelligent Autonomous Systems, February 15-19, 1993, Pittsburg PA, USA, IOS Press, Washington, Oxford, Amsterdam, Tokyo, pp. 328-337.
27. Negoita, C.V.: Expert systems and Fuzzy systems, The Benjamin/Cummings Publishing Company, Inc., Menlo Park, California, 1985.
28. Rangwala, S. S. and Dornfeld, D. A.: Learning and optimatization of machining operations using computing abilities of neural networks, IEEE Transactions on Systems, Man, and Cybernetics, vol. 19, No. 2, March/April, 1989, pp. 299-314.
29. Rummelhart, D. E. and McClelland, J. L.: Parallel distributed processing, MIT Press, Cambridge, 1986
30. Zadeh, L.A.: Fuzzy sets, Information and Control, Vol. 8, 1965, pp. 338-353.
31. Zimmermann, H.J.: Fuzzy sets, decision making, and expert systems, Kluwer Academic Publishers, Boston, Dordrecht, Lanchester, 1987.

Knowledge Based Hybrid Systems (B-11)
I. Mezgár and P. Bertók (Editors)
Elsevier Science Publishers B.V. (North–Holland)

Maintenance knowledge base design methodology and building of an Integrated Preventive MAintenance System (IPMAS).

D. MARTEEL - D. MENEXIADIS - R. SOENEN

Laboratoire de Génie Industriel et Logiciel UA n°1118, Université de Valenciennes et du Hainaut-Cambresis, BP 311, Le Mont houy, 59304 VALENCIENNES CEDEX, FRANCE

Abstract

The aim of the paper is to present an integrated preventive maintenance system, used to assist the engineer in the supervision of machines without shutting them down.

In the IPMAS system, a module deals with the automatic acquisition of real measurements and the transfer of these to a knowledge based system. This computerized maintenance system allows the modeling of the machines' working states. The acquisition module allows the analysis of phenomenon, coming from real measurements. Thereby IPMAS is able to analyse a large scale of disturbances in a realistic environement.

To obtain this system, we have defined a methodology based on an extension of Failure Modes, Effects and Criticality Analysis (FMECA) that has been called Preventive Maintenance FMECA (PM FMECA). This methodology directly provides the concepts used to build the knowledge base of the maintenance system. This approach is helpful to build and to structure this knowledge base .

This rigourous and exhaustive methodology was used to build the generic system IPMAS. This one is consecrated to the resolution of preventive maintenance problems and could be applied to a large scale of machines.

Keyword Codes : D.1.5; I.2.1; I.2.4
Keywords : Object-oriented Programming; Applications and Expert Systems; Knowledge Representation Formalisms and Methods.

INTRODUCTION

The present economic context demands a growing availibility of the production system. To face to these demands, a certain number of methods and tools allow to keep track of the state of machines, without production breakdowns. One of the most efficient methods is the vibration analysis [1].

The different methods of preventive maintenance need specific knowledge and cover various fields as signal treatment, vibration analysis associated with knowledge of working conditions, [2] etc... To face this treatment variety, we have developed a tool for the maintenance of the production systems (fig.1), the Integrated Preventive MAintenance System intended to be used more particularly in process type industries (refinery, chemical, steel industries, ...).

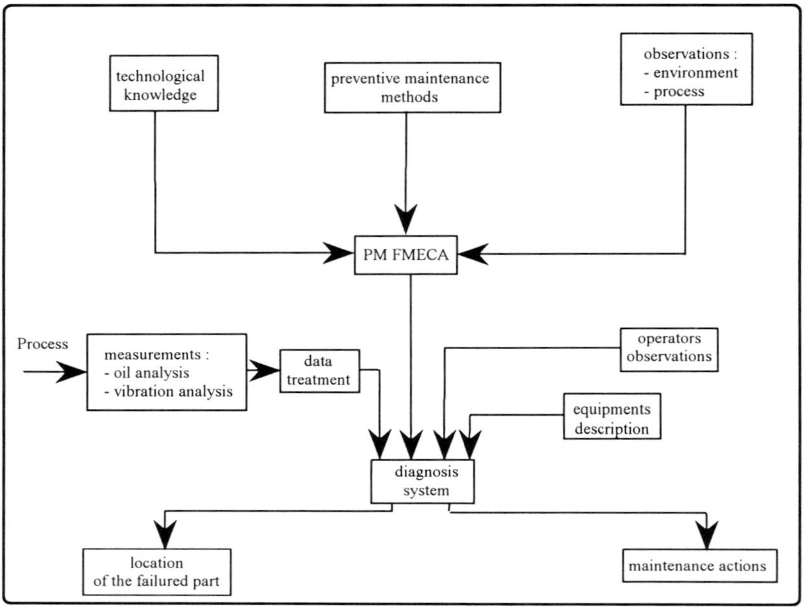

fig.1 Integrated Preventive MAintenance System

1. MEASUREMENT ACQUISITION

The first function of the system allows the automatic acquisition of real measurements, then the transfer of these to the knowledge based system. The measurement is firstly stored in an analyser (FTT) memory.

The vibration measurement (fig.2) represents a modeling of the functioning's state of machines, and constitutes one of the main concepts which are used by the maintenance system. With these spectrum, we put to highlight some failures, as for example, a bearing wear or a gearing problem [3], [4] ...

The measurement acquisition environment is build around three treatment modules :

- measurements acquisition module,
- measurements storage module,
- data formatting module.

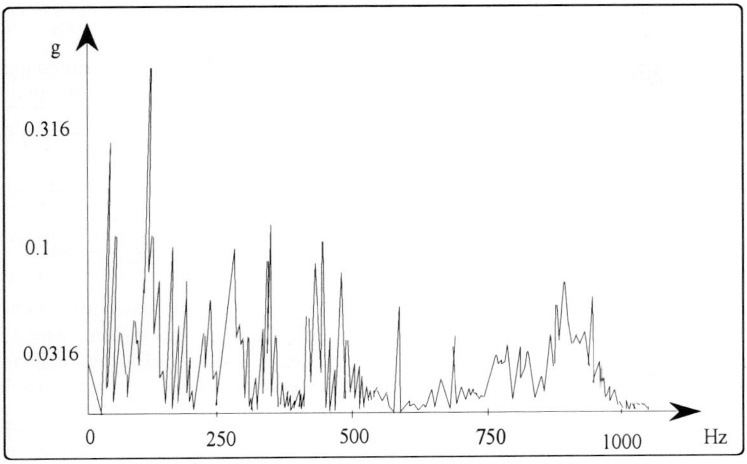

fig.2 : vibration spectrum

2 .KNOWDELGE BASE BUILDING METHODOLOGY

An essential task, within the context of the building of a maintenance system concerns the knowledge modelization. It's a matter of modeling the deductions that an expert uses in his diagnosis process, as well as the whole of the knowledge over which deductions have been based [5].

One of the major difficulties is to build and to fill up the knowledge base of the system. There is no systematic method to collect and structure this knowledge. Generally speaking they have to be extracted one by one for each situation. The knowledge base building represents then a work which is long and tedious. Moreover the knowledge base is not always built in a generic sense; so this lack of methodology implies the creation of a new knowledge base for any new application.

Preventive Maintenance FMECA (PM FMECA)

To generate the knowlegde base, we have finalized a method stemming from FMECA (Failures Modes Effects and Criticality Analysis) [6].

The FMECA method allows to list all the potential failure modes, the expected causes, the eventual means to detect them [7]. It holds some knowledge which are directly helpful for the maintenance diagnosis, while bringing a structural and functional view of the system.

The Preventive Maintenance FMECA contains the classical FMECA that is widened by some specific notions of the preventive maintenance and a set of useful concepts for the knowledge base system such as the testing notion or the priority notion in the deductive reasonings (fig .3).

PREVENTIVE MAINTENANCE FMECA

Classical FMECA	criticality	testability	certainty	strategy	- - - - - - -	maintenance actions

fig.3 : Preventive Maintenance FMECA

The aim of the PM FMECA is to connect the symptoms of a failure which are perceived as a deterioration of the function of the system, and the component or sub-system, which is at the origin of the failure. Then we take into account all the necessary resources and actions to undertake after the first symptom of failure appears (interventions, specific set of tools ...), so we define a *maintenance model for each component.*

Advantages of the PM FMECA

Besides the advantage of saving time for the knowledge base creation, the use of the PM FMECA has the benefit of sytematizing the knowledge acquisition and taking into account the availability notion by the integration of the maintenance resources. Moreover it formalizes the approach of the experts and give some help in the realization of tests and criteria for the strategy's diagnosis.

The combination of the failure modes, effects and criticality analysis and the description of one or more preventive maintenance methods and some appropriate concepts based on the expert system approach, allows to perform some maintenance actions in order to avoid the failure effects.

First of all the PM FMECA has been created in order to build the knowledge base of the IPMAS system taking into account all the necessary parameters of expert evaluation. Moreover PM FMECA thanks to its structure could be applied for all preventive analysis methods and for every type of machine.

This rigourous, and exhaustive technique allowed us to build the IPMAS system that we present in the next paragraph.

3. BUILDING OF THE INTEGRATED PREVENTIVE MAINTENANCE SYSTEM

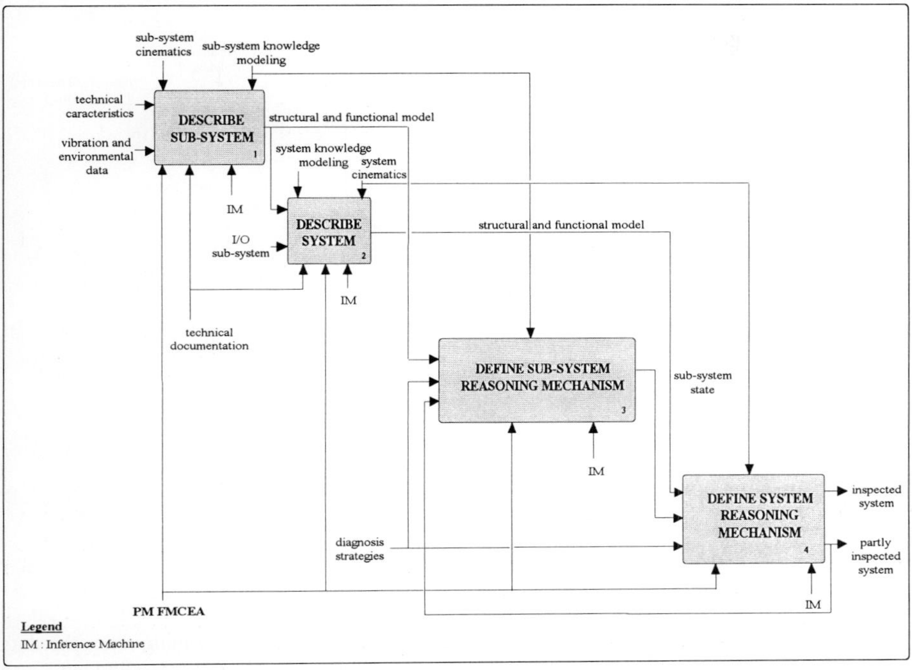

fig.4 : functional analysis of IPMAS

The evaluation of the condition of one industrial system starts on observing its behaviour in a given context, and uses, among others its structural and fonctional description, the Integrated Preventive MAintenance System conception respects this principle[8].
The system is composed of different modules. On the one hand these one turn on a structural and fonctional description of the installation and on the other hand on the reasoning mechanism (fig.4).

3.1 Components description (fig.4: "to describe sub-system")

The first stage consists in describing the components structure of the system (fig.5: "to describe sub-system structure"), thanks to an object approach. The data organisation allows to divide the system, for example, an electric engine in several tens of components according to an object hierarchical structure.
Then we describe the components behaviour with regard to the environment. Different preventive analysis methods (oil analysis, thermography, ...) allow to determine the state and the behaviour of a part, this can notice and modelize the wear of this part (fig.5 : "to describe sub-system functioning").

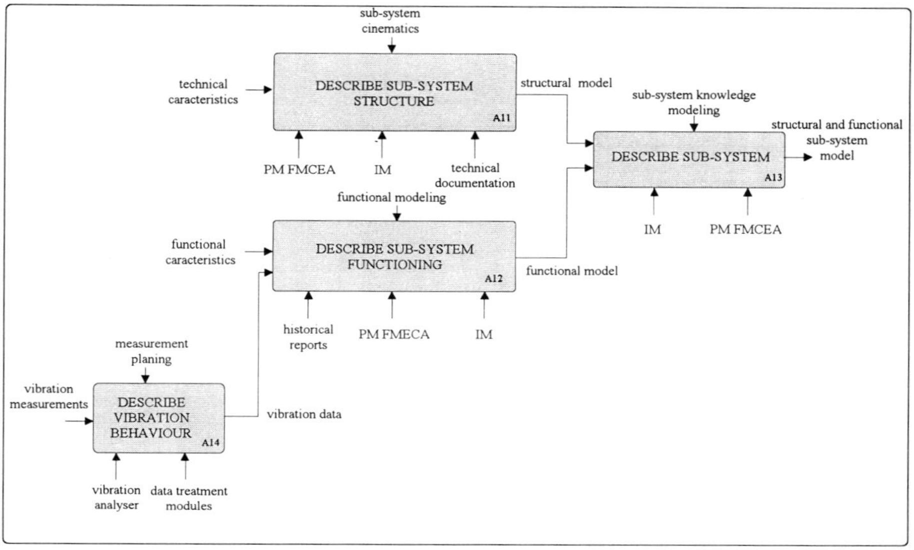

fig.5 : components description

The vibration behaviour description is performed thanks to the data coming from the data acquisition system, and relies on some vibratory and environmental concepts allowing us to modelize the state of the machine supervised.

Now we specify the type of vibration information communicated to the system. Only the frequency-amplitude couples corresponding to a measurement of a vibratory peak are dealt by the acquisition module, then the knowledge based system selects the couples useful for the reasoning.

This extraction process of information allows to structure our data of this way (fig.6).

To simplify the data extraction, we structure our data files so that one class could represent a spectrum with which we link "n" objects ("n" is varying between 30 to 255, in accordance with the measurement mode). So an object representing a frequency-amplitude couple contains as proprieties an amplitude and a frequency.

The functional description complete a structural description to describe physically the components of a machine. This components description associated to expert knowledge allows us to create standard components libraries.

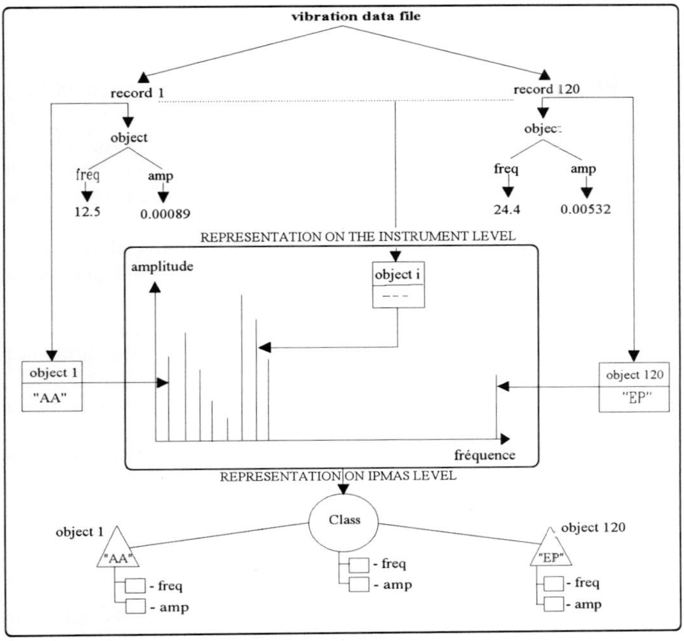

fig.6 : vibration data structure

3.2 Description of the system

The complete description of the installation is obtained by the assembly of components beforehand selected by the user. We also have the possibility to add some new properties to the assembled components.

Concerning the system level the functional description allows us to consider the installation as an entity with some specific properties due to inter-components contacts, that can be the origin of perturbations or vibration phenomenon (shock problem, misalignment,....), [9], [10].

example :

IF vibrations due to shocks in sub-system A

IF vibrations due to shocks in sub-system B to the same vibration frequency that the shocks appeared in sub-system A

IF vibration amplitude of this shocks in sub-system B is less important that in sub-system A

IF the amplitude evolution of B is proportional to A

THEN the vibrations origin in the sub-system B is a consequence of the shocks appeared in sub-system A.

3.3 The reasoning mecanism

We reason at the component level (fig.4 :" to define sub-system reasoning mechanism"). The reasoning linked with each component is based on the description presented before. It allows us to define the state of the components considering not only vibration concepts but also some other maintenance preventive techniques (flow analysis, infra red thermography, ...).

The reasoning at the system level (fig.4 : "to define system reasoning mecanism") is directly based on the reasoning made at the elementary components level and performs thus a complete inspection of the installation.

The construction of the knowledge base with the PM FMECA allows us to establish a correlation between the constraints of the maintenance method used and the conditions of production of the machine inspected.

Several campaigns of measurements have been necessary to evaluate and to analyse the quality of the expert rules which permitted us to locate, and quantify the different failures of an installation.

The realization of various test-cases have shown that IPMAS system is able to locate the failures thanks to vibation signal combined with the observation of the parameters linked to the environment. The gain to the knowledge base with PM FMECA will be submitted to a more detailed analysis of the concepts linked with the new preventive maintenance techniques.

PERSPECTIVES AND CONCLUSION

The search of reliability and availability for production equipments led us to realize the IPMAS system based on the PM FMECA methodolody in order to generate a specific maintenance knowledge base, including the principles of the expert systems approach.

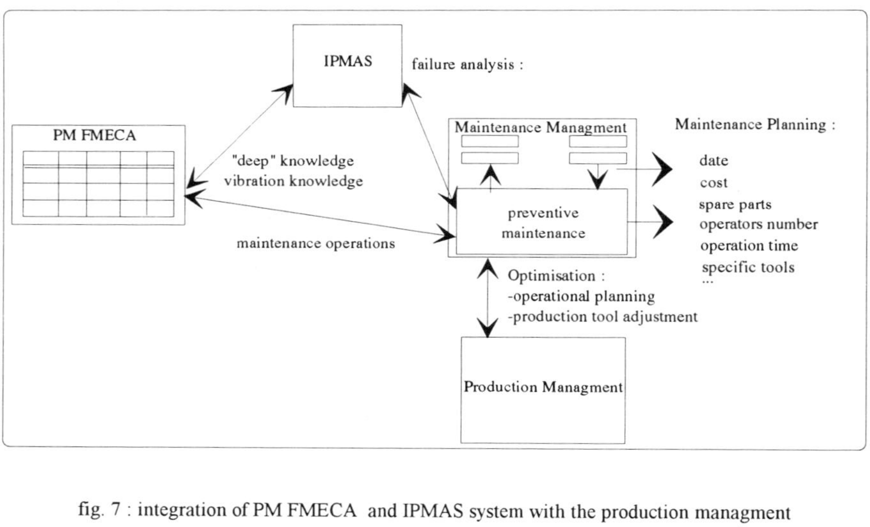

fig. 7 : integration of PM FMECA and IPMAS system with the production managment

The integration of the PM FMECA, and the IPMAS in a more global maintenance system (fig.7) will allow an improvement of the production tool availability. This new approach will help us to ameliorate the maintenance planning and therefore to optimise the human and material resources management [11], [12].

In the field of maintenance operations planning and management the integration of the preventive maintenance and diagnosis with a maintenance management information system could allow a better management of working sheets, budget, costs and operators [13], [14].

The access to the information contained in the maintenance management system, such as intervention reports, analysis of statistics, reliability, availability..., would permit the IPMAS to perform a more precise and complete diagnosis, which would take into account a great number of further data.

Besides a better equipment inspection, an application such as presented in this paper, ought to lead to the optimisation of the maintenance as a whole .

REFERENCES

[1] R. ENGELHARDT
"ARME: Système d'aide au diagnostic appliqué à l'analyse vibratoire des machines tournantes"
Doc. Laborde and Kuffer and Repelec, Groupe Alsthom, 1987

[2] M. DAVID - J.P. KRIVINE
"Utilisation de prototypes dans un système expert de diagnostic: le projet DIVA"
6èmes Journées Internationales, Avignon 86, volume 1

[3] C. PACHAUD
"Détection des défauts localisés de roulements par analyse de résonance hautes fréquences"
Rev. Matériaux Mécaniques Electriques, n°424, mars-avril 1988

[4] J.F. FLECKENSTEIN - D. LE REVEREND - P. NEAU
"La surveillance vibratoire des groupes turbo-alternateurs"
Rev. Matériaux Mécaniques Electriques, n°424, mars-avril 1988

[5] M.C. SUHNER - M. GABRIEL
"L'AMDEC pour générer la base de connaissances de systèmes d'aide au diagnostic"
L'ADMEC un atout pour les PMI, recueil de conférences CETIM 1992

[6] J.P. JEANETTE
"Analyse Fonctionnelle et AMDEC, principes et objectifs"
L'AMDEC un atout pour les PMI, Recueil de conférences CETIM 1992

[7] NORME AFNOR X60-510
Techniques d'analyse de la fiabilité des systèmes
Procéd ures d'analyse des modes de défaillance et de leur Effets (AMDE)

[8] S. MINAULT - F. JOUSSET
"Architecture temps réel pour un système de maintenance prédictive"
11èmes Journées Internationales, Avignon 91, volume 7

[9] A. BOULENGER - E. FOREL - C. PACHAUD
"l'analyse des vibrations ou le "bilan de santé" des machines tournnantes"
Journée de réflexion AIF sur la maintenance conditionnelle, Paris, février 91

[10] VIBRATIONS ACTUALITE
"Etude choc"
Rev. Campagna Varenne - année 1984 - n°3

[11] D. MENEXIADIS
"Conception d'un système expert d'aide au diagnostic pour les machines tournantes"
Thèse de Doctorat, Université de Valenciennes, Nov 1988

[12] OBSERVATOIRE FRANÇAIS DES TECHNIQUES AVANCEES
"Systèmes experts et conduite de process"
Ed. Masson, Aragon 8, 1989

[13] A. KELLY
"Maintenance planning and control"
Ed. Butterworths, 1989

[14] R. SKATTEBOE - E. LIHOUD - R. AAS HYSTAD
"Dianon: A knowledge based system for fault diagnostic and maintenance planning for rotating machinery"
8èmes Journées Internationales, Avignon 88, volume 2

MINI-SYMPOSIUM I.
Hybrid Tools for CIM System Development

nowledge Based Hybrid Systems (B-11)
Mezgár and P. Bertók (Editors)
Isevier Science Publishers B.V. (North–Holland)

CIM-CASE Demonstration Description

L.M. Camarinha-Matos; H.J. Pinheiro-Pita and A.L. Osório

Faculdade de Ciências e Tecnologia, Universidade Nova de Lisboa, Quinta da Torre, 2825 Monte Caparica - PORTUGAL

Abstract
The CIM-CASE prototype system is described, as well as the main phases of its demonstration. A set of screen hardcopies illustrates the global process of designing a new CIM software architecture and selecting an adequate set of tools to implement such architecture.

1. INTRODUCTION TO DEMONSTRATOR

This demonstrator is intended to show a prototype of an Intelligent C.A.S.E. for developing software architectures of Computer Integrated Manufacturing Systems as well as a prototype of an Engineering Information Management System.
Its baseline consists of a great emphasis in CIM Reference Models, Development of Tools, Information Integration, Integrating Infrastructure and Standards.

The demonstrator has the following objectives:
- i) To show the basic functionalities of the CIM-CASE concept
- ii) To demonstrate feasibility of different approaches to tools selection
- iii) To evaluate the use of hybrid programming paradigms
- iv) To support requirements definition of future CIM-CASE tools

The prototype presented in this demonstration was developed by UNL in the context of CIM-PLATO, an ESPRIT project of the European Community, involving 14 partners (universities and companies) from 7 countries. The main objective of CIM-PLATO was the development of an industrial toolbox prototype consisting of computer-based procedures and tools which support the design, planning and installation of FMS and FAS systems in a CIM environment.
This prototype was implemented using CRL/KC (Common Representation Language of Knowledge Craft) and its Graphical Interface was developed in C language resorting to Xlib library functions of the X Windows system. In this implementation different programming paradigms were used - frame-based programming, reactive programming, rule-based programming, concurrent programming etc.. Current prototype version runs on a SUN Sparc Station using SUN-OS Operating System.
This description shall be read together with the paper "Hybrid Programming Paradigms in CIM-CASE" of the same authors

CIM-CASE concept

CIM-CASE is an interactive system, providing functionalities to manage an electronic catalog of CIM software tools and to support the interactive generation [hierarchical interactive planning] of particular configurations of subsets of tools dedicated to specific cases, i.e., particular architectures in the CIM-OSA sense.

A **Particular Architecture** is a software configuration for a particular CIM system, comprising a set of tools (selected from the catalog) integrated via an Information System.

In close connection with the catalog of tools, another central component of this system is a metaknowledge base containing knowledge about CIM activities, application domains, reference information concepts and configuration knowledge, i.e., - **CIM reference models**.

A **Catalog Manager** module is designed to support maintenance, graphical browsing and explanation facilities associated to the catalog and reference models.

A second functional component of the CIM-CASE concept is the **Particular Architectures Planner**, an interactive support system for specification and derivation of particular architectures, including tools selection and derivation of shared IS concepts.

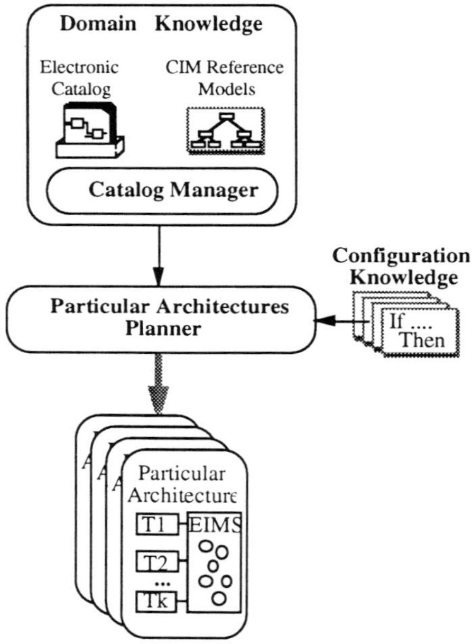

Fig. 1 - General model of CIM-CASE

The derivation of particular architectures proceeds in the following phases:

i) Design of the system's functional structure, using an IDEF0/SADT-like methodology and driven by CIM Reference Models and configuration knowledge (fig.s 6,7,8,9);

ii) Determination of a realization structure and selection of tools (fig.s 10,11,12)

iii) Derivation of basic (shared) information concepts (fig. 13)

iv) Specification of administrative policies for a specific installation (flow of tool's activation, concurrency constraints, installation parameters, version-keeping strategy, etc.).

In current version of the prototype, only phases i) to iii) are available.

In all these phases the system uses the information of its Configuration Knowledge to help the user in his decisions when proposing system solutions.

Configuration Knowledge is represented by a set of rule bases intended to assist the human expert - in a interactive way - during the various phases of the hierarchical planning. This assistance can assume various forms:

i) Criticizing or validating design options made by the human, taking into account domain knowledge. Embedded or on-line criticism (fig. 8) and explicit or on-demand criticism may be used ;

ii) Reducing the focus of attention. In each decision step, once a choice is made, the amount of domain knowledge can be appropriately "configured"/reduced for the next iterations (phases) ;

iii) Facilitating the convergence of the logical decomposition towards a realizable system driven by reference models and taking into account the characteristics of available Primitive Functionalities .

The main objective of the work is to investigate an approach to achieve a kind of specialized **CIM-CASE** system to support the interactive development of CIM systems. However, it doesn't represent a finished system but a first implementation of an emerging concept.

A more detailed presentation to this concept can be seen somewhere in this book [Hybrid Programming Paradigms in CIM-CASE].

2. DEMONSTRATION SEQUENCE

The demonstration has seven parts:

- Inspection of Domain Knowledge;
- Configuration's functional specification;
- On Demand Criticism;
- Building up the Realization Structure;
- Tool's Selection and evaluation;
- Save the chosen solution (list of chosen tools and data concepts to fill the IMS);
- Engineering Information Management System (EIMS);

Inspection of Domain Knowledge

In this phase we illustrate the taxonomies of CIM engineering activities, application domains, mechanisms and data I/O (glossary), the Electronic Catalog of software tools, historical of configurations, and exchange table. The browsing facilities as well as the flexibility to move from one data structure to another using different graphical windows are also demonstrated.

Figure 2 shows the taxonomy of CIM engineering activities and the internal model of an activity - FUNCTIONAL-MODELING activity.
There isn't a unique solution on how to decompose CIM into activities and subactivities. Therefore, the taxonomy of activities must be flexible and easily modified, to be customized by each user.
Another important aspect that this taxonomy supports is the need for relationships - other than taxonomic ones - to capture uncertain information on CIM reference models. In the model of FUNCTIONAL-MODELING the slots STRONGLY-RELATED, TYPICALLY-RELATED and MIGHT-BE-RELATED perform those capture, indicating to which other activities this one is related.
Figure 3 shows the a partial view of the taxonomy of basic information concepts. For each concept a brief description can be obtained.

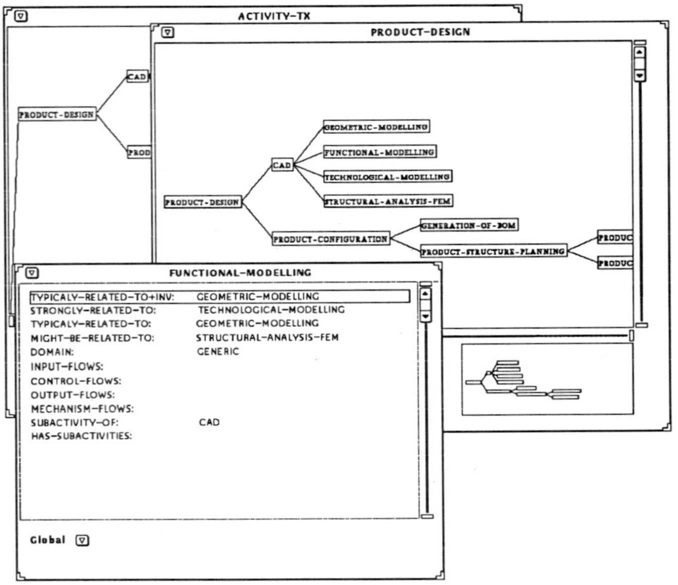

Fig. 2 - Partial view of Taxonomy of CIM engineering activities

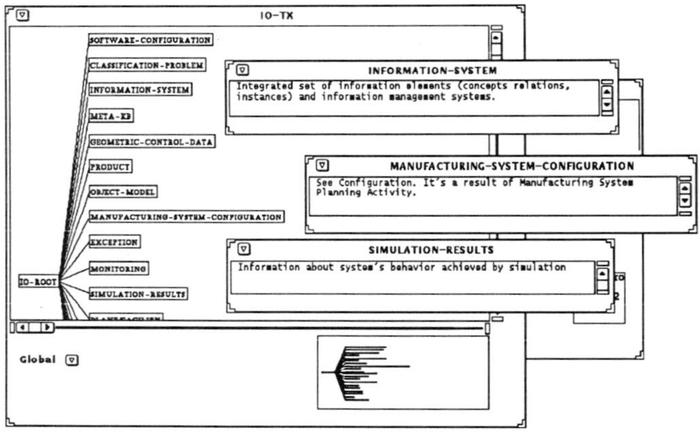

Fig. 3 - Common Glossary

Figure 4 shows the Electronic Catalog of Software Tools and the internal model a tool- APPCONV. Each tool's model has a relationship to each activity it can perform and to its application domain. In this mode, slots CAN-BE-MERGED, CAN-BE-CONNECTED-WITH, and CONNECTABLE-WITH-TRANSLATION are used to capture the knowledge about connectivity capabilities among tools.

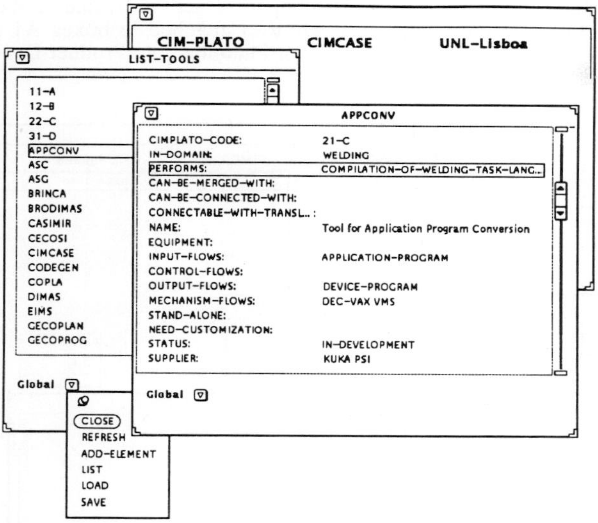

Fig. 4 - Electronic Catalog

Configuration's functional specification

At this phase a multi-level editor (SADT/IDEF0 editor) is demonstrated as well as its intelligent edition facilities: ICOM (Input Control Output & Mechanism) inheritance mechanism in both directions, embedded criticism, reduction of the focus of attention, redefinition of an ICO, etc..

Figure 5 shows the general aspect of the SADT/IDEF0 editor.

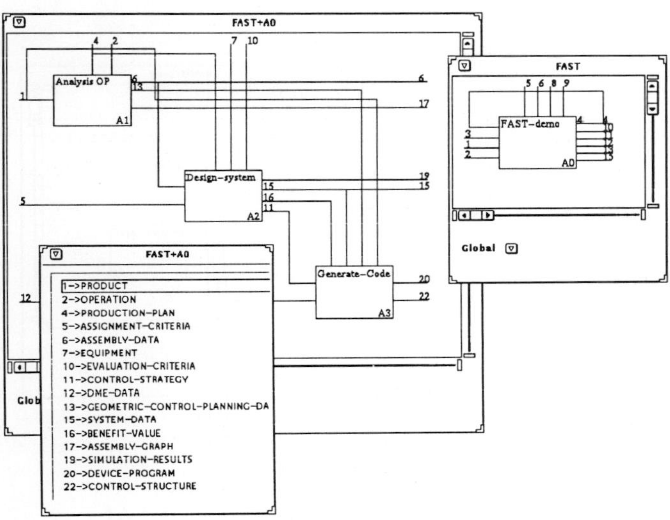

Fig. 5 - SADT/IDEF0 editor

Figure 6 illustrates the Semantic Attachment of activities, one of the steps at the functional specification phase. The activities attached to boxes A1 and A2 shall belong to the sub-taxonomies "Production-Planning" or "Product-Design", because these are de activities attached to box A0.

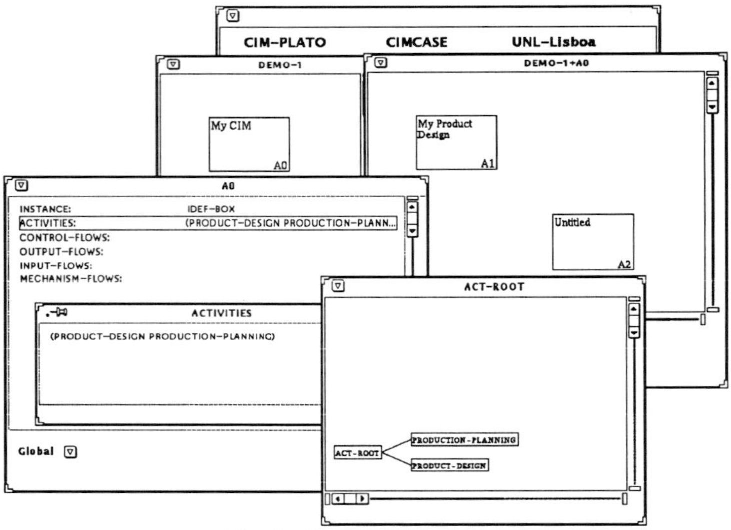

Fig. 6 - Semantic Attachment

Figure 7 illustrates the redefinition of input Assembly for box A1. The IDEF0 editor implements mechanisms of inheritance of IDEF ICOM's in both senses as a way to maintain the coherence among the differents diagrams. However, the user can use a more abstract data concept in the upper level and redefine it, in terms of its decomposition, at the lower levels.

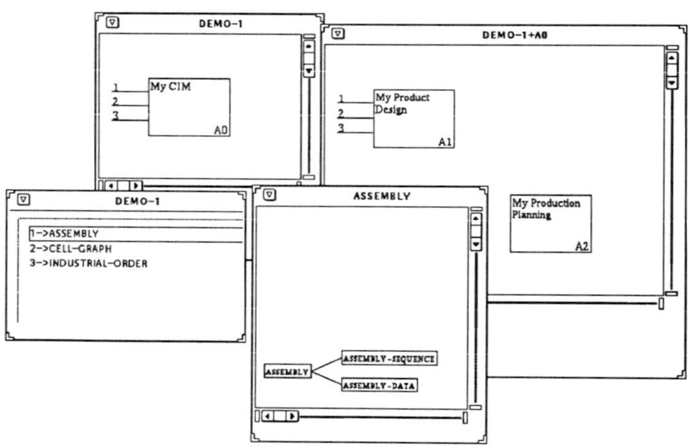

Fig. 7 Redefinition of Input Assembly for box A1

Behind the editor, and using reactive programming, there exists a set of critics that alert the user for options that don't agree with the Domain Knowledge. Figure 8 illustrates one of these critics.

Embedded Criticism

Activity in Domain?

IF
(criticism is on-line) & (activity doesn't belong to selected application domain)

THEN
alert user <activity out of domain >

Fig. 8 Example of an Embedded Criticism's rule

On Demand Criticism

Since the process of specification of the functional characteristics of a configuration doesn't follow a pre-defined skeleton, some of the coherence tests could not be done during that phase. Therefore, exists a set of critics that is activated by the user - On demand criticism.
At this phase the list of available critics is shown as well as an analysis of message reporter contents.

Figure 9 shows a partial view of the list of critics, and some examples of critics'messages.

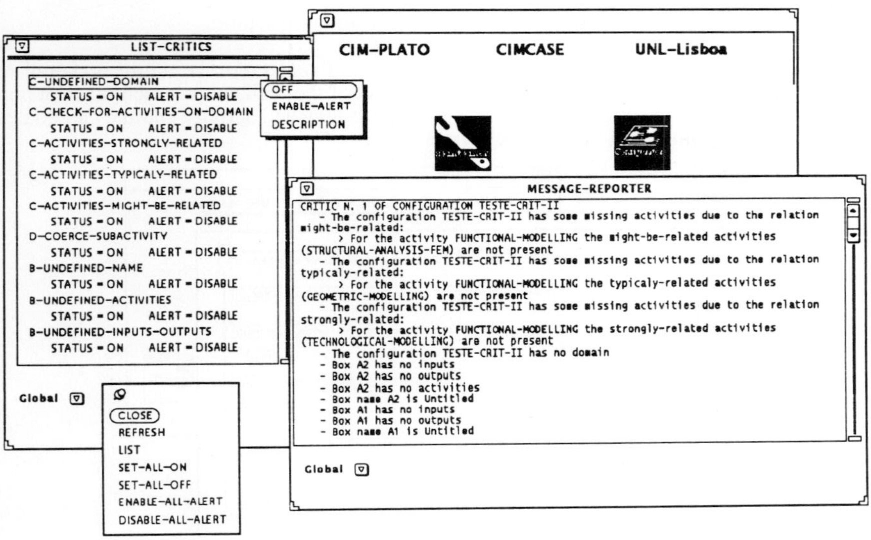

Fig.9 - Critics

Building Up the realization structure

Normally, the realization structure is equal to the functional specification, and the boxes of the deepest level are used in the Tool's selection phase. However, the user might made an overdetailed decomposition of an hierarchy branch regarding the functionalities available in the Electronic Catalog. In this case a kind of "pruning" is done; or, instead of this, the level of decomposition performed by the user, at a given branch, is so abstract that the planner couldn't select any tool. In this case, if the attached activity contains subactivities, the user is advised to perform a more detailed decomposition of that branch. Figure 10 illustrates this idea.

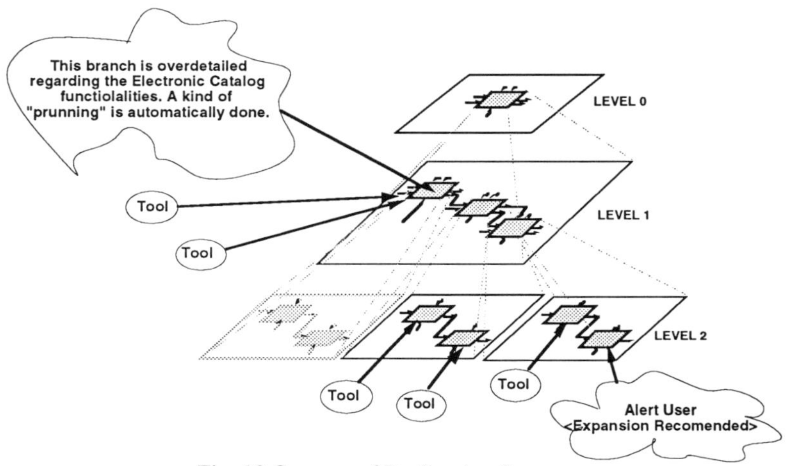

Fig. 10 Concept of Realization Structure

In this phase of the demonstration the concept of realization structure is therefore introduced. Figure 11 shows the graphical aspect of the realization structure: each box of the deeper chosen level (realization-boxes), its attached activity and the candidates tools to perform this activity are shown.

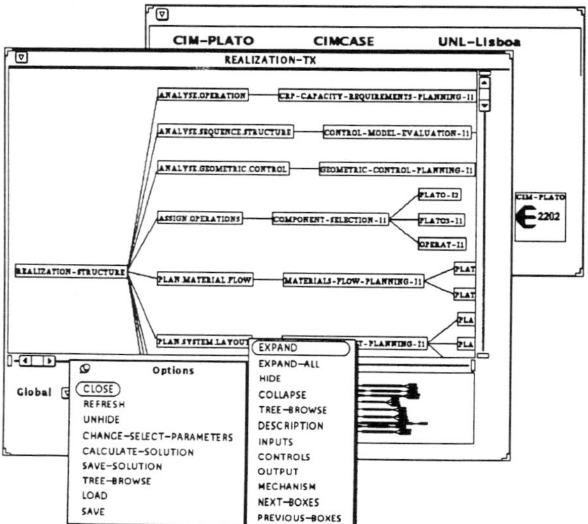

Fig. 11 - Graphical aspect of realization structure

Tool's Selection and evaluation

This phase corresponds to the selection of one of the candidates tools to perform the activity attached to the realization-boxes. A set of filters are applied in a user defined order. A kind of **what if** analysis can be performed changing the order of application of filters, enable/disable some filters or changing the filters parameters. Different solutions can be compared and evaluated. Figure 12 shows the graphical aspect of this phase.

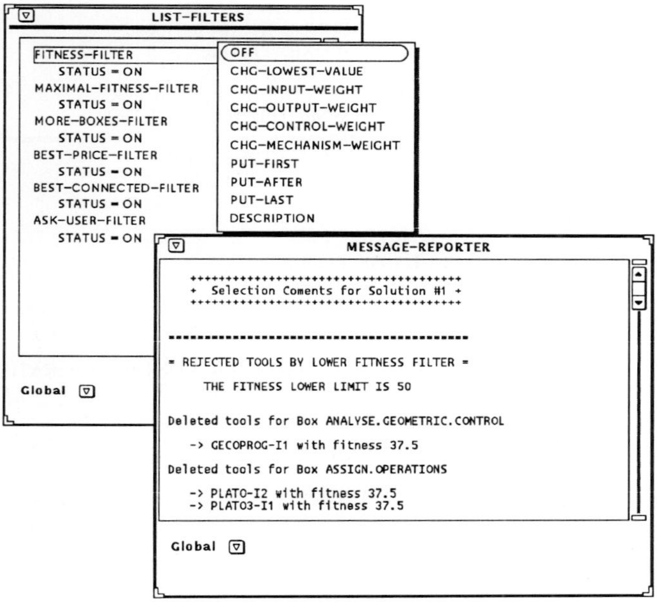

Fig. 12 List of Filters

Save the chosen solution

In this phase the user chooses one of the achieved solutions and saves it. Three files are produced: one that contains a report of the solution (Figure 13 shows a partial view of this file), other that contains the model of the tools and a third one that contains the basic data concepts. These last two files are used by the Engineering Information management system.

Engineering Information Management System (EIMS)

Finally, using the files produced by the CIM-CASE the EIMS is demonstrated. This system provides both information management functionalities and integrating infrastructure for a distributed collection of client tools.
Figure 14 illustrates the concept of EIMS, while figure 15 shows its graphical aspect.

```
SOLUTION FOR THE CONFIGURATION FAST
        1992/2/12                          11:41
        CONFIGURATION DOMAIN : MECHANICAL-ASSEMBLY

PARAMETERS
        LOWEST VALUE OF FITNESS = 0%
        FILTER SEQUENCE:
                        OUTPUT-INPUT-FILTER
                        MAXIMAL-FITN-FILTER
                        BEST-CONNECTION-FILTER
                        MORE-BOXES-FILTER
                        ASK-USER-FILTER
                        BASED-ON-EQUIPMENT-FILTER
                        NO-CONNECTED-TOOLS-FILTER
==========================================================================
        BOX : ANALYSE.SEQUENCE.STRUCTURE
        ACTIVITY: CONTROL-MODEL-EVALUATION
        TOOL: CASIMIR
INPUTS
        PRECEDENCE-GRAPH
                SUB-CLASSE: PROCESS-PLAN
                PRODUCT-NAME
                LIST-OF-OPERATIONS
CONTROLS
        OPERATION
                SUPER-CLASSE: GEOMETRIC-CONTROL-OPERATION
                              OPERATION-PART
                SUB-CLASSE: PROCESS-PLAN
                OPER-ID
                OPER-DESCRIPTION
                OPER-TIME
                RESOURCE-TYPE
                JOINING-METHOD
                PRECEDENCE-CONSTRAINT
                INSEPERABLE-CONSTRAINT
                SIMULTANEOUS-CONSTRAINT
                CONSISTENCY-CONSTRAINT
                OPERANDS-LIST
OUTPUTS
        PRECEDENCE-GRAPH
                SUB-CLASSE: PROCESS-PLAN
                PRODUCT-NAME
                LIST-OF-OPERATIONS
MECHANISM
==========================================================================
        BOX : ANALYSE.GEOMETRIC.CONTROL
        ACTIVITY: GEOMETRIC-CONTROL-PLANNING
        TOOL: GECOPROG
```

Fig.13 - Final Solution

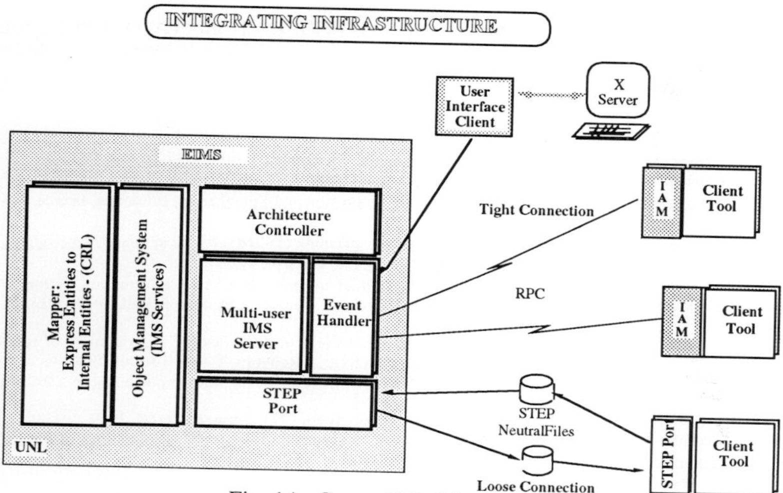

Fig. 14 - General Model of EIMS

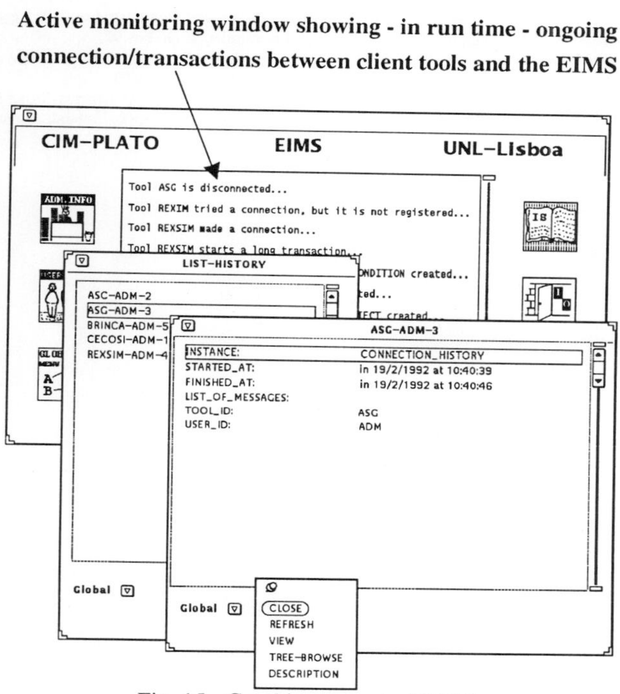

Fig. 15 - Graphical aspect of EIMS

Acknowledgments

This work was supported in part by EEC [project ESPRIT II 2202, CIM-PLATO], JNICT [project SARPIC and CIM-CASE] and INIC. The authors would like to thank Mr. João Moura-Pires, Mr. Luís Lopes and Dr. Hannu Ahonen for their contributions.

References

[1] L.M. Camarinha-Matos; H. Pinheiro-Pita, "Intelligent C.A.S.E. for CIM", Proceedings of IEEE Int. Workshop on Emerging Technologies and Factory Automation, Melbourne, Australia, 11-14 Aug. 92.

[2] L.M. Camarinha-Matos and F.Sastron, "Information Integration for CIM Planning Tools", Proceedings of CAPE'91 IFIP Conference on Computer Applications in Production and Engineering, Bordeaux, France, 10-12 Sept. 1991.

[3] L.M. Camarinha-Matos and H.J. Pinheiro-Pita "Interactive planning of CIM software systems", Proceedings of 2nd UNIDO Workshop on Industrial Robotics and CIM, Belgrade, 4-6 Sept. 91.

[4] A. Steiger-Garção; L.M. Camarinha-Matos - "Design of a knowledge-based information system", in *Integration of Robots into CIM* [Ed. Bernhardt, Dillman, Hörmann, Tierney], Chapman & Hall, 1992 - Cap. 21 & 22.

[5] L.M. Camarinha-Matos; A. Steiger-Garção "Knowledge architecture for flexible programming of robotic cells", Proceedings of the 20th International Symposium on Industrial Robotics, Tokyo, 4-6 Out. 1989.

[6] L.M. Camarinha-Matos and H.J. Pinheiro-Pita, "CIM-CASE: relatório de uma implementação", Relatório Interno, Março 1992

[7] AMICE , Esprit Consortium, [Eds.], Open System Architecture for CIM, Springer-Verlag, 1989.

[8] L.M. Camarinha-Matos "Toolbox Manager: Towards a concept of CIM-CASE", Proceedings of the CIM-PLATO Workshop on CIM Planning Tools, University of Karlsruhe, 25-26 Fev 92,

[9] Ross, D.T., Application and extensions of SADT, IEEE Computer, Apr 1985.

[10] L.M. Camarinha-Matos; H. Pinheiro-Pita; J. Barata; R. Rabelo, "Taxonomy of CIM Activities - 2nd revision", *Esprit CIM-PLATO Document, IS-UNL-04.91/1*, Apr 91.

[11] L.M. Camarinha-Matos; J. Moura-Pires; H. Pinheiro-Pita, "CIM-PLATO Common Glossary - 4th revision", *Esprit CIM-PLATO Document IS-UNL-04.91/2*, Apr 91.

[12] ISO CD 10303-11, Exchange of Product Model Data - Part II: The Express Language, Jul 1990.

[13] Rolf Bernhard "CIM Systems Planning Toolbox - Project Survey and Demonstration", Proceedings of the CIM-PLATO Workshop on CIM Planning Tools, University of Karlsruhe, 25-26 Fev 92,

[14] L.M. Camarinha-Matos; H. Pinheiro- Pita; "Interactive Planning in CIM-CASE", to be presented at 1993 IEEE Internationla Conference on Robotics and Automation, Atlanta, Georgia, May 93.

[15] L.Camarinha-Matos; A. Luis Osório; "*CIM Information Management System: An Express-Based Integration Platform* ", IFAC Workshop on CIM in Process and Manufacturing Industries, Espoo, Finland, November 23-25, 1992.

Summary of Mini-Symposium I.

Topic: Hybrid Tools for CIM System Development

Moderators: Imre Horvath (Hungary), Andrew Kusiak (USA)

Demonstration title: Hybrid programming paradigms in CIM-CASE

Contributors: L.M. Camarinha-Matos, H.J. Pinheiro-Pita (Portugal)

1. User oriented description of the demonstrated system

The system presented is an intelligent CASE tool for the development of computer integrated manufacturing systems' software architecture, and at the same time it is an engineering information management system. It was developed as part of an ESPRIT project, that included 14 universities and companies from 7 countries. It was demonstrated how the fundamental structure of a hypothetical CIM system can be defined by using an IDEF0 / SADT like methodology.

A catalogue manager was used to select the required resources. Functional modelling of the CIM system was supported by systematic classification of activities. The resources were listed in a catalogue, and a glossary contains the explanation of different terms. Based on stored rules, the system validated design decisions, checked for incoherence in the final structure and offered possible improvements.

2. Discussions

The criticising mechanism and its implementation raised considerable interest. Several questions were asked about practical aspects, such as how to fill the database, how it can be used in configuring a real flexible manufacturing system. It seemed that a detailed help system would be advantageous in handling the complex system.

The different programming methods (procedural, frame based, rule based and reactive) were successfully combined to make a complex system. The presented software verified that the design of complex information structures can be assisted effectively by specific CASE tools. The system can support the process of CIM system configuration by producing a CIM architecture out of the CIM reference model and specify its components according to the requirements of the particular application. A possible way of further development can be the connection with a real CIM system the and the use of the information feedback for reconfiguration.

3. Conclusions

Different tools with different knowledge representation forms and inference methodologies can be successfully combined in one system, which will improve the efficiency of the user's work. This was clearly demonstrated by the above hybrid CASE system.

MINI-SYMPOSIUM II.
Hybrid Systems for Product Design

Knowledge Based Hybrid Systems (B-11)
I. Mezgár and P. Bertók (Editors)
Elsevier Science Publishers B.V. (North–Holland)

FROOM - A demonstration session

O.W. Salomons, F. van Slooten, F.J.A.M. van Houten, H.J.J. Kals

Laboratory of Production and Design Engineering, Dept. of Mechanical Engineering, University of Twente, P.O. Box 217, 7500 AE Enschede, The Netherlands

Abstract

FROOM is a prototype of a feature based re-design support system. A summary is provided of the design and implementation of FROOM. Two different types of users are discerned for FROOM: end-users and system manager users. Demonstration sessions are provided from the viewpoints of both user types.

1. INTRODUCTION

FROOM is a prototype of a feature based re-design support system. FROOM stands for Features and Relations used in Object Oriented Modelling. FROOM is under development at the Laboratory of Production and Design Engineering, University of Twente, the Netherlands. The FROOM prototype is meant to provide a medium to discuss new CAD technologies, and to explore possible solutions for better support of the (re-)design process and its link with process planning. For FROOM, two different types of users are discerned: *end-users* and *system manager users*.

The *end-user* does the actual re-design work. End-users must be provided with entities that have meaning in their application domain: assemblies, components and features. End-users must also be supported in making selections from catalogues and standards tables. Not only bottom-up support must be provided to the end-user by the re-design support system, but also top-down support must be available. Ideally, it should be possible to work in a mixed top-down and bottom-up design style [1].

The *system manager user* configures the system to an application domain, to a certain company as well as to certain users. For instance, the system manager user defines the (company specific) features, components and assemblies. He also defines the catalogues and tables which can be used by the end-user. Thus, the system manager user has to acquire knowledge and store it in the FROOM system. FROOM could therefore be called a knowledge based system [1].

The organization of this paper is as follows. Section 2 gives some more details on the design and implementation of the FROOM system. Section 3 describes how an end-user will interact with the system. Section 4 shows some system manager tools. Finally, in section 5 conclusions will be drawn.

2. DESIGN AND IMPLEMENTATION OF THE FROOM SYSTEM

The conceptual and functional design of FROOM has been described in [1]. A future paper will describe the configuration and detail design of FROOM in more detail [2]. In the following we will focus somewhat on the design of the user interfaces (section 2.1) and on the implementation of FROOM (section 2.2).

2.1 Design of the user interfaces.

The user interfaces (for both end-user and system manager user) that have been designed and implemented are shown in the figures in sections 3 and 4.

The end-user interface consists of two main parts: a graphic-interactive window in which graphic interactive tasks are supported similar to those in commercial CAD systems (mainly bottom-up functionality). In the second window, the context structure is displayed, which can be considered as a conceptual graph representing objects and their relations. Objects (features, components etc.) are represented as rectangles in the context structure, while the circles represent the relations between the various objects. Objects and relations in the context window can be detailed into more detailed objects and relations in a top-down approach. Thus, the graphic interactive window supports bottom-up design tasks while the context window supports top-down design tasks. There is a bi-directional association between objects in the context window and objects in the geometry window, such that a mixed top-down and bottom-up design approach is allowed. More details on the end-user interface can be found in section 3.

In the present implementation, the end-user interface and system manager user interface are not strictly divided. Thus, a system manager user also has two main windows: a graphic-interactive window and a context window. A context structure can be pre-defined by the system manager; he can do this by the context editing facility in the context window. The graphic-interactive window could be used for tasks such as interactive geometry definition, which however is currently not implemented. More details on the system manager user interface can be found in section 4.

2.2 Implementation.

As features, components and assemblies can all be seen as objects, the object oriented paradigm has been adopted. The object oriented analysis, design and implementation methodology by Aksit and Bergmans [3] has been applied.

The tools that have been chosen for implementing the user interface are X-windows™/Motif™ and Phigs™. Further implementation was done using the C++ programming language, the Oracle™ database management system, and ACIS, an object oriented solid modeller. The reason for choosing the above mentioned tools was based on using accepted industry standards as well as robustness. The main reason for choosing ACIS for kernel modelling is that ACIS is becoming the defacto standard modeller, being built on C++. Moreover, ACIS being a non-manifold modeller, would allow for the use of incomplete geometry.

Because the implementation is realized using object oriented design methods, and an object oriented programming language, the term 'object' will also refer to objects within the software system, as instances of classes. Regarding this distinction, within the software system, also objects with 'management' tasks can be identified. For example, the context structure object acts as a kind of container of the various objects and relations, and handles constraint management and other inter-object behaviour. There are also separate objects that manage tasks like catalogue selection, database interaction, and user - interface facilities.

Objects within the software system communicate by means of message passing. Messages usually are generated by the user, by performing an action on the user interface, e.g. selecting from a menu. For example, when the object managing the user interface receives such a message, it determines to which object this message should be sent. For example, when the user presses the menu button File-Open-Assembly (figure 1), the object managing the user interface will tell the object that manages the context structure to create a new context structure. This object in turn will invoke a message to the object managing the database to retrieve necessary information, and invoke a message to the assembly to display the geometry that is associated with the new context structure.

3. END-USER DEMONSTRATION SESSION

A demonstration session of an end-user, re-designing a gearbox in FROOM, will be given. The standard (CAD) functionality of FROOM, like scaling, zooming, graphical editing etc., will not be discussed here.

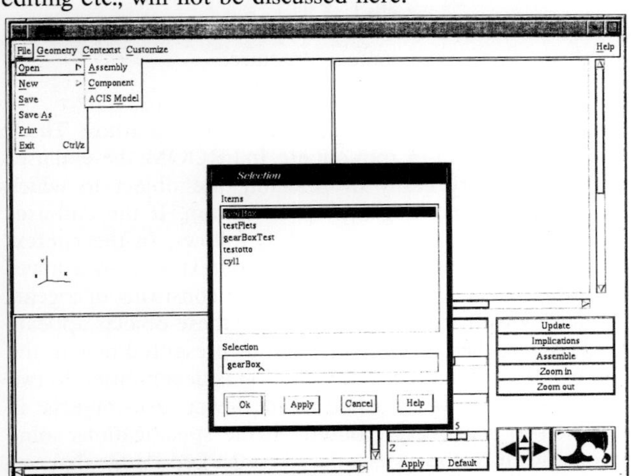

Figure 1. Opening an assembly for detailing.

Our designer could choose to work either in a top-down or bottom-up fashion. If he chooses to work top-down, he could choose a template of a design (e.g. a gearbox consisting of 2 shafts, 2 gearwheels, bearings and a housing). In the designer prefers to work in a bottom up fashion, he could choose an existing design that has been detailed completely. Suppose that our designer chooses to detail a template of a gearbox. How this is done is shown in figure 1; the user opens an already existing, pre-defined template of a gearbox assembly. In FROOM the gearbox template has been pre-defined by the system manager. Further detailing of the template is left to the designer. After opening the gearbox template, a single block, representing the gearbox, will appear in the context window.

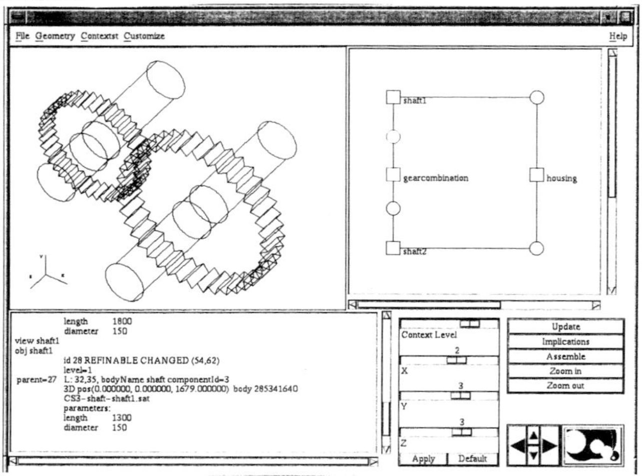

Figure 2. The second level of abstraction of the context structure.

Usually the end-user will want to attach some specifications to the gearbox. These specifications could be: power transmitted, rpm-in, rpm-out etc. In FROOM, the-end user can attach specifications to an object very easily by selecting the object to which specifications have to be attached and by pushing the 'specs' button. If the end-user would like to detail the gearbox further, he could proceed as follows. In the context window he can select the box representing the gearbox. In the context window a lower abstraction level of the gearbox will appear. In our case a gearbox consisting of a gear-combination, two shafts and a housing with the relations between these objects appears in the context window (figure 2). Again, the objects and relations presented now in the context window can be detailed further, e.g. the gear-combination is decomposed in two related gearwheels. Also, specifications can be attached to the lower level objects; in figure 3 specifications are attached to gearwheel1. Based on the specifications some geometry might be inferred automatically by the system, based on the knowledge that was implemented previously by the system manager. In figure 2 the assembly geometry is instantiated in the graphic interactive window (upper left). In figure 4 only the gearwheel geometry according to the specifications in figure 3 is derived.

Figure 3 shows that parameters within a specification can be defined to have dependencies on parameters of other objects. For example, the hole diameter of the gearwheel depends on the diameter of the shaft. At the moment tolerances cannot be defined.

After having detailed some components to a certain extent, the end-user might want to

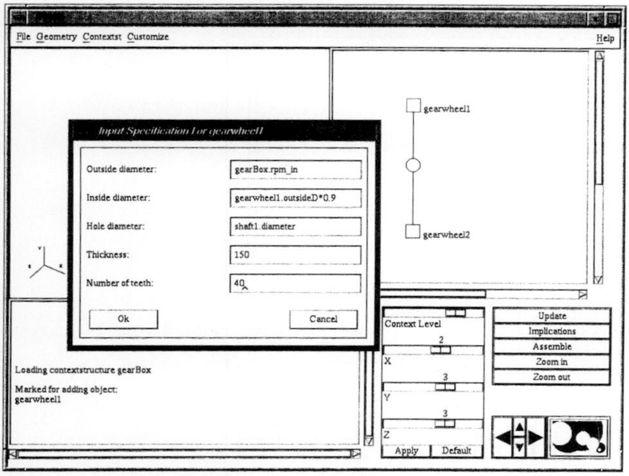

Figure 3. Adding specifications to object gearwheel.

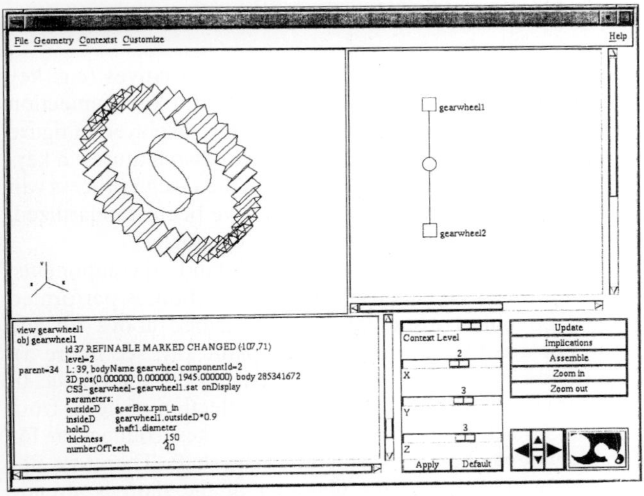

Figure 4. Instantiation of a single object's geometry.

determine what kind of connection to choose for connecting these components. In this case, shafts and gearwheels could have been detailed to some extent after which the designer determines the way the shafts connect with the geawheels.This can be done by further detailing the shaft-gearwheel relation in the context structure. Some pre-defined alternatives could be offered to the end-user to choose from. Of course these alternatives

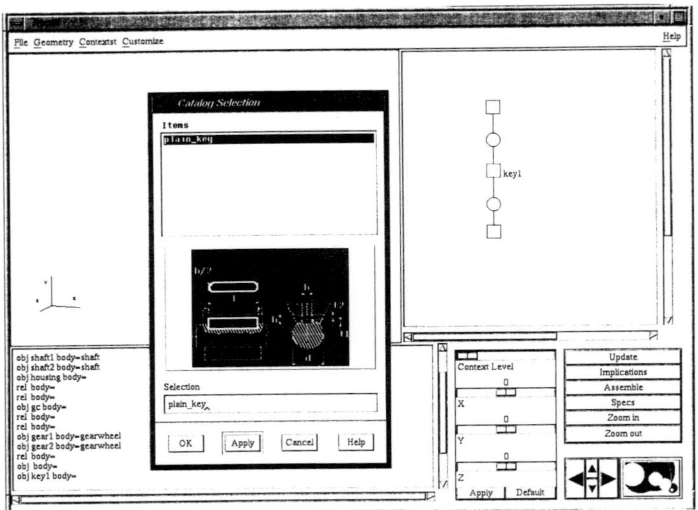

Figure 5. Selection of a plain key from a catalogue.

have been defined by the system-manager (see section 4). These alternatives (e.g. key connection, weld connection) are presented to the user. Suppose that a key connection will be chosen by our end-user. The corresponding context structure is displayed in figure 5. The context structure also shows that a key component comes with two features: a key-way on the shaft and a key-way on the gearwheel. The next step in the design process will be to choose the most appropriate key (and key-ways). As keys have been standardized, suppose it is obligatory for the end-user to choose a key from a catalogue.

The user is supported by a catalogue selection tool to instantiate standard components, like bolts, nuts, keys etc. Instantiation of an object by catalogue selection, is performed in the same manner as the general object instantiation as described above for the gearwheel instantiation. The user selects the object type (here plain_key, see figure 5), and selects the catalogue-field (figure 6) to determine the parameters of the object. In the ideal case, the system could provide the end-user with a pre-selection of items from which the end-user can make his final choice. In the case of the key connection for example, the parameters of the key depend on the diameter of the shaft. Therefore not all keys in the table are suitable for selection as the diameter of the shaft is already known. However this option has not been implemented yet. After a selection of key parameters has been made, the object(s) will be instantiated automatically. In this case, both a key and two key-ways will be instantiated automatically. In this way errors can be prevented. Time savings can be made as well. Objects, like the key and key-ways behave 'intelligently', they know how to position themselves. An object can search the context structure for referenced object parameters (dependencies).

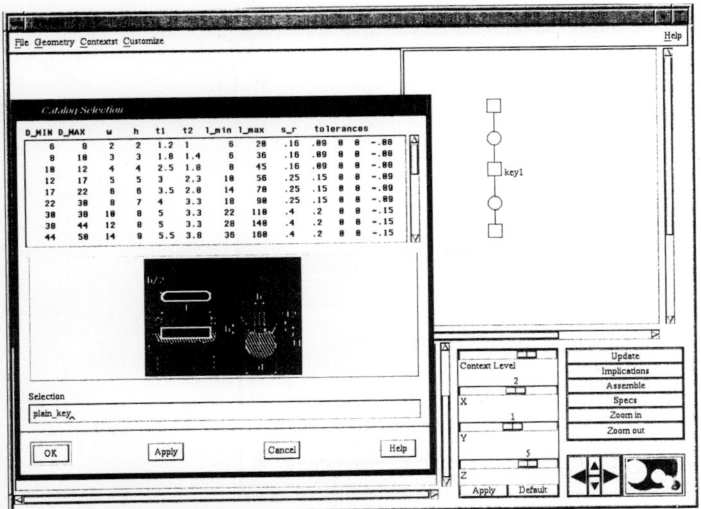

Figure 6. Selection from the catalog table of key parameters.

4. SYSTEM MANAGER SUPPORT TOOLS

The main functionality that is presently offered to the system manager is: interactive object definition, context structure editing, database management and system customization.

Interactive Object Definition (IOD) is a means to provide the end-user with pre-defined templates of objects (features, components, assemblies), of which an instance can be made by interactively specifying the parameters of such an object. The system-manager uses the IOD tool to define the template object, its type (generic or standard), its parameters and the geometric shape. In our gearbox case, a generic object could be a gearwheel, which has a generic shape. A standard object is a component that can be purchased off-the-shelf, e.g. keys, bolts and nuts. For this kind of objects the system-manager has to enter the corresponding catalogue data of the object. The IOD tool has a module to support this interactively. How the interactive object definition functionality presently looks like is shown in fig. 7.

The system-manager uses the context structure editing tool (figure 8) to define the context structure that is to be further detailed by the end-user. The system-manager can leave some decisions about the detail design to the end-user by offering alternative context structures for certain relations or objects to the end-user. In section 3 this was already mentioned when alternatives for the shaft-gearwheel relation where considered.

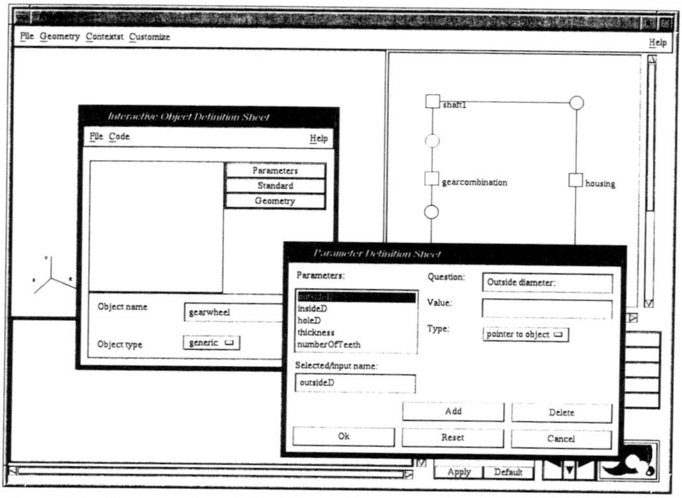

Figure 7. Interactive definition of a gearwheel object.

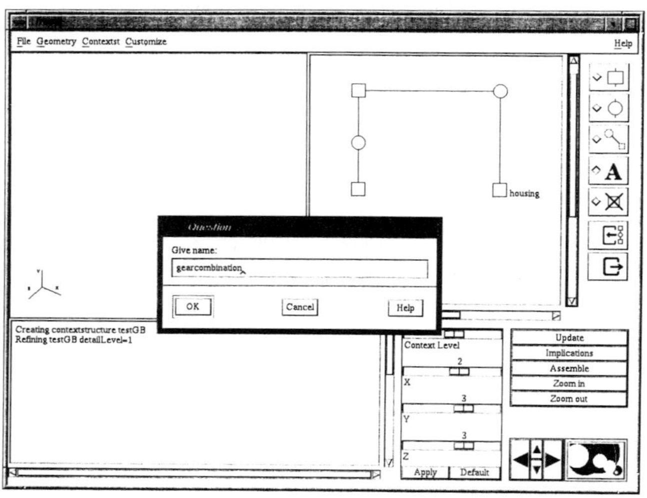

Figure 8. Interactive editing of a context structure.

5. CONCLUSIONS

The description as presented in this paper is a preliminary one. The FROOM prototype is only a first version, which needs further extension. Future development will focus on: -

- The Interactive object definition tool.
  Interactive definition of object parameters and catalogue tables is now possible. The definition of object geometry is not very user-friendly yet; although it can be done from within the system manager user interface, it merely comes down to programming geometry using ACIS functionality. More has to be done on providing the system-manager with some interactive tool to define the geometry of an object.
- Context editing tool.
  This tool now provides context editing functionality, but needs also validity checking of context structures and a way to present alternative context structures to the end-user.
- Design history.
  To capture design intent, and to track decisions regarding chosen alternatives and other design decisions, a design history tool would be necessary. This is a topic of future research.
- Standard feature based CAD functionality.
  To be able to support a bottom-up design style, more standard CAD functionality has to be developed.
- Constraint management.
  At the moment only some relatively simple parameter dependency constraints are supported. In the future more advanced constraint management mechanisms have to be supported. Tolerancing has to be supported as well.

References

1 Salomons O.W., Kappert J.H., Slooten F. van, Houten F.J.A.M. van, Kals H.J.J., Computer support in the (re)design of mechanical products, a new approach in feature based design, focusing on the link with CAPP, accepted for publication in the, Proceedings of KNOWHSEM '93, Budapest, April 1993.

2 Salomons O.W., Slooten F. van, Houten F.J.A.M. van, Kals H.J.J., A computer support tool for re-design, a prototype system resulting from applying a methodic design approach, submitted to the International Conference on Engineering Design, ICED'93, August, The Hague, 1993.

3 Aksit M., Bergmans L., Object oriented analysis, design and implementation, course notes, University of Twente, Enschede, 1991.

Summary of Mini-Symposium II.

Topic: Hybrid Systems for Product Design

Moderators: Andrew Kusiak (USA), Imre Horvath (Hungary)

Demonstration title: Computer support in the (re)design of mechanical products

Contributors: O.W. Salomons, J.H. Kappert (Netherlands)

1. User oriented description of the demonstrated system

The system supports redesign of mechanical parts by feature based modelling in applications where shape design and assembly descriptions are linked together. It is a quasi-expert system, as it has a knowledge base about the available mechanical components, but it does not perform automatic problem solving. The system has two levels of access: a system manager maintains the knowledge base and the end user applies it to the redesign work.

The implementation is based on well known commercial software tools, such as XWindows, Motif, Oracle data base and ACIS 1.4 object oriented solid modeller. The extensive use of these tools, especially of the window management, makes it very efficient. The demonstration, although performed on a simple set of shafts and cogwheels, showed some important trends in CAD system development, such as continuous increase in the level of interactivity, object conceptions, sets of modelling entities with dependency relations, reasoning with abstract entities, constraint management, different forms of context description.

2. Discussions

The importance of the raised issues was emphasised by the fact, that other participants also reported similar efforts, although in less advanced stages in implementation. Several questions were raised on the implementation details, such as managing libraries of standard parts, feature management and representation, and on further applications, such as designing new parts, tolerance analysis, i.e. linking tolerance model with product model.

The system being a prototype, developers are still identifying places of improvement, especially in the system's functionality.

3. Conclusion

The application of AI techniques in product design became a topic of research interest recently. On the other hand system integration is combining geometric models with other models describing the product from other aspects, and a clear distinction has to be made between hybrid and integrated systems.

MINI-SYMPOSIUM III.
Layout Design

Knowledge Based Hybrid Systems (B-11)
I. Mezgár and P. Bertók (Editors)
Elsevier Science Publishers B.V. (North–Holland)

LayoutManager: A Decision Support System for Facilities Planning

L.R. Foulds

Department of Management Systems, University of Waikato, Private Bag, Hamilton 3105, New Zealand.

Abstract

Most facilities layout techniques embrace a single objective and do not allow for user interaction or experimentation. We describe a user-friendly, menu-driven decision support system for facilities layout. It accommodates a variety of objectives, including the maximization of adjacency scores and the minimization of transportation costs. Based on various graph theoretic layout heuristics described elsewhere in these proceedings, the system is written in Pascal, within the Microsoft Windows environment for an IBM-compatible personal computer. A successful case study based on the system is reported.

Keyword Codes: G.2.2; I.2.1; J.6
Keywords: Graph Theory; Applications and Expert Systems; Computer-aided Engineering.

1. INTRODUCTION

Here we present a decision support system for facilities layout which is user-friendly and menu-driven. It is based on various graph theoretic heuristics, which are described in Foulds (1993), elsewhere in these proceedings. The decision support system, called LayoutManager is detailed in Section 2. We conclude with a description of a case study, in which LayoutManager was successfully used, in Section 3.

The author is currently exploring the possibility of other strategies, including simulated annealing and tabu search. Further graph theoretic methods for facilities layout have been described by Leung (1992 a,b), Hammouche and Webster (1985), Hassan and Hogg (1987) and by Boswell (1992).

2. LAYOUT MANAGER

LayoutManager is a user-friendly menu-driven decision support system for facilities layout which can accommodate any of the objective functions mentioned in Foulds (1993). It employs the graph theoretic heuristics described by Foulds (1993) and implements routines to output a colour-coded, labelled block plan. It is written in Pascal, within the Microsoft Windows environment, for an IBM-compatible personal computer. A simple interactive computer program for designing a layout with the objective of minimizing transportation costs only has been reported by Wascher and Chamoni (1987).

The user provides LayoutManager with a particular problem to solve by loading a standard ASCII text file that provides all the information necessary to specify the problem. Input files may be created and edited with any suitable text editor or they can be created and edited with the built-in editor. The input file must specify the objective function to be adopted and the number of facilities to be considered, including the exterior facility. There is a soft limit imposed by the program of 40 facilities. This limit may be increased through recompilation if necessary, provided the additional computer memory requirements can be met.

For the objective functions of the Deltahedron and N-Boundary heuristics, as described by Foulds (1993), relationship charts of closeness ratings between every pair of facilities is required. For the N-Boundary heuristic, relationship charts are required for each distance from 1 to $[(n - 2)/3] + 1$ where n is the number of facilities.

For any given pair of facilities the closeness rating can be specified as a floating point number allowing the user to define exactly the input parameters; or when the measure is less precise and only a rough estimate of closeness is available the user can specify the closeness rating using a restricted set of values represented by the letters A,E,I,O,U and X.

The letters A,E,I,O,U,X are representative of actual floating point values in decreasing order where A represents the highest value and closeness rating and X is the lowest. The user may specify, in the input file or interactively, the equivalent floating point value for each letter. Default values of 64.0, 32.0, 16.0, 4.0, 1.0, 0.0 and -1024.0 will be assigned to A,E,I,O,U,X respectively if no values are specified. The large default negative value fo the letter X is designed to be used for facility pairs whose adjacency is infeasible. The letter representation is loaded in by LayoutManager and is internally converted to the equivalent floating point value, but is still viewed as letter values.

To generate a block plan, and for use in the Super Deltahedron heuristic, as described by Foulds (1993), we require the area of each facility. Areas are expressed as positive floating point values. A unique name must be assigned to each facility so that it can be identified.

An optional property of facility colour can also be assigned to each facility. The facility colour is used to distinguish the facilities in the block plan and as a key to the facility names. Each facility colour is specified as a background colour code, a foreground colour code and a pattern index. The values for the background and foreground colour are limited to the capability of the output display device. The pattern index specifies one of the following patterns: no pattern (background colour only), horizontal lines, vertical lines, left and right diagonal lines, crosshatching and diagonal crosshatching. If no facility colour is specified LayoutManager will assign a unique colour combination.

2.1 THE EDITOR

To aid the user in defining a valid problem in the format required the built-in editor can be used to create a new problem data set or to edit any of the details of an existing problem.

The editor allows

- facilities to be added and deleted,
- facility names, areas and colours to be altered, and
- relationship weights between any two facilities to be changed.

The editor thus allows a user to change interactively various problem parameters. The affect on the solution of editing can be observed by reapplying the relevant heuristic. Corrections and experimental changes can be quickly made and their effect noted immediately. Any changes to the problem data can then be saved as a file so that they may be recalled as required.

2.2 THE BLOCK PLAN

Once a problem has been specified, the appropriate heuristic can then be applied to the input data to produce a planar adjacency graph and a score. The score is displayed and, in the case of the Deltahedron heuristic, the score is also displayed as a percentage of a calculated upper bound to provide some indication of how close to optimality the solution is.

Once the layout heuristic has constructed a deltahedron, the user may apply the edge replacement and vertex relocation improvement heuristics as described by Foulds (1993). For maximum flexibility, these improvement heuristics can be applied in several ways. Either improvement heuristic can be chosen and one instance will be applied to the current deltahedron. For the edge replacement heuristic, LayoutManager will choose the edge which, when replaced, will give the best improvement. Once chosen, the deltahedron is updated and the block plan is redrawn to represent the change. The new score is also displayed. If no edge will improve the deltahedron, LayoutManager will not make any change. Similarly, for vertex relocation, the vertex which will give the best improvement will be relocated, the block plan redrawn and the new score given. If no vertex relocation will improve the deltahedron score no change is made. LayoutManager also gives the option of applying one instance of the heuristic which gives the best improvement. There is also the option of repeatedly applying the heuristic which gives the best improvement until no further improvement is possible.

The adjacency graph produced by the heuristic is internally represented as a tree structure and is not directly accessible to the user. Instead LayoutManager will generate and display a block plan based on the resulting adjacency graph. Because the adjacency graph is invariably planar, we know that it is possible to construct an equivalent block plan from it. The block plan is constructed such that only the adjacencies represented by edges in the deltahedron are present between the facilities in the block plan.

The block plan is constructed by inserting facilities in much the same order as was used to construct the deltahedron. Each facility is inserted adjacent to exactly three already inserted facilities. Let us call these three facilities the parent facilities of the child facility to be inserted.

When inserting a child facility into the chosen parent facility the area encompassed by the parent facility is increased to match the size of both facilities. The area is then divided so that each facility is of the correct size. Because a block plan introduces the added requirement of area, there are several issues that the deltahedron does not address concerning the shape and positioning of facilities. Care must be taken to insert facilities into the correct parent facility to avoid the possibility that its insertion could remove an existing adjacency. If facilities are inserted into the parent facility that was the last of the parents to be inserted, we can guarantee no unwanted adjacencies are created and that existing adjacencies are retained. Inserting in this manner implies that at most three child facilities may be inserted into any given parent facility.

For the initial four facilities, the exterior facility is assumed to be inserted first. The remaining three facilities are then ordered based on their areas, starting with the facility with smallest area. Assigning insertions in this order implies that no facilities will be inserted into the exterior facility, as both the other two parents of a child facility must have been inserted later than the exterior. Similarly, no facility will be inserted into the smallest initial facility as at least one other parent must have been inserted later. The third initial facility inserted can only have the child facility inserted whose parents are exactly the first, second and third initial facilities. The other facilities that can be inserted at this stage, must be inserted into the fourth initial facility which has been arranged to be the largest. Choosing most insertions to be made into the larger facility is important in producing a useful block plan which has few areas whose shapes have been elongated by large insertions to the extent that their area has become unusable.

Once the facilities in the initial tetrahedron have been constructed, there are three points in each facility where other facilities can be inserted; these correspond to three corners of the facility. Each of these corners intersect with exactly two other facilities. The remaining corner of the rectangular facility does not intersect with other facilities and is redundant in terms of the insertion process.

The presence of the redundant corner means a facility can be inserted in one of two ways. Facilities can be inserted by *boxing* that is, creating a box at the appropriate corner of the enlarged parent facility. Boxing in this fashion does not remove any existing adjacency and ensures that the inserted facility is adjacent to only its three parents. The second insertion method, known as *carving*, extends an inserted facility from its insertion corner to the redundant corner, carving a complete slice out of the parent facility. The redundant corner can be used in this fashion because it does not represent any specific adjacency in the deltahedron.

The advantage of carving over boxing is seen by the effect of insertion on the parent facility. With boxing, the originally rectangular parent facility assumes an L-shape, and if two child facilities are inserted into the same parent by boxing, the parent has a T-shape. If all three possible insertions into a parent facility are boxed the parent facility's shape becomes even more irregular. An irregular shaped facility may be unusable for its intended purpose and should be avoided. When carving however, the shape of the parent facility remains rectangular, and for this reason is preferred.

As there is only one redundant corner in each facility, only one child facility can be inserted by carving into each parent facility, and this child facility cannot be carved if its insertion corner is diagonally opposite the redundant corner. If there is a choice between facilities to carve, LayoutManager will choose the larger of the facilities as the one to carve. It is preferable to carve larger facilities, as carved facilities have a dimension in common with their parent which will tend to be larger than the child because of the insertion order used in the heuristic. If carving is used where possible, all facilities in the final plan will either be rectangular, L-shaped or T-shaped.

The final block plan shape is assumed to be rectangular and the ratio of width to height can be specified by the user. The shape can be adjusted dynamically although the total area must always equal the sum of the facility areas. by default, the final block plan shape is assumed to be square.

The shape of an individual facility is controlled by the shape of its parent facility and the insertion method used. For carved facilities, one dimension matches the dimension of its parent and the other is determined by the area of the child facility. For boxed facilities, thee is a degree of freedom in their shape. The width and height of a boxed facility may be chosen, provided the dimensions of the facility correspond to the required area, the facility remains completely inside the parent, and an adequate length of boundary is retained by the parent and its adjacent facilities. When more than one box is added to a facility care is taken that the child facilities do not overlap.

LayoutManager uses the ratio of a parent facility dimension to fix the shape of the box. Using this ratio gives some similarity to the shapes used in the block plan, and the length of the adjacencies between the parent and its parents are maintained as a ratio of their original lengths. The box ratio is moderated so that the box does not become too elongated as square facilities are assumed to be the most desirable shape. Provided the box remains inside its parent, the ratio of its dimension is adjusted to be in the range between 1:1 and 1:2.

A minimum width of facility and of adjacent boundary is also used to adjust the shape of facilities. LayoutManager initially calculates the minimum width to be

$$\frac{1}{2}\sqrt{A/n}$$

where A is the total area of the problem and n is the number of facilities. LayoutManager attempts to shape facilities so that this minimum is maintained.

There are, however, situations where this minimum cannot be satisfied using the insertion methods available. For example, if a sufficiently small facility is carved into a large facility, one dimension of the small facility is forced to be large causing the other dimension to be smaller than any minimum width that can be chosen. Similarly, boxing a sufficiently large facility into a small facility will cause the adjacencies of the small facility to be less than any minimum width chosen. If the minimum width criteria cannot be satisfied, LayoutManager reduces the minimum width until a solution can be found and the block plan drawn. If the minimum width was reduced, the user is informed of the required minimum width and of the facility which forced the reduction.

The block plan is drawn to scale within a window using the final block plan shape specified and the appropriate colours and patterns. A key to the colour patterns is also displayed so the user can relate the shapes back to the names provided. Having produced a block plan the user can evaluate the results and return to the input data to make any necessary modifications using the built-in editor. The changes can then be reapplied to the heuristic and a new block plan will be drawn.

When the user is satisfied with the results or wishes to save a particular block plan for further study, the block plan can be printed out or saved as a file. Saved block plans, when reloaded, are displayed in separate windows so that several instances of a problem, with minor modifications applied, can be viewed at once. One intended use for this option is to observe the changes that are produced when the edge replacement and vertex relocation improvement heuristics are successively applied to the problem. We now go on to describe a case study in which LayoutManager was used to redesign successfully a manufacturing system.

3. A CASE STUDY

Computer Automation (Ireland) Limited is a manufacturer of printed circuit boards and commercial and industrial computers, located in Dublin. Recently, the company began to experience difficulty in maintaining efficiency in the highly competitive computer manufacturing market. Upon examination, one of the problems appeared to be a reduction in efficiency of the mechanical assembly area, where the final products were actually put together. The layout of the area was deemed to be quite inappropriate. As a result, LayoutManager was developed and ultimately implemented in order to achieve the necessary efficiency in the flow of materials and people in this area. Twelve activities were identified that take place concerning this region, one of them being the exterior region itself. This was included because it is important that certain facilities are placed adjacent to the perimeter of the overall area. Relationship charts were developed and the area required for each facility was estimated. This information was used as part of the input data for the system. Eventually a workable plan was produced, which was ultimately introduced and is now in use. It took some time to convince people of the need to change but eventually all of those involved saw that the recommendation was simple, logical, represented common sense, and was in no way threatening. Indeed those involved became convinced that the new layout actually made their lives more pleasant. As a result of the designed process inefficient assembly sequences were redesigned and worker and goods travel was reduced significantly.

The following lessons were learnt in the overall process. The total commitment of the top management of the company was vital. It was essential to involve the workers on the shop floor very early and to gain their ideas, many of which were quite innovative and valuable. This early involvement meant that the workers had ownership of the ideas and thus were later far more receptive to the changes that were ultimately recommended. It took a great deal of team work to achieve the change and of course, mistakes were made during the process. Another lesson learned was that of gradualism: that is, it was clear that a sudden and wholesale change in the layout, with little explanation, would have resulted in workers attempting to revert to the old design. One thing that is clear from most successful applications is that one must have a great deal of patience. There was an initial dip in productivity in the short term, because of the loss of productivity during the change, when people were getting used to the new system. It was important that the top management had faith and confidence that this dip was a worthwhile investment, which it turned out to be.

4. SUMMARY AND CONCLUSIONS

We have described a new decision support system for facilities layout. It will accommodate various objectives, such as the maximization of adjacency scores and the minimization of transportation costs, and is based on a number of graph theoretic heuristics. The system was used in practice in the design of a computer manufacturing factory. As a result of numerical experimentation and practical implementation of the system, we are confident that it represents a useful addition to the facility planner's toolkit.

REFERENCES

Boswell, S.G. 1992. TESSA - A new greedy heuristic for facilities layout planning, *IJPR*. **30**, 1957-1968.

Foulds, l.R. 1993. Graph theoretic-based decision support for facilities layout. These proceedings.

Hammouche, A. and D.B. Webster. 1985. Evaluation of an application of graph theory to the layout problem. *IJPR* **23**, 987-1000.

Hassan, K.M.D., and G.L. Hogg. 1987. A review of graph theory application to the facilities layout problem. *OMEGA* **15**, 291-300.

Leung, J. 1992a. A graph-theoretic heuristic for designing loop-layout manufacturing systems. *EJOR* **57**, 243-252.

Leung, J. 1992b. A new graph-theoretic heuristic for facility layout. *Mgmt. Sci.* **38**, 594-605.

Wascher G. and P. Chamoni. 1987. MICROLAY: An interactive computer program for factory layout planning on microcomputers. *EJOR* **31**, 185-193.

Summary of Mini-Symposium III.

Topic: Layout Design

Moderators: E.D. Adamides (Switzerland), H.J. Pinheiro-Pita (Portugal)

Demonstration title: LayoutManager: A Decision Support System for Facilities Planning

Contributor: L.R. Foulds (New-Zealand)

1. User oriented description of the demonstrated system

LayoutManager is a menu driven software system made for helping factory layout designers. It can be used in the designer's office as well as in the shop floor. It was implemented in the Microsoft Windows environment and its hardware requirements are modest.

The system considers adjacency, near adjacency and physical distance criteria in the process of finding the best layout, and produces a graphical image of the proposed layout. Due to various limitations and constraints imposed by the system on the geometry of facilities the result should be treated as indicative. Even so, the system has already proven its usefulness in industrial projects.

2. Discussions

The interactive user interface may need improvement, especially in the problem definition phase. Presenting the outcome of the consultation session as alternatives in a hierarchy would also be beneficial, especially when having a large number of facilities, whose dimensions vary considerably. Further improvement could be achieved by allowing the user to take active part in the decision making process, e.g. offering selective changes or keeping one part of the layout fixed while optimising the rest. An analysis of how sensitive the system is to input data variations remains to be done.

3. Conclusions

The advantages of an interactive system over automatic problem solvers have clearly been demonstrated, especially, that co-operation of human experts and machines is more efficient and productive.

MINI-SYMPOSIUM IV.
Management of Production Systems

Knowledge Based Hybrid Systems (B-11)
I. Mezgár and P. Bertók (Editors)
Elsevier Science Publishers B.V. (North–Holland)

MASTer-For Production Systems That Run Themselves

John E. Lenz
CMS Research, Inc.
Oshkosh, Wisconsin, U.S.A.

LEVEL I
CAPACITY PLANNING LEVEL

MASTer is a day-to-day management tool for flexible cells. It provides three levels of essential tasks: Level I) Capacity Planning, Level II) Data Sharing With Real Time Control, and Level III) Performance Monitoring. With MASTer, you can take full advantage of the flexibility of your cell. Put flexibility to strategic use with MASTer.

MASTer allows you to allocate flexible capacity in response to all of the following:

Determine Actual Cost of Production
Raw material cost, labor cost, machine burden, tooling, and other costs are used to determine the actual cost of producing a product. Value added is used to establish the cost versus value to justify the cell as an efficient production system.

Establish Production Requirements
Daily production requirements can be evaluated. If new material shortages occur, reallocate the capacity to another product without wasting capacity.

Introduce New Product
Allocate appropriate amount of capacity for prototype parts. Bring new product manufacturing on-line with minimal effect upon daily requirements.

Respond to Changes in Capacity
When machine is not of service, reallocate the available capacity to match production requirements. Then when the machine returns to operations, allocate parts to prevent uncontrolled pallet surges.

Establish Correct Number of Pallets
Given a production requirement, identify the correct number or types of pallet fixtures. Activate-Deactivate pallets which produce to production requirements without bottlenecking the operation.

Determine Appropriate Number of Operators
Establish number of operations required for load/unload and machine respondence to failure.

These decisions are supported with the following "tools."

Production Scheduler
Enter and list production requirements for day, week, month, or year time basis.

Route Editor
List operations for each product and its corresponding cycle times. Identify specific stations for each operation to provide as much or as little flexibility as needed.

Cost Data
The costs of all components are used to establish the burden rate for the cell. Then the value added to a product is compared to its cost (in terms of burden rate) to support daily decisions. Cost information can be used as an alternative to performance to decide production schedule, routing, new products, and pallet mix.

Pallet Mix Definition
Establish a where-used relation between products and pallet fixture types. Determine the correct number of pallets for a production schedule.

Physical Layout
The exact description of a cell layout is included in MASTer. This includes number/type of each station, type of material handling, distances between all steps, and speeds/shuttling time. No further information is required.

Exact Control Decisions
MASTer uses the exact same control decisions as used in the real time control. Algorithms will be selected from your explicit cell configuration of the control.

History of Alternative Routings
Maintain a history of route configurations. Then simply recall alternative routing for a product with a minimum of route editing.

Hardware Platform
IBM RS6000
16MB memory
320MB hard disk
AIX V3.2 operating system
PC simulator software
15" high resolution color monitor

LEVEL II
DATA SHARING LEVEL

MASTer is a day-to-day management tool for flexible cells. It provides three levels of essential tasks: Level I) Capacity Planning, Level II) Data Sharing With Real Time Control, and Level III) Performance Monitoring. With MASTer, you can take full advantage of the flexibility of your cell. Put flexibility to strategic use with MASTer.

MASTer Level II includes all the capability and features of Level I with the addition of data sharing with real time control. This includes an upload of data from the control of the following data:

Actual Operation Durations
The actual cycle time for each product type at each station will be logged. This data will be automatically filled into the Route Editor for the specific part/operation. As engineering changes are performed in the part programs, MASTer will automatically be updated for accurate capacity planning.

Station Availability History
The pattern of station available/unavailable will be logged. This will provide average time between failure, average repair time, and average availability. This data can then be included into the capacity planning to provide realistic allowance for system efficiency.

Current Status
MASTer Level II will obtain an up-to-the-minute report of current machine status, pallet positions, vehicle location, and times to complete for all operations. This will be used as the initial conditions for capacity planning. That is, the capacity planning can begin from current location of all pallets, and in cycle machines will include remaining time left for the operation.

LEVEL III
ON-LINE MONITORING LEVEL

MASTer is a day-to-day management tool for flexible cells. It provides three levels of essential tasks: Level I) Capacity Planning, Level II) Data Sharing With Real Time Control, and Level III) Performance Monitoring. With MASTer, you can take full advantage of the flexibility of your cell. Put flexibility to strategic use with MASTer.

Level III closes the decision support loop between capacity planning and actual operations. It uses the capacity allocation of Level I with the data sharing of Level II to provide the performance monitoring of cell efficiency. It accomplishes this through the following features:

Flow Time Control (FLOTIM) Chart for Each Product Type

The flow time (pallet cycle time) is used as an indicator of how well the FMS is running. The results of Level I--Capacity Planning, establish a target flow time and this is used to generate a control chart. Then whenever a part is completed, its flow time is recorded on the control chart. Trends in cell efficiency can be observed and responded to in a timely manner.

Identify with Products Running Well

Using the Flow Time Control Chart, it is easy to identify those products which are being produced at the required rate. Their observed flow time will be above the target level. When this occurs, it will be possible to reallocate some capacity away from this product to help out those which might be below required production rates.

Identify Which Product Is Running Behind Schedule

Anticipate/allocate capacity to avoid product shortages by using the Flow Time Control FLOTIM Chart. Those products which have observed flow times longer than planned will be producing below required rates. They can be identified from the FLOTIM Chart and more flexible capacity can be given to these products.

Download Route (Job Control) Data

Use MASTer to send Route Data to control. After capacity planning, a specific route configuration is defined along with pallet quantities. This information is defined as the Routing Table and can be downloaded directly into the control. In this sense, the decision of how best to allocate flexible capacity is automatically implemented in the real time control.

Download Product Requirements

MASTer can download the daily production schedule from its capacity planning results. This data along with the Routing Table provides management control to ensure that the cell capacity is configured and allocated precisely as management decides.

Summary of Mini-Symposium IV.

Topic: Management of Production Systems

Moderators: H.J. Pinheiro-Pita (Portugal), E.D. Adamides (Switzerland)

Demonstration title: MASTer-For Production Systems That Run Themselves

Contributor: John E. Lenz (USA)

1. User oriented description of the demonstrated system

MASTer is a multi-purpose software tool to help management in many areas from capacity planning to real time monitoring of equipment. It can identify the current state of the system, such as products running well or behind schedule, machine failure, material shortage etc., and come up with resource allocation / reallocation plans. Thus, it helps in making better use of flexible facilities.

The system has a user friendly interface with graphics and menus, that is well tailored to the application.

2. Discussions

Enhancements are planned to make the system more open towards integration with other systems, such as a process planning system or an enterprise wide database. Issues of data communication, database support for MASTer itself, e.g. for history data, and access of these data with multiple keys will have to be addressed as well.

An important merit of the system is that it includes tool handling, which is omitted in many cases. It may be better to develop MASTer towards being a decision support tool for the production supervisor than making it part of a fully automated production system.

The basic performance measurement could be more comprehensible for the end users, as they have to supply the selection and evaluation criteria for the decisions made by the system. A more easily comprehensible model would facilitate the integration of the system into the enterprise environment.

3. Conclusions

The complex approach to evaluation of manufacturing systems is highly important. That means that the inclusion of different subsystems, e.g. tool handling, as well as using comprehensive evaluation methods, e.g. flow time analysis describes the operation better than utilisation rate of machines. A balance between tasks performed by human experts and tasks performed by automated systems has to be carefully maintained.

Author Index

IFIP

The INTERNATIONAL FEDERATION FOR INFORMATION PROCESSING is a multinational federation of professional and technical organisations (or national groupings of such organisations) concerned with information processing. From any one country, only one such organisation – which must be representative of the national activities in the field of information processing – can be admitted as a Full Member. In addition a regional group of developing countries can be admitted as a Full Member. On 1 October 1992, 45 organisations were Full Members of the Federation, representing 62 countries.

The aims of IFIP are to promote information science and technology by:

- fostering international co-operation in the field of information processing;
- stimulating research, development and the application of information processing in science and human activity;
- furthering the dissemination and exchange of information about the subject;
- encouraging education in information processing.

IFIP is dedicated to improving worldwide communication and increased understanding among practitioners of all nations about the role information processing can play in all walks of life.

Information technology is a potent instrument in today's world, affecting people in everything from their education and work to their leisure and in their homes. It is a powerful tool in science and engineering, in commerce and industry, in education and adminstration. It is truly international in its scope and offers a significant opportunity for developing countries. IFIP helps to bring together workers at the leading edge of the technology to share their knowledge and experience, and acts as a catalyst to advance the state of the art.

IFIP came into official existence in January, 1960. It was established to meet a need identified at the first International Conference on Information Processing which was held in Paris in June, 1959, under the sponsorship of UNESCO.

Organisational Structure
The Federation is governed by a GENERAL ASSEMBLY, which meets once every year and consists of one representative from each Member organisation. The General Assembly decides on all important matters, such as general policy, the programme of activities, admissions, elections and budget.

IFIP ORGANISATION CHART

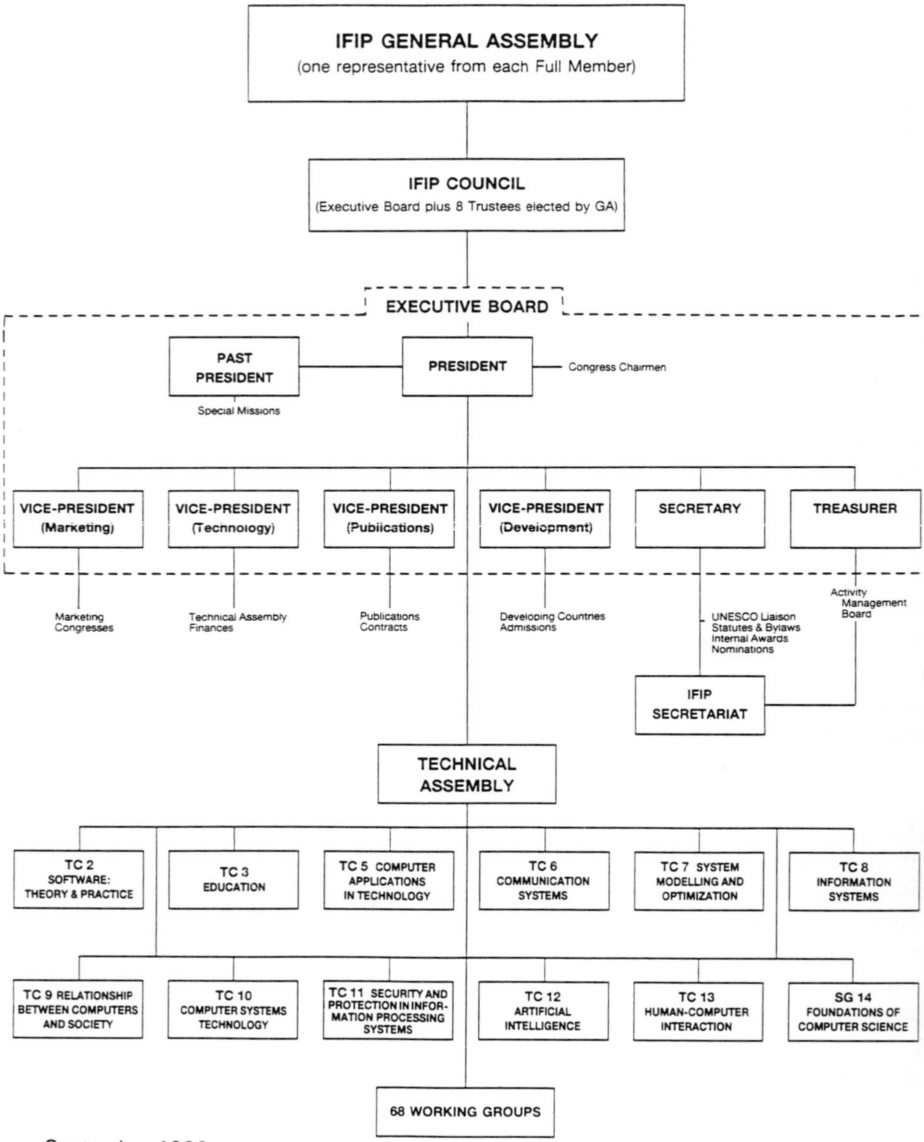

September 1992

The day-to-day work of IFIP is directed by its Officers: the President, Vice-Presidents, Secretary and Treasurer, who are elected by the General Assembly and together constitute the EXECUTIVE BOARD.

The COUNCIL, consisting of the Officers and up to eight Trustees elected from the General Assembly, meets twice a year and takes decisions which become necessary between General Assembly meetings.

The headquarters of the Federation are in Geneva, Switzerland where the IFIP Secretariat administers its affairs.

For further information please contact:

IFIP Secretariat
attn. Mme. GWYNETH ROBERTS
16 Place Longemalle
CH-1204 Geneva, Switzerland
telephone: 41 (22) 28 26 49
facsimile: 41 (22) 781 23 22
Bitnet: ifip@cgeuge51

IFIP's MISSION STATEMENT

IFIP's mission is to be the leading, truly international, apolitical organisation which encourages and assists in the development, exploitation and application of Information Technology for the benefit of all people.

Principal Elements

1. To stimulate, encourage and participate in research, development and application of Information Technology (IT) and to foster international co-operation in these activities.
2. To provide a meeting place where national IT Societies can discuss and plan courses of action on issues in our field which are of international significance and thereby to forge increasingly strong links between them and with IFIP.
3. To promote international co-operation directly and through national IT Societies in a free environment between individuals, national and international governmental bodies and kindred scientific and professional organisations.
4. To pay special attention to the needs of developing countries and to assist them in appropriate ways to secure the optimum benefit from the application of IT.
5. To promote professionalism, incorporating high standards of ethics and conduct, among all IT practitioners.
6. To provide a forum for assessing the social consequences of IT applications; to campaign for the safe and beneficial development and use of IT and the protection of people from abuse through its improper application.

7. To foster and facilitate co-operation between academics, the IT industry and governmental bodies and to seek to represent the interest of users.
8. To provide a vehicle for work on the international aspects of IT development and application including the necessary preparatory work for the generation of international standards.
9. To contribute to the formulation of the education and training needed by IT practitioners, users and the public at large.

Note to Conference Organizers

Organizers of upcoming IFIP Working Conferences are urged to contact the Publisher. Please send full details of the Conference to:

Mrs. STEPHANIE SMIT
Administrative Editor – IFIP Publications
ELSEVIER SCIENCE PUBLISHERS
P.O. Box 103, 1000 AC Amsterdam
The Netherlands
telephone: 31 (20) 5862481
facsimile: 31 (20) 5862616
email: s.smit@elsevier.nl

IFIP TRANSACTIONS

IFIP TRANSACTIONS is a serial consisting of 15,000 pages of valuable scientific information from leading researchers, published in 35 volumes per year. The serial includes contributed volumes, proceedings of the IFIP World Conferences, and conferences at Technical Committee and Working Group level. Mainstream areas in the IFIP TRANSACTIONS can be found in Computer Science and Technology, Computer Applications in Technology, and Communication Systems.

From 1993 onwards the IFIP TRANSACTIONS are only available as a full set.

IFIP TRANSACTIONS A:
Computer Science and Technology
1992: Volumes A1–A19
1993: Volumes A20–A40
ISSN 0926-5473

IFIP Technical Committees that are involved in IFIP TRANSACTIONS A

Software: Theory and Practice (TC2)
Education (TC3)
System Modelling and Optimization (TC7)
Information Systems (TC8)
Relationship Between Computers and Society (TC9)

Computer Systems Technology (TC10)
Security and Protection in Information Processing Systems (TC11)
Artificial Intelligence (TC12)
Human-Computer Interaction (TC13)
Foundations of Computer Science (SG14)

IFIP TRANSACTIONS B:
Applications in Technology
1992: Volumes B1–B8
1993: Volumes B9–B14
ISSN 0926-5481

IFIP Technical Committee that is involved in IFIP TRANSACTIONS B
Computer Applications in Technology (TC5)

IFIP TRANSACTIONS C:
Communication Systems
1992: Volumes C1–C8
1993: Volumes C9–C16
ISSN 0926-549X

IFIP Technical Committee that is involved in IFIP TRANSACTIONS C
Communication Systems (TC6)